CAMBRIDGE TRACTS IN
MATHEMATICS

General Editors

B. BOLLOBAS, P. SARNAK, C. T. C. WALL

107 Duality and Perturbation Methods in Critical Point Theory

T0276130

NASSIF GHOUSSOUB

University of British Columbia

Duality and perturbation methods in critical point theory

CAMBRIDGE UNIVERSITY PRESS
Cambridge, New York, Melbourne, Madrid, Cape Town, Singapore, São Paulo

Cambridge University Press
The Edinburgh Building, Cambridge CB2 8RU, UK

Published in the United States of America by Cambridge University Press, New York

www.cambridge.org
Information on this title: www.cambridge.org/9780521440257

© Cambridge University Press 1993

First published 1993
This digitally printed version 2008

A catalogue record for this publication is available from the British Library

ISBN 978-0-521-44025-7 hardback
ISBN 978-0-521-07195-6 paperback

To LOUISE

CONTENTS

PREFACE

The aim of these notes is to give a self-contained presentation of the min-max approach to critical point theory while emphasizing the role of *duality and perturbation* methods. Actually, this monograph originated in a project where we set out to show that *duality* is a fundamental concept that underlies many aspects of critical point theory. The goal was to try to reprove and improve selected results in min-max theory by exploiting the notion of *dual* families of sets and its ramifications. It turned out that, by adopting this point of view, the whole theory can be nicely developed and vastly enriched.

On the other hand, by *perturbation methods*, we mean the aspect of infinite dimensional critical point theory where, in order to deal with the possible lack of compactness or with the presence of degeneracy, one tries to modify the functional or the problem under study to a neighboring one that can be more manageable. We shall adhere to this methodology throughout these notes.

This monograph owes its existence to a very dear friend, Ivar Ekeland, who introduced me to non-linear analysis and, more importantly, influenced greatly my global vision of mathematics. Special thanks go to another dear friend, Bernard Maurey, for all the years of collaboration and from whom I learned so much. Many of the relevant examples included here are due to Gabriella Tarantello. I am very grateful to her for permitting me to include her results, some of which have not yet appeared in print. Many thanks to my graduate students, Guangcai Fang and David Robinson for their important contributions to the mathematical content as well as to the pedagogical aspect of this monograph. I would like to thank Rita McIlwaine from the staff of the department of mathematics at the University of British Columbia for helping me TEX some of this! Many thanks go to the senior editor of Cambridge University Press, David Tranah who makes book publishing look so easy! My deep gratitude goes to my wife Louise for coping with all that and much more ...

N. Ghoussoub
Vancouver, B.C.

INTRODUCTION

Eigenvectors, geodesics, minimal surfaces, harmonic maps, conformal metrics with prescribed curvature, subharmonics of Hamiltonian systems, solutions of semilinear elliptic partial differential equations and Yang-Mills fields are all critical points of some functional on an appropriate manifold. This is not surprising since many of the laws of mathematics and physics can be formulated in terms of *extremum principles*.

Finding such points by minimization is as ancient as the least action principle of Fermat and Maupertuis, and the calculus of variations has been an active field of mathematics for almost three centuries. For more general, unstable extrema, the methods have a more recent history. Two, not unrelated, theories are available for dealing with the existence of such points: *Morse theory* and *the min-max* methods (or the calculus of variations in the large) introduced by G. Birkhoff and later developed by Ljusternik and Schnirelmann in the first half of this century. Currently, both theories are being actively refined and extended in order to overcome the limitations to their applicability in the theory of partial differential equations: limitations induced by the infinite dimensional nature of the problems and by the prohibitive regularity and non-degeneracy conditions that are not satisfied by present-day variational problems.

In this monograph, we will be concerned with the old problems of existence, location, multiplicity and structure of critical points via the methods of the calculus of variations in the large, but we shall present a new point of view that might help in dealing with some of the difficulties mentioned above. Actually, we shall build upon the well known min-max methods that are presently used in non-linear differential equations and which are well documented in the lecture notes of Rabinowitz [R 1] and the recent books of Ekeland [Ek 2], Mawhin-Willem [M-W] and Struwe [St]. However, unlike the above mentioned monographs, our emphasis will be on trying to develop, in a systematic way, a "general theory" for variationally generated critical points, that can be applied in a variety of situations, under a minimal set of hypotheses. But our real goal is to advertise a whole array of newly discovered *duality and perturbation methods* that are instrumental in carrying out the refinements of that theory.

One of the central questions we address is the following: what can one say about the *critical structure* of a functional defined on an infinite dimensional domain without imposing the usual compactness conditions *à la Palais-Smale*, or the non-degeneracy conditions *à la Fredholm*? We shall present various new variational principles which, besides ensuring the existence of critical points under minimal hypotheses, give valuable information about their location, their multiplicity and their structure. These principles will yield most of the classical results. However we shall emphasize those applications where the standard methods do not apply but the ones offered in this monograph do.

Here is an overview of our approach. Consider a C^1-functional φ on a smooth infinite dimensional manifold X. If φ is bounded below on X, then one can try to find a critical point at the level $c = \inf_X \varphi$ by checking whether the infimum is attained. If φ is indefinite or if one is looking for *unstable* critical points, one then considers a family $\mathcal{F}$ of compact subsets of X that is stable under a certain class of homotopies and then shows that φ has a critical point at the level

$$c = c(\varphi, \mathcal{F}) = \inf_{A \in \mathcal{F}} \max_{x \in A} \varphi(x).$$

In both cases, one can easily find an *almost critical sequence* at the level c; that is a sequence $(x_n)_n$ in X satisfying:

$$\lim_n \varphi(x_n) = c \quad \text{and} \quad \lim_n \|d\varphi(x_n)\| = 0. \tag{1}$$

The main problem of existence reduces to proving that such a sequence is convergent. This is usually where the hard analysis is needed. Any function that possesses such a property is said to satisfy the *Palais-Smale condition at the level* c (in short, $(PS)_c$). However, various interesting functionals originating in partial differential equations and in differential geometry do not satisfy such a property, or they may only satisfy it for certain levels c. These problems usually occur in situations involving the critical exponent in the Sobolev embedding theorems, or in the cases where *scale or gauge invariance* requirements give rise to non-compact group actions.

Our main purpose in this direction is to try to find almost critical sequences for φ that possess some extra properties which might help to ensure their convergence. Here is the first idea that comes to mind: one can find an almost critical sequence that is arbitrarily close to any *min-maxing sequence* $(A_n)_n$ in $\mathcal{F}$. That is, if $(A_n)_n$ in $\mathcal{F}$ is such that $\lim_n \max_{A_n} \varphi = c$, then one can find a sequence $(x_n)_n$ that satisfies (1) as well as

$$\lim_n \operatorname{dist}(x_n, A_n) = 0. \tag{2}$$

What is needed then, is *a Palais-Smale condition along one min-maxing sequence* $(A_n)_n$ *in* $\mathcal{F}$. As we shall see in Chapter 3, this weakening of the Palais-Smale condition turns out to be relevant for the solution of a *resonance problem.*

Another point of view consists of considering a family $\mathcal{F}^*$ of closed sets that is *dual to* $\mathcal{F}$ (i.e., satisfying $F \cap A \neq \emptyset$ for all $F \in \mathcal{F}^*$ and $A \in \mathcal{F}$) and such that

$$\sup_{F \in \mathcal{F}^*} \inf_{x \in F} \varphi(x) = \inf_{A \in \mathcal{F}} \max_{x \in A} \varphi(x) = c.$$

One of the main results of this monograph (Theorem 4.5) shows that one can then construct an almost critical sequence that is also arbitrarily close to any *max-mining sequence* $(F_n)_n$ *in* $\mathcal{F}^*$. In other words, if $(F_n)_n$ in $\mathcal{F}^*$ is such that $\lim_n \inf_{F_n} \varphi = c$, then one can find a sequence $(x_n)_n$ that satisfies (1), (2) and

$$\lim_n \operatorname{dist}(x_n, F_n) = 0. \tag{3}$$

There are three important features to this new min-max principle. First, we could expect that the topology, which provides the dual families, can sometimes help the analysis involved in proving the convergence of such pseudo-critical sequences. We shall see in Chapter 8, that in various examples, one can indeed push back the threshold of noncompactness by ensuring that an almost critical sequence is arbitrarily close to a certain dual set.

Another aspect of that refinement is that it helps relax the boundary conditions: Indeed, the homotopies that preserve the elements of the family $\mathcal{F}$ are usually required to leave a certain boundary B invariant. In that case, the standard existence results require that $\sup \varphi(B) < c = c(\varphi, \mathcal{F})$. In Chapters 4 and 5, we exhibit situations where the boundary condition can be relaxed to $\sup \varphi(B) = c$, especially in the presence of dual sets.

The third and most important feature of this principle is that it locates *critical points* whereas the classical min-max principle only identifies possible *critical levels*. Indeed, as mentioned in (3) above, we could find, under adequate compactness conditions, a critical point on any dual set F. The largest dual set is clearly $\{\varphi \geq c\}$. The trick is to find proper subsets F of $\{\varphi \geq c\}$ that still intersect all the members in $\mathcal{F}$, since the smaller F is, the more information we have about the critical points it contains. By appropriate choices of dual sets, we manage to prove various old and new results concerning the multiplicity and the Morse indices of critical points generated by this procedure. In Chapter 6, we use this method to classify the critical points generated by the mountain pass theorem in a setting that is not covered by Morse theory. In Chapter 7, we show how this same information about the location of critical points on various dual sets gives easy proofs of the standard multiplicity results, while it leads naturally to new and unexpected ones. In Chapter 9, it helps us relate the Morse index of a critical point to the *topological dimension* (homotopic, cohomotopic or homological) of the family $\mathcal{F}$ used to obtain it in the case of non-degenerate C^2-functionals on Riemannian manifolds. In Chapter 10, we study the degenerate case and find that the same principle gives new types of multiplicity results: these concern the size of sets of critical points having a common estimate on their Morse indices.

Let us return to the possibility of "improving" the Palais-Smale sequences in order to ensure their convergence. Suppose now that φ is a C^2-functional. One can then try to get an almost critical sequence $(x_n)_n$ with some information about the second derivatives $d^2\varphi(x_n)$. For example, this can be done in minimization problems: one can then construct an almost critical sequence $(x_n)_n$ that is minimizing and that satisfies for each $n \in \mathbb{N}$:

$$\langle d^2\varphi(x_n)v, v\rangle \geq -\|v\|^2/n \quad \text{for any } v \in X. \tag{4}$$

One way to obtain such sequences consists of establishing "perturbed variational principles". (A typical example would be Ekeland's Theorem). If, due to the lack of compactness, we cannot find critical points for a given function, can one then perturb it so that the new functional has critical points of the kind that is expected for the original one. This type of result is much stronger than finding almost critical points for the original function. Indeed, if the perturbation is C^2-small, then knowing that the almost critical point is a true critical point for the new functional allows us to use Morse theory and therefore to get some information about the Hessian of the original function at that point.

For minimization problems, several results of this kind have recently been established with various degrees of generality. In Chapter 1, we include a general variational principle where the perturbations can be taken to be as smooth as the norm on the domain of the functional under study. As an illustration, we give a result of P. L. Lions about the solutions of the *Hartree-Fock equations* as well as an application to the existence of *viscosity solutions* for first-order *Hamilton-Jacobi equations*. We also give, in Chapter 2, two recent results of a similar nature: In the first, which can be applied in reflexive Banach spaces, the perturbations can be taken to be linear while the second covers spaces as "topologically and geometrically bad" as L^1, provided one settles for plurisubharmonic perturbations. This will be used in Chapter 2 to get generic minimization results in the case of critical exponents and non-zero data.

Perturbed variational principles for problems not involving minimization are more involved and one should expect to need *hyperbolic perturbations*. Results of this type are very recent and are not yet in their final form. However, we shall present in Chapter 11, a new method devised by Fang and Ghoussoub, for constructing directly –without establishing the perturbation result– an almost critical sequence with the appropriate second order information, provided one has an additional assumption of Hölder continuity on the first and second derivatives. For example, in the context of *the mountain pass theorem*, we obtain an almost critical sequence $(x_n)_n$ and a sequence of subspaces $(E_n)_n$ of codimension one such that the sequence $(x_n)_n$ satisfies, in addition to (1), (2) and (3) above, the following condition:

$$\langle d^2\varphi(x_n)v,v\rangle \geq -\|v\|^2/n \quad \text{for any } v \in E_n. \tag{5}$$

In other words, for each $n \in \mathbb{N}$, $d^2\varphi(x_n)$ has *at most one* eigenvalue below $-1/n$, which clearly implies that any potential cluster point for $(x_n)_n$ will be a critical point of Morse index at most one. More importantly, and as already noted by P. L. Lions in his study of the Hartree-Fock equations, this additional information is sometimes crucial for proving the convergence of such sequences, especially when the standard Palais-Smale condition is not satisfied. We shall include this example as an application of our results in the homotopic case.

Let us mention that another weakening of the Palais-Smale condition also turned out to be relevant in some geometric problems. It is the (PS) condition along a fixed orbit of the *negative gradient flow* of the functional: that is when compactness holds for those pseudo-critical sequences of the form

$$x_n = \sigma(t_n, x) \tag{6}$$

where σ is the solution of the Cauchy problem

$$\dot{\sigma} = -\nabla\varphi(\sigma), \quad \sigma(0,x) = x.$$

We shall not discuss this phenomenon here and we refer the reader to the book of Bahri [Ba 2] for a detailed analysis of several important examples.

Throughout this monograph, we have also emphasized the role of various *local perturbation methods* that are now available and which are interesting in their own right. These techniques consist of changing a problem appropriately to a neighboring one that can be more manageable. Here is a sample: In Theorem 4.5, we perturb the function and the variational setting to reduce the new refined min-max principle to the old one. In Theorem 5.2, we use yet another perturbation to reduce the new relaxed boundary condition to the classical one. In chapter 10, we use the Marino-Prodi perturbation method to change a degenerate case to a non-degenerate one. A method for restricting a homotopy-stable class of sets to submanifolds is implemented, while another kind of perturbation is needed to isolate certain subsets of critical points.

We have tried to make this monograph as self-contained as possible, especially for the part that deals with "the general theory". It was hard to do the same with the examples and therefore they may seem sketchy at times. However, the necessary results and concepts are included – some without proofs – in an appendix prepared skillfully by David Robinson. This is definitely not a comprehensive study of critical point theory, nor of any subset of the theory of partial differential equations. It is merely a collection of some recently formulated variational principles, followed by some examples justifying their relevance. Our main goal is to make these new methods accessible to non-linear analysts, hoping that more interesting applications will follow. To illustrate these principles, we chose the simplest examples known to us, and often did not include the strongest known versions, in order to avoid the complications that were irrelevant to our discussion.

Before closing this introduction, we should mention that at present, there are at least two other approaches that have proved very successful in dealing with the difficulties of infinite dimensional critical point theory and which are unfortunately missing from this monograph. We have in mind Conley's and Floer's approaches to Morse theory ([Co], [Fl 1,2]) and the machinery involving *critical points at infinity* that was mainly developed by A. Bahri [Ba 2]. The reason for these omissions is simply that both of these theories are substantialy more involved than the elementary approach adopted in this monograph since they require a much heavier backgroung from algebraic topology and geometry.

1

LIPSCHITZ AND SMOOTH

PERTURBED MINIMIZATION PRINCIPLES

We start this chapter by stating and proving Ekeland's well known variational principle since it will be used frequently throughout this monograph. We also give some of its lesser known applications to constrained minimization problems that eventually yield *global* critical points for the functional in question. We introduce the *Palais-Smale condition around a set* and we present the first of many examples which show its relevance. We give an existence result for nonhomogeneous elliptic equations involving the critical Sobolev exponent, due to Tarantello. We then establish the more recent *smooth variational principle* of Borwein-Preiss and we apply it to the study of Hartree-Fock equations for Coulomb systems as was done by P.L. Lions. Finally, we follow the ideas of Ghoussoub and Maurey to deal with the more general problem of identifying those classes of functions that can serve as *perturbation spaces* in an appropriate minimization principle. As an application, we give a result of Deville et al, stating that the perturbations can be taken to be as smooth as the norm of the Banach space involved. We then apply this result to the problem of existence and uniqueness of *viscosity solutions* for first order Hamilton-Jacobi equations on general Banach spaces.

1.1 Ekeland's variational principle

The following theorem will be of constant use throughout this monograph. The applications of this principle to non-linear analysis are numerous and well documented in several books ([A-E], [Ek 1], [De]). We shall only be concerned with those that are relevant for this monograph.

Theorem 1.1 (Fig 1.1): *Let (X, d) be a complete metric space and* consider a function $\varphi : X \rightarrow (-\infty, +\infty]$ that is *lower semi-continuous, bounded from below and not identical to* $+\infty$. Let $\varepsilon > 0$ and $\lambda > 0$ be given and let $u \in X$ be such that $\varphi(u) \leq \inf_X \varphi + \varepsilon$. Then there exists $v_\varepsilon \in X$ such that

(i) $\varphi(v_\varepsilon) \leq \varphi(u)$

(ii) $d(u, v_\varepsilon) \leq 1/\lambda$

(iii) For each $w \neq v_\varepsilon$ in X, $\varphi(w) > \varphi(v_\varepsilon) - \varepsilon\lambda d(v_\varepsilon, w)$.

Proof: We shall show the existence of $v_\varepsilon \in X$ such that

$$\varphi(v_\varepsilon) \leq \varphi(u) \tag{1}$$

$$d(u, v_\varepsilon) \leq 1 \tag{2}$$

and, for each $w \neq v_\varepsilon$ in X,

$$\varphi(w) > \varphi(v_\varepsilon) - \varepsilon d(v_\varepsilon, w). \tag{3}$$

The theorem will then follow by replacing the metric d by λd.

Introduce the following partial order on X

$$w \prec v \quad \text{provided } \varphi(w) + \varepsilon d(v, w) \leq \varphi(v).$$

We construct inductively a sequence (u_n) as follows: Start with $u_0 = u$. Once u_n is known, let $S_n = \{w \in X : w \prec u_n\}$ and choose $u_{n+1} \in S_n$ such that

$$\varphi(u_{n+1}) \leq \inf_{S_n} \varphi + \frac{1}{n+1}.$$

Clearly, $S_{n+1} \subset S_n$, as $u_{n+1} \prec u_n$, and since φ is lower semi-continuous, S_n is closed. Now, if $w \in S_{n+1}$, $w \prec u_{n+1} \prec u_n$ and hence

$$\varepsilon d(w, u_{n+1}) \leq \varphi(u_{n+1}) - \varphi(w) \leq \inf_{S_n} \varphi + \frac{1}{n+1} - \inf_{S_n} \varphi = \frac{1}{n+1}$$

so that $\text{diam}(S_{n+1}) \leq \frac{2}{\varepsilon(n+1)}$. Since X is complete, this implies that

$$\bigcap_n S_n = \{v_\varepsilon\} \tag{4}$$

for some $v_\varepsilon \in X$. In particular, $v_\varepsilon \in S_0$, which means that

$$\varphi(v_\varepsilon) \leq \varphi(u) - \varepsilon d(u, v_\varepsilon) \leq \varphi(u)$$

and

$$d(u, v_\varepsilon) \leq \varepsilon^{-1}(\varphi(u) - \varphi(v_\varepsilon)) \leq \varepsilon^{-1}(\inf_X \varphi + \varepsilon - \inf_X \varphi) = 1.$$

To obtain (iii), it suffices to notice that if $w \prec v_\varepsilon$, then for each $n \in \mathbf{N}$, $w \prec u_n$ so that $w \in \cap_n S_n$ and, by (4), $w = v_\varepsilon$.

Our first application is to the existence of minimizing sequences that are also almost critical. Throughout this monograph and unless stated otherwise (as in sections 1.8 and 1.9), we shall say that a real-valued function φ on a Banach space X is *differentiable* if it is *Fréchet differentiable*; that is for any $x \in X$, there is $p \in X^*$ (denoted $\varphi'(x)$) such

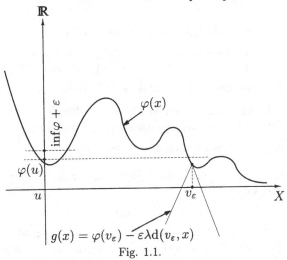

$$g(x) = \varphi(v_\varepsilon) - \varepsilon\lambda\mathrm{d}(v_\varepsilon, x)$$

Fig. 1.1.

that

$$\lim_{t \to 0} t^{-1}[\varphi(x + th) - \varphi(x) - \langle p, th \rangle] = 0$$

uniformly for $h \in X$ with $\|h\| = 1$.

We shall say that φ is a C^1-functional if $x \to \varphi'(x)$ is a continuous map from X to its dual X^*.

Corollary 1.2: *Let X be a Banach space and let $\varphi : X \to \mathbb{R}$ be a function that is bounded from below and differentiable on X. Then, for each $\varepsilon > 0$ and for each $u \in X$ such that $\varphi(u) \leq \inf_X \varphi + \varepsilon^2$, there exists $v_\varepsilon \in X$ such that*

  (i) $\varphi(v_\varepsilon) \leq \varphi(u)$.
  (ii) $\|u - v_\varepsilon\| \leq \varepsilon$.
  (iii) $\|\varphi'(v_\varepsilon)\| \leq \varepsilon$.

Proof: For $\varepsilon > 0$ given, use Theorem 1.1 with $\lambda = 1/\varepsilon$ to find $v_\varepsilon \in X$ such that (i), (ii) hold as well as

$$\varphi(w) > \varphi(v_\varepsilon) - \varepsilon\|v_\varepsilon - w\| \tag{$*$}$$

for all $w \neq v_\varepsilon$ in X. By applying $(*)$ to $w = v_\varepsilon + th$ with $t > 0, h \in X, \|h\| = 1$, we get $\varphi(v_\varepsilon + th) - \varphi(v_\varepsilon) > -\varepsilon t$. Dividing both sides by t and letting $t \to 0$, we obtain $-\varepsilon \leq \langle \varphi'(v_\varepsilon), h \rangle$ for all $h \in X$ with $\|h\| = 1$, and hence (iii) is verified.

Corollary 1.3: *Let X be a Banach space and let $\varphi : X \to \mathbb{R}$ be a function that is bounded from below and differentiable on X. Then, for each minimizing sequence (u_k) for φ, there exists a minimizing sequence (v_k) for φ such that $\varphi(v_k) \leq \varphi(u_k)$, $\lim_k \|u_k - v_k\| = 0$ and $\lim_k \varphi'(v_k) = 0$.*

Proof: If (u_k) is a minimizing sequence for φ, take

$$\varepsilon_k = \begin{cases} \varphi(u_k) - \inf_X \varphi & \text{if } \varphi(u_k) - \inf_X \varphi > 0 \\ 1/k & \text{if } \varphi(u_k) - \inf_X \varphi = 0 \end{cases}$$

and then take v_k associated to u_k and ε_k in Corollary 1.2.

Definition 1.4: Say that φ verifies the *Palais-Smale condition at the level* c (in short $(PS)_c$), if any sequence $(x_n)_n$ satisfying $\lim_n \varphi(x_n) = c$ and $\lim_n \|\varphi'(x_n)\| = 0$ has a convergent subsequence.

Throughout this monograph, we shall denote by K_c the set of critical points at level c, i.e.,

$$K_c = \{x \in X; \varphi(x) = c, d\varphi(x) = 0\}.$$

The following two corollaries show that the (PS) condition is quite restrictive. In particular, it forces the function to be *coercive*, i.e., $\liminf_{\|u\| \to \infty} \varphi(u) = \infty$.

Corollary 1.5: *Let φ be C^1-function on a Banach space X.*
  (i) *If φ is bounded below and verifies $(PS)_c$ with $c = \inf_X \varphi$, then every minimizing sequence for φ is relatively compact. In particular, φ achieves its minimum at a point in K_c.*
 (ii) *If $d = \liminf_{\|u\| \to \infty} \varphi(u)$ is finite , then φ does not verify $(PS)_d$.*

Proof: (i) follows immediately from Corollary 1.3. For (ii), we shall show the existence of a sequence $(u_n)_n$ in X such that $\|u_n\| \to \infty$, $\varphi(u_n) \to d$ and $\|\varphi'(u_n)\| \to 0$.
  For that, define for $r \geq 0$ the function

$$m(r) = \inf_{\|u\| \geq r} \varphi(u).$$

Clearly $m(r)$ is nondecreasing and $\lim_{r \to \infty} m(r) = d$. For $\varepsilon < 1/2$, find $r_0 \geq 1/\varepsilon$ such that $d - \varepsilon^2 \leq m(r)$ for $r \geq r_0$, then choose u_0 with $\|u_0\| \geq 2r_0$ such that

$$\varphi(u_0) < m(2r_0) + \varepsilon^2 \leq d + \varepsilon^2.$$

Apply now Ekeland's theorem in the region $D = \{\|u\| \geq r_0\}$, to find v_0 with $\|v_0\| \geq r_0$ such that

$$\varphi(v_0) \leq \varphi(u) - \varepsilon\|u - v_0\| \quad \text{for all } u \in D.$$

It follows that

$$d - \varepsilon^2 \leq m(r_0) \leq \varphi(v_0) \leq \varphi(u_0) - \varepsilon\|u_0 - v_0\|.$$

Hence $\|u_0 - v_0\| \leq 2\varepsilon$ and $\|v_0\| > r_0$. Since v_0 belongs to the interior of the region D, the argument in Corollary 1.2 gives that $\|\varphi'(v_0)\| \leq \varepsilon$.

Corollary 1.6: *Let φ be C^1-function satisfying the (PS) condition on a Banach space X. If u_0 is a local minimum for φ, then there is $\varepsilon > 0$ such that the following alternative holds:*

(i) Either $\varphi(u_0) < \inf\{\varphi(u) : \|u - u_0\| = \alpha\}$ for some $0 < \alpha < \varepsilon$,

(ii) or for each α with $0 < \alpha < \varepsilon$, φ has a local minimum at a point u_α with $\|u_\alpha - u_0\| = \alpha$ and $\varphi(u_\alpha) = \varphi(u_0)$.

Proof: Let $\varepsilon > 0$ be such that $\varphi(u_0) \leq \varphi(u)$ for $\|u - u_0\| \leq \varepsilon$. If (i) does not hold, then for any given α with $0 < \alpha < \varepsilon$, we have

$$\varphi(u_0) = \inf\{\varphi(u) : \|u - u_0\| = \alpha\} \tag{1}$$

Let $\delta > 0$ be such that $0 < \alpha - \delta < \alpha + \delta < \varepsilon$ and consider the restriction of φ to the ring $R = \{u \in X : \alpha - \delta \leq \|u - u_0\| \leq \alpha + \delta\}$. From (1), we can find a sequence $(u_n)_n$ in R such that

$$\|u_n - u_0\| = \alpha \text{ and } \varphi(u_n) \leq \varphi(u_0). \tag{2}$$

By Ekeland's theorem, we can then find $(v_n)_n$ in R such that

$$\varphi(v_n) \leq \varphi(u_n), \quad \|u_n - v_n\| \leq \frac{1}{n} \tag{3}$$

and

$$\varphi(v_n) \leq \varphi(u) + \frac{1}{n}\|u - u_n\| \text{ for all } u \in R. \tag{4}$$

If n is large enough, we get from (3) that v_n is in the interior of R, which then implies that $\|\varphi'(v_n)\| \leq \frac{1}{n}$. The (PS) condition now ensures that a subsequence of the $(v_n)_n$ converges to a point u_α. It is clear that φ has a local minimum at u_α and that $\|u_\alpha - u_0\| = \alpha$ while $\varphi(u_\alpha) = \varphi(u_0)$.

Remark 1.7: If u_0 is a strict local minimum, then clearly alternative (ii) cannot hold, and we then obtain an $\alpha > 0$ such that

$$\varphi(u_0) < \inf\{\varphi(u) : \|u - u_0\| = \alpha\}.$$

1.2 Constrained minimization and global critical points

The (PS) condition being so restrictive, one can try to find pseudo-critical sequences with some additional properties that might help in proving their convergence. To do that, we introduce the following concept which will play a central role throughout this monograph.

Definition 1.8: Say that φ verifies the *Palais-Smale condition at the level c and around the set F* (in short, $(PS)_{F,c}$) if any sequence $(x_n)_n$ in X verifying $\lim_n \varphi(x_n) = c$, $\lim_n \|\varphi'(x_n)\| = 0$ and $\lim_n \text{dist}(x_n, F) = 0$ has a convergent subsequence.

The most naive approach for finding such points consists of minimizing the functional φ on the submanifold F and to check whether such a relative minimum is a global critical point for φ. We have observed that this is the case if, for instance, the points obtained via Ekeland's theorem are in the interior of the constraint set. In the sequel, we shall present other settings where this method can apply. On the other hand, in Chapter 4, we shall implement a global (unconstrained) variational principle that will yield almost critical points that are close to certain *dual* sets. These sets can have an empty interior, and do not need to have any differentiable structure. We refer to Chapter 8 for some examples.

For now, we shall only consider situations where constraint minimization yields global critical points. We denote by $B_\varepsilon(X)$ the ball in the Banach space X centered at 0 and of radius ε. We start with the following:

Lemma 1.9: *Let φ be a C^1-functional on a Hilbert space H and let F be a subset of H verifying the following property:*

For any $u \in F$ with $\varphi'(u) \neq 0$, there exists, for a small enough $\varepsilon > 0$, a Fréchet differentiable function $s_u : B_\varepsilon(H) \to \mathbb{R}$ such that, by setting $t_u(\delta) = s_u\left(\delta\frac{\varphi'(u)}{\|\varphi'(u)\|}\right)$ for $0 \leq \delta \leq \varepsilon$, we have

$$t_u(0) = 1 \text{ and } t_u(\delta)(u - \delta\frac{\varphi'(u)}{\|\varphi'(u)\|}) \in F.$$

If φ is bounded below on F, then for every minimizing sequence (v_n) in F for φ, there exists $(u_n)_n$ in F such that

$$\varphi(u_n) \leq \varphi(v_n), \quad \lim_n \|u_n - v_n\| = 0 \text{ and}$$

$$\|\varphi'(u_n)\| \leq \frac{1}{n}(1 + \|u_n\|\, |t'_{u_n}(0)|) + |t'_{u_n}(0)|\, |\langle\varphi'(u_n), u_n\rangle|. \quad (*)$$

Proof: Let $c = \inf \varphi(F)$. Use Ekeland's theorem to get a minimizing sequence $(u_n)_n$ in F with the following properties

(i) $\varphi(u_n) \leq \varphi(v_n) < c + 1/n$.

(ii) $\lim_n \|u_n - v_n\| = 0$.

(iii) $\varphi(w) \geq \varphi(u_n) - \frac{1}{n}\|w - u_n\|$ for all $w \in F$.

Let us assume $\|\varphi'(u_n)\| > 0$ for n large, since otherwise we are done. Apply the hypothesis on the set F with $u = u_n$ to find $t_n(\delta) := s_{u_n}\left(\delta\frac{\varphi'(u_n)}{\|\varphi'(u_n)\|}\right)$ such that $w_\delta = t_n(\delta)\left(u_n - \delta\frac{\varphi'(u_n)}{\|\varphi'(u_n)\|}\right) \in F$ for all small enough δ.

Use now the mean value theorem to get

$$\frac{1}{n}\|w_\delta - u_n\| \geq \varphi(u_n) - \varphi(w_\delta)$$

$$= (1 - t_n(\delta))\langle\varphi'(w_\delta), u_n\rangle$$

$$+ \delta t_n(\delta)\langle\varphi'(w_\delta), \frac{\varphi'(u_n)}{\|\varphi'(u_n)\|}\rangle + o(\delta)$$

where $o(\delta)/\delta \to 0$ as $\delta \to 0$. Dividing by $\delta > 0$ and passing to the limit as $\delta \to 0$ we derive

$$\frac{1}{n}(1 + |t'_n(0)|\, \|u_n\|) \geq -t'_n(0)\langle\varphi'(u_n), u_n\rangle + \|\varphi'(u_n)\|,$$

which is our claim $(*)$.

In order to obtain *global* critical points via the above constrained minimization procedure, we need to insure that the right hand side of formula $(*)$ above approaches 0. A step in that direction would be to consider the submanifold

$$M_\varphi = \{u \in H; \ u \neq 0 \text{ and } \psi(u) := \langle \varphi'(u), u \rangle = 0\}.$$

However, an additional (second order) condition is needed. Indeed, if we assume that φ actually attains its minimum at a point u_0 of M_φ, then there exists a Lagrange multiplier λ such that

$$\varphi'(u_0) = \lambda \psi'(u_0).$$

Applying now u_0 to both sides of this equation and using the fact that $\langle \varphi'(u_0), u_0 \rangle = 0$, we obtain that $\lambda = 0$ provided, of course that $\langle \psi'(u_0), u_0 \rangle \neq 0$. In the case where the minimum is not supposed to be a priori attained, the asymptotic property $\limsup_n |\langle \psi'(u_n), u_n \rangle| > 0$ seems to be the natural hypothesis as the following proof illustrates.

We can now prove the following:

Corollary 1.10: *Let φ be a C^2-functional on a Hilbert space H that is coercive and bounded below on the set*

$$M_\varphi = \{u \in H; \ u \neq 0 \text{ and } \psi(u) := \langle \varphi'(u), u \rangle = 0\}.$$

Suppose $\langle \psi'(u), u \rangle \neq 0$ for any $u \in M_\varphi$ and that for any sequence $(u_n)_n$ in M_φ that is minimizing for φ on M_φ, we have that $(\psi'(u_n))_n$ is bounded in H^* and $\limsup_n |\langle \psi'(u_n), u_n \rangle| > 0$.

Then, for every minimizing sequence (v_n) for φ on M_φ, there exists a sequence $(u_n)_n$ in M_φ such that $\varphi(u_n) \leq \varphi(v_n)$, $\lim_n \|u_n - v_n\| = 0$ and $\lim_n \|\varphi'(u_n)\| = 0$.

In particular, if φ verifies $(PS)_{M_\varphi,c}$ where $c = \inf \varphi(M_\varphi)$, then K_c is nonempty.

Proof: We first show that M_φ verifies the hypothesis of Lemma 1.9. Indeed, for a fixed $u \in M_\varphi$, define $G : \mathbb{R} \times H \to \mathbb{R}$ by $G(s,w) = \psi(s(u - w))$. Since $G(1,0) = 0$ and $G_s(1,0) \neq 0$, we can apply the implicit function theorem at the point (1,0) and get that for $\varepsilon > 0$ small enough, there exists a differentiable function $s_u : B_\varepsilon(H) \to \mathbb{R}^+$ such that

(i) $s_u(0) = 1$, $s_u(w)(u - w) \in M_\varphi$ for $w \in B_\varepsilon(H)$, and

(ii) $\langle s'_u(0), w \rangle = \frac{\langle \psi'(u), w \rangle}{\langle \psi'(u), u \rangle}$.

Apply now Lemma 1.9 to obtain a sequence $(u_n)_n$ in M_φ verifying $\varphi(u_n) \leq \varphi(v_n)$, $\lim_n \|u_n - v_n\| = 0$ and $(*)$. Since φ is coercive, we can assume that $(u_n)_n$ is bounded by a constant C, so that we have

$$\|\varphi'(u_n)\| \leq \frac{1}{n}(1 + C|t'_n(0)|) + t'_n(0) \langle \varphi'(u_n), u_n \rangle.$$

Note now that $\langle \varphi'(u_n), u_n \rangle = 0$ for each n and that

$$t'_n(0) = \langle s'_{u_n}(0), \frac{\varphi'(u_n)}{\|\varphi'(u_n)\|} \rangle = \frac{\langle \psi'(u_n), \frac{\varphi'(u_n)}{\|\varphi'(u_n)\|} \rangle}{\langle \psi'(u_n), u_n \rangle}$$

which is uniformly bounded. Hence $\|\varphi'(u_n)\| \to 0$.

Corollary 1.11: *Let φ be a C^1-functional on a Hilbert space H. Let F be a closed convex subset of H such that $(I - \varphi')F \subset F$. If φ is bounded below on F, then for every minimizing sequence (v_k) in F, there exists a sequence (u_k) in F such that $\varphi(u_k) \leq \varphi(v_k)$, $\lim_k \|u_k - v_k\| = 0$ and $\lim_k \|\varphi'(u_k)\| = 0$.*

In particular, if φ verifies $(PS)_{F,c}$ where $c = \inf_F \varphi$, then $K_c \cap F$ is nonempty.

Proof: Let t be identically equal to 1 on H. Assume $u \in F$ with $\varphi'(u) \neq 0$. Then by the convexity of F, we get for any $0 < \delta < \|\varphi'(u)\|$ that

$$u - \delta \frac{\varphi'(u)}{\|\varphi'(u)\|} = (1 - \frac{\delta}{\|\varphi'(u)\|})u + \frac{\delta}{\|\varphi'(u)\|}(I - \varphi')u \in F.$$

Hence F verifies the hypothesis of Lemma 1.9. The conclusion now follows from $(*)$ and the fact that $t'_{u_n}(0) = 0$.

1.3 Nonhomogeneous elliptic equations involving the critical Sobolev exponent

Consider the Dirichlet problem

$$\begin{cases} -\Delta u = |u|^{p-2}u + f & \text{on } \Omega \\ u = 0 & \text{on } \partial\Omega. \end{cases} \tag{1}$$

where $\Omega \subset \mathbb{R}^N$ ($N \geq 3$) is a bounded domain, $H = H^1_0(\Omega)$, $f \in H^{-1}$ and $p = 2^* = \frac{2N}{N-2}$ is the limiting exponent in the Sobolev embedding. (Appendix A).

Weak solutions for this problem are the critical points of the functional

$$\varphi(u) = \frac{1}{2} \int_\Omega |\nabla u|^2 \, dx - \frac{1}{p} \int_\Omega |u|^p \, dx - \int_\Omega f u \, dx$$

defined on H. (Appendix B)

If $f = 0$, then (1) does not generally have non-trivial solutions. (See for instance Struwe [St] for a discussion of this phenomenon.) In the following, we shall present a result of Tarantello [T1], showing that solutions may exist if $f \neq 0$. The first solution will be obtained via Corollary 1.10 and will turn out to be a local minimizer for φ. We shall come back to this problem in Chapter 8, in order to find another solution which is not a local minimum for φ.

Consider the manifold $M_\varphi = \{u \in H : u \neq 0, \langle \varphi'(u), u \rangle = 0\}$. We now prove the following result:

Theorem 1.12: *Suppose $f \neq 0$ and satisfies the following condition*

$$\inf_{\|u\|_p=1} (K_N \|\nabla u\|_2^{\frac{N+2}{2}} - \int_\Omega fu) =: \mu_0 > 0 \tag{2}$$

where $K_N = \frac{4}{N-2} \left(\frac{N-2}{N+2}\right)^{\frac{N+2}{4}}$. Then $\inf \varphi(M_\varphi) = c_1$ is finite and is achieved at a point $u \in M_\varphi$ which is a critical point for φ.

Remark 1.13: Actually, G. Tarantello [T1] proved the above result (and more) assuming only that $\mu_0 \geq 0$ in (2). Notice that this assumption holds if

$$\|f\|_{H^{-1}} \leq K_N S^{N/2}$$

where S is the best Sobolev constant (see Chapter 8 and Appendix A). We shall not include this refinement since it is not relevant to our discussion.

Proof of Theorem 1.12: It consists of verifying that the set M_φ satisfies the hypothesis of Corollary 1.10. First, we show the following:

Claim (i): φ is coercive and bounded below on M_φ.

Indeed, for $u \in M_\varphi$, we have

$$\psi(u) := \langle \varphi'(u), u \rangle = \int_\Omega |\nabla u|^2 - \int_\Omega |u|^p - \int_\Omega fu = 0.$$

Thus

$$\begin{aligned}
\varphi(u) &= \frac{1}{2}\int_\Omega |\nabla u|^2 - \frac{1}{p}\int_\Omega |u|^p - \int_\Omega fu \\
&= \frac{1}{N}\int_\Omega |\nabla u|^2 - (1 - \frac{1}{p})\int_\Omega fu \\
&\geq \frac{1}{N}\|\nabla u\|_2^2 - \frac{N+2}{2N}\|f\|_{H^{-1}}\|\nabla u\|_2
\end{aligned}$$

which yields our first claim.

Claim (ii): For every $u \in M_\varphi$, we have

$$\langle \psi'(u), u \rangle = \|\nabla u\|_2^2 - (p-1)\|u\|_p^p \neq 0.$$

To do that define for $u \neq 0$ the function

$$\chi(u) = K_N \frac{\|\nabla u\|_2^{\frac{N+2}{2}}}{\|u\|_p^{\frac{N}{2}}} - \int_\Omega fu.$$

For any $t > 0$, $\|u\|_p = 1$ we have

$$\chi(tu) = t\left[K_N \|\nabla u\|_2^{\frac{N+2}{2}} - \int_\Omega fu\right].$$

For any $\gamma > 0$, we derive from assumption (2) that

$$\inf_{\|u\|\geq\gamma} \chi(u) \geq \gamma\mu_0. \tag{3}$$

Suppose now that for some $u \in M_\varphi$ we have

$$\|\nabla u\|_2^2 - (p-1)\|u\|_p^p = 0. \tag{4}$$

Then the definition of M_φ and (4) yield

$$0 = \|\nabla u\|_2^2 - \|u\|_p^p - \int_\Omega fu = (p-2)\|u\|_p^p - \int_\Omega fu. \tag{5}$$

Condition (4) and the Sobolev inequality imply

$$\|u\|_p \geq \left(\frac{S}{p-1}\right)^{\frac{1}{p-2}} =: \gamma.$$

From (3) and (5) we obtain

$$0 < \mu_0 \gamma \leq \chi(u) = \left(\frac{1}{p-1}\right)^{\frac{p-1}{p-2}} (p-2) \left(\frac{\|\nabla u\|_2^{2(p-1)}}{\|u\|_p^p}\right)^{\frac{1}{p-2}} - \int_\Omega fu$$

$$= (p-2) \left(\left(\frac{1}{p-1}\right)^{\frac{p-1}{p-2}} \left(\frac{\|\nabla u\|_2^{2(p-1)}}{\|u\|_p^p}\right)^{\frac{1}{p-2}} - \|u\|_p^p \right)$$

$$= (p-2)\|u\|_p^p \left(\left(\frac{\|\nabla u\|_2^2}{(p-1)\|u\|_p^p}\right)^{\frac{p-1}{p-2}} - 1 \right) = 0$$

which is a contradiction.

Claim (iii): $c_1 = \inf \varphi(M_\varphi) < 0$.

Let $v \in H$ be the unique solution for $-\Delta u = f$. Since $f \neq 0$, then $\int_\Omega fv = \|\nabla v\|_2^2 > 0$. We claim that

$$\text{there exists } t_0 > 0 \text{ such that } t_0 v \in M_\varphi. \tag{6}$$

Indeed, the function $\ell(t) = t\|\nabla v\|_2^2 - t^{p-1}\|v\|_p^p$ is concave and achieves its maximum at

$$t_{\max} = \left(\frac{\|\nabla v\|_2^2}{(p-1)\|v\|_p^p}\right)^{\frac{1}{p-2}}.$$

Also

$$\ell(t_{\max}) = \left(\frac{1}{p-1}\right)^{\frac{p-1}{p-2}} (p-2) \left(\frac{\|\nabla v\|_2^2}{(p-1)\|v\|_p^p}\right)^{\frac{1}{p-2}} = K_N \frac{\|\nabla v\|_2^{\frac{N+2}{2}}}{\|v\|_p^{N/2}}.$$

Since by (2) we have

$$\ell(0) = 0 < \int_\Omega fv < K_N \frac{\|\nabla v\|_2^{\frac{N+2}{2}}}{\|v\|_p^{N/2}} = \ell(t_{\max}),$$

we may deduce the existence of a (unique!) $0 < t_0 < t_{\max}$ such that

$\ell(t_0) = \int_\Omega fv$ and therefore $t_0 v \in M_\varphi$. Consequently,

$$\varphi(t_0 v) = \frac{t_0^2}{2}\|\nabla v\|_2^2 - \frac{t_0^p}{p}\|v\|_p^p - t_0\|\nabla v\|_2^2$$

$$= -\frac{t_0^2}{2}\|\nabla v\|_2^2 + \frac{p-1}{p}t_0^p\|v\|_p^p$$

$$< -\frac{t_0^2}{N}\|\nabla v\|_2^2 = -\frac{t_0^2}{N}\|f\|_{H^{-1}}^2$$

This yields $c_1 < -\frac{t_0^2}{N}\|f\|_{H^{-1}}^2 < 0$ and Claim (iii) is proved.

Claim (iv): If $(u_n)_n$ is a minimizing sequence for φ in M_φ, then
$$\liminf_n |\langle \psi'(u_n), u_n \rangle| > 0.$$

Since φ is coercive, we can assume $(u_n)_n$ uniformly bounded. Moreover, since $\varphi(0) = 0$ and $c_1 < 0$, we can assume, modulo passing to a subsequence, that for some $\delta > 0$,

$$\|\nabla u_n\|_2 \geq \delta. \tag{7}$$

Suppose now that

$$\langle \psi'(u_n), u_n \rangle = \|\nabla u_n\|_2^2 - (p-1)\|u_n\|_p^p = o(1). \tag{8}$$

By combining (7) and (8) we get that for a suitable constant $\gamma > 0$, $\|u_n\|_p \geq \gamma$ and

$$\left(\frac{\|\nabla u_n\|_2^2}{p-1}\right)^{\frac{p-1}{p-2}} - \left(\|u_n\|_p^p\right)^{\frac{p-1}{p-2}} = o(1).$$

The fact that $u_n \in M_\varphi$, combined with (8), also gives

$$\int_\Omega fu_n = (p-2)\|u_n\|_p^p + o(1).$$

This, together with (3) implies

$$\mu_0\gamma^{\frac{N+2}{2}} \leq \|u_n\|_p^{\frac{p}{p-2}}\chi(u_n) = (p-2)\left[\left(\frac{\|\nabla u_n\|_2^2}{p-1}\right)^{\frac{p-1}{p-2}} - \left(\|u_n\|_p^p\right)^{\frac{p-1}{p-2}}\right]$$

which is $o(1)$. This is impossible and hence Claim (iv) is verified.

It now remains to prove the following

Claim (v): The functional φ verifies $(PS)_{M_\varphi, c_1}$ on the space $H_0^1(\Omega)$.

Indeed, suppose $(u_n)_n$ is a sequence in M_φ such that $\lim_n \varphi(u_n) = c_1$ and $\lim_n \|\varphi'(u_n)\| = 0$, then $(u_n)_n$ is necessarily bounded and we can find a weak cluster point u in $H_0^1(\Omega)$. We can assume that $(u_n)_n$ converges weakly to u and, by the Rellich-Kondrakov theorem (Appendix A), that $u_n \to u$ strongly in $L^p(\Omega)$ for all $p < 2^*$. In particular, for any $v \in C_0^\infty(\Omega)$

$$\langle v, \varphi'(u_n) \rangle = \int_\Omega (\nabla u_n \nabla v - fv - u_n|u_n|^{2^*-2}v)\, dx$$

which converges as $n \to \infty$ to

$$\int_\Omega (\nabla u \nabla v - fv - u|u|^{2^*-2}v) \, dx = \langle v, \varphi'(u) \rangle$$

Hence $\langle \varphi'(u), v \rangle = 0$ for all $v \in H^{-1}$ which means that u is a weak solution for (1) and therefore $u \neq 0$. In particular, $u \in M_\varphi$. Since φ is weakly lower semi-continuous, we get

$$c_1 \leq \varphi(u) = \frac{1}{N}\|\nabla u\|_2^2 - (1 - \frac{1}{p})\int_\Omega fu \leq \lim_{n \to +\infty} \varphi(u_n) = c_1.$$

It follows that $\varphi(u) = c_1 = \inf \varphi(M_\varphi)$ and that $\|u_n\| \to \|u\|$ which implies that $u_n \to u$ strongly in $H_0^1(\Omega)$.

Remark 1.14: We shall prove in Chapter 8 that u is actually a local minimum for φ on the whole space H and that $\|\nabla u\|_2^2 - (p-1)\|u\|_p^p \geq 0$. If we consider the following splitting for M_φ

$$M_\varphi^+ = \{u \in M_\varphi : \|\nabla u\|_2^2 - (p-1)\|u\|_p^p > 0\}$$
$$M_\varphi^0 = \{u \in M_\varphi : \|\nabla u\|_2^2 - (p-1)\|u\|_p^p = 0\}$$
$$M_\varphi^- = \{u \in M_\varphi : \|\nabla u\|_2^2 - (p-1)\|u\|_p^p < 0\}.$$

we get that $c_1 = \inf \varphi(M_\varphi) = \inf \varphi(M_\varphi^+)$ and from Claim (ii) above that $M_\varphi^0 = \emptyset$. In order to find another critical point, we can investigate the minimization problem

$$c_2 = \inf \varphi(M_\varphi^-).$$

The same reasoning as above shows that M_φ^- verifies the hypothesis of Corollary 1.9, and therefore there exists a sequence $(v_n)_n$ in M_φ^- such that $\varphi(v_n) \to c_2$ and $\varphi'(v_n) \to 0$. However, the Palais-Smale condition at the level c_2 is a more subtle problem and we shall come back to it in Chapter 8.

1.4 Weak sub-solutions and super-solutions

Let Ω be a smooth domain in $\mathbb{R}^N$ with $N \geq 3$ and consider the equation

$$-\Delta u = f(x, u) \quad \text{in } \Omega \quad \text{and} \quad u = u_0 \text{ on } \partial\Omega. \tag{1}$$

where $u_0 \in H^1(\Omega)$ and $f : \Omega \times \mathbb{R} \to \mathbb{R}$ is a Caratheodory function: that is a function satisfying
  (i) for each $s \in \mathbb{R}$, $x \to f(x, s)$ is measurable on Ω, and
  (ii) for almost every $x \in \Omega$, $s \to f(x, s)$ is continuous.

Definition 1.15: A function $u \in H^1(\Omega)$ is said to be a (weak) *sub-solution* to (1) if $u \leq u_0$ on $\partial\Omega$ and if

$$\int_\Omega \nabla u \nabla \ell \, dx - \int_\Omega g(\cdot, u)\ell \, dx \leq 0 \quad \text{for all } \ell \in C_0^\infty(\Omega), \, \ell \geq 0.$$

Similarly $u \in H^1(\Omega)$ is a (weak) *super-solution* to (1) if the reverse inequalities hold.

Theorem 1.16: *Let f be a Caratheodory function on $\Omega \times \mathbb{R}$, and suppose $u_1 \in H^1(\Omega)$ is a bounded sub-solution while $u_2 \in H^1(\Omega)$ is a bounded super-solution to problem (1) such that $u_1 \leq u_2$ almost everywhere (a.e.) in Ω. Then there exists a weak solution $u \in H^1(\Omega)$ of (1), such that $u_1 \leq u \leq u_2$ a.e. in Ω.*

Proof: Even though the theorem holds as stated, we shall prove it under the following additional conditions on f since they are usually satisfied in most applications. (See also Remark 1.17).

(iii) $|f(x,s)| \leq C|s|^{p-1} + b(x)$ where $C > 0$, $1 \leq p \leq \frac{2N}{N-2}$ and $b \in L^{p^*}$ ($\frac{1}{p} + \frac{1}{p^*} = 1$), and

(iv) there exists $L > 0$ such that the function $g(x,s) = f(x,s) + Ls$ is increasing in s on $[a,b]$ where $a = \inf_\Omega u_1$ and $b = \sup_\Omega u_2$.

We shall also assume with no loss of generality that $u_0 = 0$. Define now the following equivalent norm on $H_0^1(\Omega)$

$$\|u\|^2 = L\|u\|_2^2 + \|\nabla u\|_2^2$$

and let $\langle .,. \rangle$ be the corresponding inner product. Let φ be the functional defined by

$$\varphi(u) = \frac{1}{2}\langle u, u \rangle - \int_\Omega G(x,u) \, dx$$

where $G(x,s) = \int_0^s g(x,t) \, dt$ is a primitive of g. It is clear that φ is a C^1-functional on $H_0^1(\Omega)$ whose critical points are weak solutions for equation (1). Consider

$$F = \{u \in H_0^1(\Omega); u_1 \leq u \leq u_2 \text{ almost everywhere}\}.$$

This set is closed and convex in $H_0^1(\Omega)$. To show that $(I - \varphi')F \subset F$, consider $u \in F$ and let $v = (I - \varphi')u$. For any $\ell \in C_0^\infty(\Omega)$, $\ell \geq 0$, we have

$$\langle v, \ell \rangle = \langle u, \ell \rangle - \langle u, \ell \rangle + \int_\Omega g(x,u)\ell(x) \, dx$$

and hence

$$\langle v - u_1, \ell \rangle \geq \int_\Omega [g(x,u) - g(x,u_1)]\ell(x) \, dx \geq 0$$

which, in view of the maximum principle, implies that $u_1 \leq v$ almost everywhere. The same reasoning gives that $v \leq u_2$ almost everywhere. Therefore $(I - \varphi')F \subset F$.

On the other hand, $\varphi(u) \geq \frac{1}{2}\langle u, u \rangle - K_1$ on F, where K_1 is some constant. Hence φ is bounded below on F and Corollary 1.11 applies to yield a minimizing sequence $(u_n)_n$ in F, such that $\varphi'(u_n) \to 0$.

To prove that φ is weakly lower semi-continuous on F, suppose that $v_m \rightharpoonup v$ weakly in $H_0^1(\Omega)$, where $v_m, v \in F$. By passing to a subsequence, if necessary, we may assume that $v_m \to v$ pointwise almost everywhere and since $|G(x, v_m(x))| \leq K_2$ uniformly, we get from Lebesgue's dominated convergence theorem that

$$\int_\Omega G(x, v_m) \, dx \to \int_\Omega G(x, v) \, dx.$$

To show that φ verifies $(PS)_{F,c}$ where $c = \inf_F \varphi$, it is enough to consider a minimizing sequence $(u_n)_n$ for φ in F, such that $\varphi'(u_n) \to 0$. If u is a weak cluster point for $(u_n)_n$, it is clear that $u \in F$ and, since φ is weakly lower semi-continuous on F, we have

$$\inf_F \varphi \leq \varphi(u) = \frac{1}{2}\langle u, u \rangle - \int_\Omega G(x, u) \, dx \leq \lim_{n \to \infty} \varphi(u_n) = \inf_F \varphi$$

It follows that $\|u_n\| \to \|u\|$ and therefore $u_n \to u$ strongly in $H_0^1(\Omega)$. Consequently u is a critical point for φ and hence a solution in F for (1).

Remark 1.17: Without assumption (iii) on the function $f(x, s)$, the functional φ is not necessarily defined on all of $H^1(\Omega)$ but φ is defined on F since the latter is a bounded subset of L^∞. Corollary 1.11 cannot be applied (as stated above) and one has to verify directly that a minimum for φ on F still gives a weak solution for (1).

On the other hand, if u_1 and u_2 are supposed to be in C^2, then one can avoid assumption (iii) by altering the function f in the following way: define

$$\tilde{f}(x, s) = \begin{cases} \min\{f(x, u_2(x)), -\Delta u_2(x)\} & \text{if } s > u_2(x) \\ f(x, s) & \text{if } u_1(x) \leq s \leq u_2(x) \\ \max\{f(x, u_1(x)), -\Delta u_1(x)\} & \text{if } s < u_1(x). \end{cases}$$

It is clear that $\tilde{f}$ verifies the hypothesis of Theorem 1.16. Moreover, a standard application of the maximum principle yields that any minimum for the functional associated to $\tilde{f}$ is necessarily a solution for (1) in the set F. The details are left to the reader.

As an application of the above, consider the following equation on a smooth bounded domain Ω in $\mathbb{R}^N$, $(N \geq 3)$,

$$-\Delta u = \alpha(x)u - u|u|^{p-2}, \quad u = u_0 \text{ on } \partial\Omega \tag{2}$$

where $p = \frac{2N}{N-2}$, and where α is a continuous function such that $1 \leq \alpha(x) \leq K < \infty$ on Ω. Suppose $u_0 \in C^1(\bar{\Omega})$ satisfies $u_0 \geq 1$ on $\partial\Omega$. Then $u_1 \equiv 1$ is a sub-solution while $u_2 \equiv M$ for large $M > 1$ is a super-solution to equation (2). Consequently, Theorem 1.16 applies and we get that (2) admits a solution u larger than 1.

1.5 A minimization principle with quadratic perturbations

Let (X, d) be a complete metric space and consider the class $\mathcal{Q}$ of all

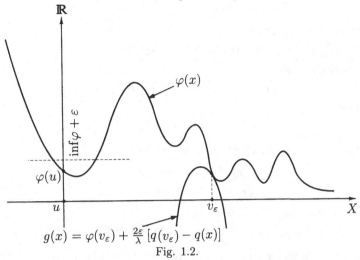

$$g(x) = \varphi(v_\varepsilon) + \tfrac{2\varepsilon}{\lambda}\,[q(v_\varepsilon) - q(x)]$$

Fig. 1.2.

real-valued functions on X of the form

$$q(x) = \frac{1}{2}\sum_{n=1}^{\infty}\mu_n d(x, v_n)^2$$

where $\mu_n \geq 0$, $\sum_{n=1}^{\infty}\mu_n = 1$ and where $(v_n)_n$ is some convergent sequence in X.

We shall now present a result of Borwein and Preiss stating that the Lipschitz perturbations obtained in Ekeland's theorem can be replaced by *quadratic* perturbations in the class $\mathcal{Q}$. The relevance of this improvement becomes clear in the case where X is a Banach space admitting a differentiable norm (away from the origin), since the functions in the class $\mathcal{Q}$ are then differentiable everywhere.

Theorem 1.18 (Fig 1.2): *Let (X, d) be a complete metric space and let $\varphi : X \to (-\infty, +\infty]$ be a lower semi-continuous function that is bounded from below and not identical to $+\infty$.. Fix $\varepsilon > 0$, $\lambda > 0$ and assume that $u \in X$ satisfies $\varphi(u) < \inf_X \varphi + \varepsilon$. Then there exist $q \in \mathcal{Q}$ and $v_\varepsilon \in X$ such that*

(i) $\varphi(v_\varepsilon) < \inf_X \varphi + \varepsilon$.
(ii) $d(u, v_\varepsilon) < \lambda$.
(iii) For all $x \in X$, $\varphi(x) + 2\tfrac{\varepsilon}{\lambda^2}q(x) \geq \varphi(v_\varepsilon) + 2\tfrac{\varepsilon}{\lambda^2}q(v_\varepsilon)$.

Moreover, if X is a Banach space with a smooth norm (off the origin) and if φ is differentiable, then $\|\varphi'(v_\varepsilon)\| \leq \tfrac{2\varepsilon}{\lambda}$.

Proof: We construct q iteratively, as follows. Fix constants ε_1 and ε_2 with

$$\varphi(u) - \inf_X \varphi < \varepsilon_2 < \varepsilon_1 < \varepsilon \tag{1}$$

and then select μ with

$$0 < \mu < 1 - \varepsilon_1/\varepsilon. \qquad (2)$$

Next, choose ν with

$$0 < \nu/\mu < [1 - (\varepsilon_2/\varepsilon_1)^{1/2}]^2, \qquad (3)$$

and finally let

$$\delta = (1 - \mu)\varepsilon/\lambda^2.$$

We define sequences $\{\varphi_n\}, \{v_n\}$ as follows

$$\varphi_0 = \varphi, \qquad v_0 = u$$

$$\varphi_{n+1}(x) = \varphi_n(x) + \delta\mu^n d(x, v_n)^2, \qquad (4)$$

and then choose v_{n+1} in X so that

$$\varphi_{n+1}(v_{n+1}) \le \nu\varphi_n(v_n) + (1 - \nu)\inf_X \varphi_{n+1}. \qquad (5)$$

To see that this latter choice is possible, let $s_n = \inf_X \varphi_n, a_n = \varphi_n(v_n)$ and note that $a_n = \varphi_{n+1}(v_n)$. We want $\varphi_{n+1}(v_{n+1}) \le \nu a_n + (1 - \nu)s_n$. Either $a_n = \varphi_n(v_n) > s_n$, in which case the right-hand side above is greater than s_n, or $\varphi_{n+1}(v_n) = s_n$, in which case we can take $v_{n+1} = v_n$. Note that $\varphi_n \le \varphi_{n+1}$ so that $s_n \le s_{n+1} \le a_{n+1} \le \nu a_n + (1 - \nu)s_{n+1}$ and hence

$$0 \le a_{n+1} - s_{n+1} \le \nu(a_n - s_{n+1}) \le \nu(a_n - s_n) \le \dots \le \nu^{n+1}(a_0 - s_0). \qquad (6)$$

From (4), with $x = v_{n+1}$, we obtain

$$a_{n+1} = \varphi_n(v_{n+1}) + \delta\mu^n d(v_{n+1}, v_n)^2 \ge s_n + \delta\mu^n d(v_{n+1}, v_n)^2$$

and from (5)

$$a_{n+1} = \varphi_{n+1}(v_{n+1}) \le \varphi_n(v_n) = a_n.$$

It follows that

$$\delta\mu^n d(v_{n+1}, v_n)^2 \le a_n - s_n \le \nu^n(a_0 - s_0) < \nu^n \varepsilon_2$$

so that

$$d(v_{n+1}, v_n)^2 < (\nu^n \varepsilon_2)/(\delta\mu^n) \quad \text{and} \quad d(v_{n+1}, v_n) < (\nu/\mu)^{n/2}(\varepsilon_2/\delta)^{1/2}.$$

Using the fact that $\nu/\mu < 1$, we see that $\{v_n\}$ is a Cauchy sequence in X and for all n, m

$$d(v_m, v_n) \le (\varepsilon_2/\delta)^{1/2}/[1 - (\nu/\mu)^{1/2}]^{-1}. \qquad (7)$$

From (2) and (3) it follows that the right hand side of (7) is less that λ. Thus, if we let $v_\varepsilon = \lim_m v_m$, then $d(v_\varepsilon, v_n) < \lambda$ for all n; in particular, we obtain assertion (ii). The desired member of $\mathcal{Q}$ is defined by

$$q(x) = (1/2)\sum_n \mu_n d(x, v_n)^2, \quad \text{where} \quad \mu_n = \mu^n(1 - \mu). \qquad (8)$$

We now establish (iii). For x in the domain of φ we have, in view of (6) and $\varepsilon/\lambda^2 = \delta(1 - \mu)^{-1}$, that $\lim_n s_n = \lim_n a_n$,

$$\varphi(x) + (\varepsilon/\lambda^2)2q(x) = \sup_n \varphi_n(x) \ge \lim_n s_n = \lim_n a_n = \lim_n \varphi_n(v_n).$$

If $n \geq m$, then $\varphi_n \geq \varphi_m$, so $\varphi_n(v_n) \geq \varphi_m(v_n)$ and therefore

$$\varphi(x) + (\varepsilon/\lambda^2)2q(x) \geq \sup_m \liminf_{n \to \infty} \varphi_m(v_n) \geq \varphi_m(v_n).$$

Now, each φ_m is lower semi-continuous, so $\liminf_{n \to \infty} \varphi_m(v_n) \geq \varphi_m(v_\varepsilon)$ and therefore

$$\varphi(x) + (\varepsilon/\lambda^2)2q(x) \geq \sup_m \varphi_m(v_\varepsilon) = \varphi(v_\varepsilon) + (\varepsilon/\lambda^2)2q(v_\varepsilon), \qquad (9)$$

which is (iii). To establish (i), we first estimate $q(u)$. Since $u = v_0$ and $d(v_\varepsilon, v_n) < \lambda$ for all n, (8) then implies that $2q(u) < \mu\lambda^2$. We next use (9) to get

$$\varphi(u) + \varepsilon\mu > \varphi(u) + (\varepsilon/\lambda^2)2q(u) \geq \varphi(v_\varepsilon) + (\varepsilon/\lambda^2)2q(v) \geq \varphi(v_\varepsilon).$$

From (1) and (2) it follows that (i) holds:

$$\varphi(v_\varepsilon) \leq \varepsilon_1 + \mu\varepsilon + \inf_X \varphi < \varepsilon + \inf_X \varphi.$$

Suppose now that the norm is smooth. The function q is then differentiable at v_ε. Let $x^* = -(2\varepsilon/\lambda^2)q'(v_\varepsilon)$. It follows from (iii) that x^* is the derivative of φ at v_ε. It is also easy to see that $\|x^*\| \leq 2\varepsilon/\lambda$. This finishes the proof.

Remark 1.19: It is easy to see that if X is a Hilbert space or more generally an L^p space ($1 < p < \infty$), then the perturbation q can be taken to be of the form $q(x) = \frac{1}{2}\|x - w\|^2$ for some w usually not equal to v_ε.

The following corollary is now immediate:

Corollary 1.20: *Let φ be a C^2-functional that is bounded below on a Hilbert space X. Then, for each minimizing sequence $(u_k)_k$ of φ, there exists a minimizing sequence $(v_k)_k$ of φ such that*

(i) $\lim_k \|u_k - v_k\| = 0$

(ii) $\lim_k \|\varphi'(v_k)\| = 0$

(iii) $\liminf_k \langle \varphi''(v_k)w, w \rangle \geq 0$ *for any $w \in X$.*

1.6 The Hartree-Fock equations for Coulomb systems

We now give an example due to P.L Lions [L] where second order information on the almost critical sequences helps in proving their convergence. We only give a weak version of his result. It will be enough to illustrate the point we are stressing, without unnecessary complications.

Consider the purely Coulombic N-body Hamiltonian

$$H = -\sum_{i=1}^{N} \Delta_{x_i} + \sum_{i=1}^{N} V(x_i) + \sum_{i<j} \frac{1}{|x_i - x_j|} \qquad (1)$$

where $V(x) = -\sum_{j=1}^{m} z_j |x - \bar{x}_j|^{-1}$, $m \geq 1, z_j > 0, \bar{x}_j \in \mathbb{R}^3$ are fixed. We write $Z = \sum_{j=1}^{m} z_j$ for *the total charge of the nucleii*.

We consider the following minimization problem
$$\text{Inf } \{\varphi(u_1, ..., u_N); (u_1, ..., u_N) \in M\}, \qquad (2)$$
where φ, M are given by

$$\varphi(u_1, ..., u_N) = \sum_{i=1}^{N} \int_{\mathbb{R}^3} |\nabla u_i|^2 + V|u_i|^2 \, dx$$

$$+ \frac{1}{2} \iint_{\mathbb{R}^3 \times \mathbb{R}^3} \varrho(x) \frac{1}{|x-y|} \varrho(y) \, dx \, dy \qquad (3)$$

$$- \frac{1}{2} \iint_{\mathbb{R}^3 \times \mathbb{R}^3} \frac{1}{|x-y|} |\varrho(x,y)|^2 \, dx \, dy,$$

and

$$M = \{(u_1, ..., u_N) \in H^1(\mathbb{R}^3)^N; \int_{\mathbb{R}^3} u_i u_j^* \, dx = \delta_{ij}\}, \qquad (4)$$

where z^* denotes the conjugate of z, $\varrho(x) = \sum_{i=1}^{N} |u_i|^2(x)$ is the density and $\varrho(x,y) = \sum_{i=1}^{N} u_i(x)u_i^*(y)$ is the density matrix.

The Euler-Lagrange equations corresponding to the above problem (after a suitable diagonalization) are the following so-called Hartree-Fock equations: For $1 \le i \le N$,

$$-\Delta u_i + V u_i + (\rho * \frac{1}{|x|})u_i - \int_{\mathbb{R}^3} \rho(x,y) \frac{1}{|x-y|} u_i(y) \, dy + \varepsilon_i u_i = 0 \quad (5)$$

where $\lambda_i = -\varepsilon_i$ is a Lagrange multiplier and $(u_1, ..., u_N) \in M$.

Theorem 1.21: *Assume $Z > N$. Then, every minimizing sequence of the problem (2)–(4) is relatively compact in $(H^1(\mathbb{R}^3))^N$. In particular, there exists a minimum $(u_1, ..., u_N)$ that is (up to a unitary transformation) a solution to the Hartree-Fock equation (5).*

We shall need the following two lemmas.

Lemma 1.22: *Let A be a self-adjoint operator on a Hilbert space H and let H_1, H_2 be two subspaces such that $H = H_1 \oplus H_2, \dim H_1 = k < \infty$ and $P_2 A P_2 \ge 0$, where P_1, P_2 denote the orthogonal projections onto H_1, H_2 respectively. Then A has at most k negative eigenvalues.*

Proof: For simplicity, assume that A is bounded. Multiplying, if necessary, A by a positive constant, we may assume that $P_1 A P_1 \ge -P_1$. Set $\tilde{A} = -P_1 + P_2 A P_1 + P_1 A P_2$ and note that it is also self-adjoint, bounded and $\tilde{A} \le A$. It is therefore enough to show that $\tilde{A}$ has at most k negative eigenvalues. Now, if λ is an eigenvalue of $\tilde{A}$ different from 0 and if x is a corresponding eigenvector, one has

$$P_2 x = \frac{1}{\lambda} P_2 A P_1 x, \text{ and } P_1 A P_2 A P_1 (P_1 x) = \lambda(\lambda + 1) P_1 x.$$

Observe that $P_1 A P_2 A P_1$ is a nonnegative, self-adjoint operator on H_1.

Moreover, to each of its eigenvalues corresponds only one negative eigenvalue of A.

Lemma 1.23: *Let μ be a bounded non-negative measure on $\mathbb{R}^3$ such that $\mu(\mathbb{R}^3) < Z$. Let H_μ be the Hamiltonian given by $H_\mu = -\Delta + V + \mu * \frac{1}{|x|}$. Then, for each integer k, there is $\delta > 0$ such that H_μ admits at least k eigenvalues strictly below $-\delta$.*

Proof: We shall find for each integer k, a subspace E_k of dimension k such that
$$\max\{\langle H_\mu u, u\rangle; u \in E_k, \|u\|_2 = 1\} < 0.$$
Indeed, consider an arbitrary normalized $u \in \mathcal{D}(\mathbb{R}^3)$, and set for $\sigma > 0$, $u_\sigma(x) = \sigma^{-3/2} u(x/\sigma)$. We have

$$\langle H_\mu u_\sigma, u_\sigma\rangle = \frac{1}{\sigma^2} \int_{\mathbb{R}^3} |\nabla u|^2 \, dx + \frac{1}{\sigma} \int_{\mathbb{R}^3} V_\sigma(x) |u(x)|^2 \, dx$$
$$+ \frac{1}{\sigma} \int_{\mathbb{R}^3} (\mu_\sigma * \frac{1}{|x|}) |u|^2 \, dx$$

where

$$V_\sigma(x) = -\sum_{j=1}^m \frac{z_j}{|x - \frac{\bar{x}_j}{\sigma}|} \quad \text{and} \quad \mu_\sigma(dx) = \sigma^3 \mu(\sigma dx).$$

In particular, if we choose φ to be radially symmetric, we may write the last term as

$$\int_{\mathbb{R}^3} (\mu_\sigma * \frac{1}{|x|}) |u|^2 \, dx = \int_{\mathbb{R}^3} (|u|^2 * \frac{1}{|x|}) \, d\mu_\sigma$$
$$= \int \int_{\mathbb{R}^3 \times \mathbb{R}^3} |u(y)|^2 \max(|x|, |y|)^{-1} \, d\mu_\sigma \, dy$$
$$\leq \mu_\sigma(\mathbb{R}^3) \int_{\mathbb{R}^3} \frac{|u(y)|^2}{|y|} \, dy = \mu(\mathbb{R}^3) \int_{\mathbb{R}^3} \frac{|u(y)|^2}{|y|} \, dy.$$

It is now enough to choose any k-dimensional space of radially symmetric functions in $\mathcal{D}(\mathbb{R}^3)$ and then to let E_k be the space obtained by rescaling them as above ($u \to u_\sigma$). The claim will follow by taking σ large enough.

Proof of Theorem 1.21: Let $(\tilde{u}_1^n, ..., \tilde{u}_N^n)$ be a minimizing sequence of (2)–(4). By the smooth variational principle (Theorem 1.18), we can find another minimizing sequence $(u_1^n, ..., u_N^n)$ of (2)–(4) such that
$$\lim_n \|u_i^n - \tilde{u}_i^n\|_{H^1(\mathbb{R}^3)} = 0, \quad \text{for all } i = 1, ..., N, \tag{7}$$
and $(u_1^n, ..., u_N^n)$ is the minimum on M of
$$\varphi(u_1, ..., u_N) + \frac{1}{2}\gamma_n \sum_{i=1}^N \|u_i - v_i^n\|_{H^1(\mathbb{R}^3)}^2 \tag{8}$$

for some $(v_1^n, ..., v_N^n) \in M$ and where $\gamma_n > 0$ while $\lim_n \gamma_n = 0$.

This yields the existence of $(\varepsilon_1^n, ..., \varepsilon_N^n)$ in $\mathbb{R}^N$ such that the following holds in $L^2(\mathbb{R}^3)$:

$$
\lim_n \left(-\Delta u_i^n + V u_i^n + \left(\varrho^n * \frac{1}{|x|} \right) u_i^n \right.
$$
$$
\left. - \int_{\mathbb{R}^3} \varrho^n(x, y) u_i^n(y) \frac{1}{|x-y|} \, dy + \varepsilon_i^n u_i^n \right) = 0 \tag{9}
$$

and

$$
\sum_{i=1}^N \int_{\mathbb{R}^3} |\nabla w_i|^2 + V|w_i|^2 + \left(\varrho^n * \frac{1}{|x|} \right) |w_i|^2 + (\varepsilon_i^n + \gamma^n)|w_i|^2 \, dx
$$
$$
- \iint_{\mathbb{R}^3 \times \mathbb{R}^3} \varrho^n(x, y) \frac{1}{|x-y|} w_i(x) w_i(y) \, dx \, dy \tag{10}
$$
$$
- \frac{1}{2} \iint_{\mathbb{R}^3 \times \mathbb{R}^3} [K^n(x, y) - K^n(x) K^n(y)] \frac{1}{|x-y|} \, dx \, dy \geq 0
$$

for all $(w_1, ..., w_N) \in H^1(\mathbb{R}^3)$ satisfying

$$
\int_{\mathbb{R}^3} w_i u_j^n \, dx = 0 \quad \forall i, j; \qquad \int_{\mathbb{R}^3} w_i w_j \, dx = 0 \quad \forall i \neq j, \tag{11}
$$

where $K^n(x, y) = \sum_i u_i^n(x) w_i(y) + w_i(x) u_i^n(y)$, $K^n(x) = K^n(x, x)$.

We shall now use the second order information in (10) to find lower bounds for ε_i^n. Observe first that (10) implies in particular, that we have for each fixed i,

$$
\int_{\mathbb{R}^3} (|\nabla w|^2 + V|w|^2 + \left(\varrho^n * \frac{1}{|x|} \right) |w|^2) \, dx + (\varepsilon_i^n + \gamma^n) \int_{\mathbb{R}^3} |w|^2 \, dx \geq 0 \tag{10'}
$$

for all $w \in H^1(\mathbb{R}^3)$ such that for all j,

$$
\int_{\mathbb{R}^3} w u_j^n \, dx = 0. \tag{11'}
$$

By Lemma 1.22, this implies that the Schrödinger operator $H_n = -\Delta + V + \varrho^n * \frac{1}{|x|}$ has at most N eigenvalues strictly less than $-(\varepsilon_i^n + \gamma^n)$.

On the other hand, since $Z > N$, we may use Lemma 1.23 to find $\delta > 0$ such that H_n admits at least N eigenvalues strictly below $(-\delta)$. It follows that $\varepsilon_i^n + \gamma^n \geq \delta$. Since $\lim_n \gamma^n = 0$, we deduce that for n large enough, we have for every i,

$$
\varepsilon_i^n \geq \delta/2 > 0. \tag{12}
$$

To finish the proof, we note first that minimizing sequences are necessarily bounded in $H^1(\mathbb{R}^3)$. Indeed, by the Cauchy-Schwarz inequalities we have

$$
|\rho(x, y)|^2 \leq \rho(x)\rho(y) \quad \text{on } \mathbb{R}^3 \times \mathbb{R}^3. \tag{13}
$$

On the other hand, we have the following classical inequality which holds for any $\bar{x} \in \mathbb{R}^3$ and any $u \in H^1(\mathbb{R}^3)$

$$\int_{\mathbb{R}^3} \frac{1}{|x-\bar{x}|} |u(x)|^2 \, dx \le C \|u\|_2 \|\nabla u\|_2. \tag{14}$$

for some C independent of $\bar{x}$ and u. Combining (13) and (14) with the information that $\varphi(u_1^n, ..., u_N^n)$ is bounded above, we can easily deduce the H^1- bound on the minimizing sequence. We can also deduce that ε_i^n is bounded, and thus we may assume – by extracting subsequences if necessary – that u_i^n converges weakly in $H^1(\mathbb{R}^3)$ (and a.e. in $\mathbb{R}^3$) to some u_i and that ε_i^n converges to ε_i which satisfies $\varepsilon_i \ge \delta/2 > 0$ because of (12). Passing to the limit in (9), we get

$$-\Delta u_i + V u_i + \left(\varrho * \frac{1}{|x|}\right) u_i + \varepsilon_i \varphi_i - \int_{\mathbb{R}^3} \varrho(x,y) \frac{1}{|x-y|} u_i(y) dy = 0. \tag{15}$$

In particular, we find that

$$\limsup_n \sum_i \varepsilon_i^n \int_{\mathbb{R}^3} |u_i^n|^2 \, dx$$

$$= -\liminf_n \left\{ \sum_i \int_{\mathbb{R}^3} |\nabla u_i^n|^2 + V |u_i^n|^2 \, dx \right.$$

$$\left. + \iint_{\mathbb{R}^3 \times \mathbb{R}^3} \{\varrho^n(x)\varrho^n(y) - |\varrho^n(x,y)|^2\} \frac{1}{|x-y|} \, dx \, dy \right\}$$

$$\le -\left\{ \sum_i \int_{\mathbb{R}^3} |\nabla u_i|^2 + V |u_i|^2 \, dx \right.$$

$$\left. + \iint_{\mathbb{R}^3 \times \mathbb{R}^3} \{\varrho(x)\varrho(y) - |\varrho(x,y)|^2\} \frac{1}{|x-y|} \, dx \, dy \right\}$$

$$= \sum_i \varepsilon_i \int_{\mathbb{R}^3} |u_i|^2 \, dx.$$

Hence $\lim_n \|u_i^n\|_2 = \|u_i\|_2$ for every i, and consequently, u_i^n converges strongly in $L^2(\mathbb{R}^3)$ to u_i. The proof of Theorem 1.21 is complete.

1.7 A general perturbed minimization principle

In view of the two *perturbed variational principles* discussed above, it is natural to inquire about the classes of functions on a complete metric space X that are suitable to be *perturbation spaces* for a minimization principle. In this section, we shall isolate a fairly general condition on a class $\mathcal{A}$ of continuous functions on X that makes it eligible to be a perturbation space. We shall then say that $\mathcal{A}$ is an *admissible class*. We will check that the perturbations used in Ekeland's theorem and the one of Borwein-Preiss readily satisfy our condition. The same will hold for various spaces of smooth functions defined on a suitable Banach space. However, the proofs of the facts that, on certain Banach spaces, the spaces of *linear functionals* or cones of *plurisubharmonic functions* are admissible classes, are more involved and we shall deal with these cases in the next chapter.

Let (X, d) be a metric space and let $(\mathcal{A}, \delta)$ be a metric space of real valued functions defined on X. For any subset F of X, we shall denote by $\mathcal{A}_F$ the class of functions in $\mathcal{A}$ that are bounded above on F. For $f \in \mathcal{A}_F$, and $t > 0$, we denote by $S(F, f, t)$ the following *slice* of F

$$S(F, f, t) = \{x \in F; f(x) > \sup f(F) - t\}.$$

Definition 1.24: The space (X, d) is said to be *uniformly $\mathcal{A}$-dentable* if for every non-empty set $F \subset X$, every $f \in \mathcal{A}_F$, and every $\varepsilon > 0$, there exists $g \in \mathcal{A}_F$ such that $\delta(f, g) \leq \varepsilon$ and a non-empty slice $S(F, g, t)$ of F such that $\mathrm{diam} S(F, g, t) \leq \varepsilon$.

In the sequel, we shall associate to the metric space (X, d), the space $\tilde{X} = X \times \mathbb{R}$ equipped with the pseudo-metric

$$\tilde{d}((x, \lambda), (y, \mu)) = d(x, y).$$

On the other hand, we associate to $(\mathcal{A}, \delta)$ the class $\tilde{\mathcal{A}}$ of functions of the form $\tilde{f} := (f, -1)$ which act on $\tilde{X}$ via $(f, -1)(x, \lambda) = f(x) - \lambda$. We equip $\tilde{\mathcal{A}}$ with the distance $\tilde{\delta}(\tilde{f}, \tilde{g}) = \delta(f, g)$.

Definition 1.25: Let $\mathcal{A}$ be a class of bounded and continuous functions on a metric space (X, d) equipped with a metric δ. We shall say that $\mathcal{A}$ is an *admissible class* if the following conditions hold:
(i) There exists $K > 0$ such that $\delta(f, g) \geq K \sup\{|f(x) - g(x)|; x \in X\}$ for all $f, g \in \mathcal{A}$.
(ii) $(\mathcal{A}, \delta)$ is a complete metric space.
(iii) The product space $(\tilde{X}, \tilde{d})$ is uniformly $\tilde{\mathcal{A}}$-dentable.

Definition 1.26: Say that a function $\varphi : X \to \mathbb{R} \cup \{+\infty\}$ *strongly exposes* y in X if
(i) $\varphi(y) = \inf\{\varphi(x); x \in X\}$ and
(ii) $d(y, y_n) \to 0$ whenever $\varphi(y_n) \to \varphi(y)$.

We then say that y is *a strong minimum* for φ. Note that, in particular, y is a unique minimum for φ.

Denote by $D(\varphi)$ the set $\{x \in X; \varphi(x) < +\infty\}$. We can now state the following *general variational principle*.

Theorem 1.27: Let $(\mathcal{A}, \delta)$ be an admissible class of functions on a complete metric space (X, d) and let $\varphi : X \to \mathbb{R} \cup \{+\infty\}$ be a bounded below, lower semi-continuous function with $D(\varphi) \neq \emptyset$. Then the set

$$\{g \in \mathcal{A}; \ \varphi - g \text{ attains a strong minimum on } X\}$$

contains a dense G_δ subset $\mathcal{G}$ of $\mathcal{A}$.

Proof: Indeed, let

$$\mathcal{U}_n = \{g \in \mathcal{A}; \exists x_n \in X \text{ with } (\varphi - g)(x_n) < \inf\{(\varphi - g)(x) : d(x, x_n) \geq \frac{1}{n}\}$$

We claim that $\mathcal{U}_n$ is an open dense subset of $\mathcal{A}$. Indeed, $\mathcal{U}_n$ is open because of assumption (i) in the admissibility condition. To see that $\mathcal{U}_n$ is dense, let $g \in \mathcal{A}$ and $\varepsilon > 0$. We need to find $h \in \mathcal{A}$, $\delta(h, g) < \varepsilon$, and $x_n \in X$ such that

$$(\varphi - h)(x_n) < \inf\{(\varphi - h)(x); d(x, x_n) \geq \frac{1}{n}\}. \tag{*}$$

To do that, note that the functional $(g, -1)$ in $\tilde{\mathcal{A}}$ is bounded above on the epigraph F of φ in $\tilde{X}$, hence for any $\varepsilon' > 0$, there exists a non-empty slice $S = S(\tilde{h}, F, t)$ of F with diameter less than ε' and determined by a function $\tilde{h} = (h, -1) \in \tilde{\mathcal{A}}$ verifying $\delta(h, g) < \varepsilon'$.

Take $\varepsilon' < \min\{\varepsilon, 1/n\}$ and pick any $(x_n, \lambda_n) \in S$. For any $x \in X$ such that $d(x, x_n) \geq 1/n$, we have that $(x, \varphi(x)) \in F \setminus S$, so that

$$h(x) - \varphi(x) \leq \sup_F \tilde{h} - t < h(x_n) - \lambda_n.$$

Since $\lambda_n \geq \varphi(x_n)$, we obtain that the function h verifies (*).

Since $\mathcal{A}$ is a complete metric space, $\mathcal{G} = \bigcap_{n \geq 1} \mathcal{U}_n$ is a dense G_δ–subset of $\mathcal{A}$, by the Baire category theorem. We claim now that if $g \in \mathcal{G}$, then $\varphi - g$ attains a strong minimum on X. Indeed for each $n \geq 1$, there exists $x_n \in X$ such that

$$(\varphi - g)(x_n) < \inf\{(\varphi - g)(x); d(x, x_n) \geq \frac{1}{n}\}$$

We necessarily have that $d(x_p, x_n) < \frac{1}{n}$ for each $p > n$. Indeed, if not then by the definition of x_n, we have $(\varphi - g)(x_p) > (\varphi - g)(x_n)$. On the other hand, we will have $d(x_n, x_p) \geq \frac{1}{n} \geq \frac{1}{p}$ which gives, by the definition of x_p, that $(\varphi - g)(x_n) > (\varphi - g)(x_p)$. This is clearly a contradiction. Thus $(x_n)_n$ is a Cauchy sequence converging to some $x_\infty \in X$ and we claim that x_∞ is a strong minimum for $\varphi - g$. Indeed, since φ is lower semi-continuous,

$$(\varphi - g)(x_\infty) \leq \liminf_n (\varphi - g)(x_n)$$

$$\leq \liminf_n \inf\{(\varphi - g)(x); d(x, x_n) \geq \frac{1}{n}\}$$

$$\leq \inf\{(\varphi - g)(x); x \in X \setminus \{x_\infty\}\}$$

Moreover, let (y_n) be a sequence in X such that $(\varphi - g)(y_n)$ converges to $(\varphi - g)(x_\infty)$. Let us assume that (y_n) does not converge to x_∞. Extracting, if necessary, a subsequence, we can assume that there exists $\varepsilon > 0$ such that for all n, $d(y_n, x_\infty) \geq \varepsilon$. Thus there exists an integer p such that $d(x_p, y_n) \geq \frac{1}{p}$ for all n. Consequently

$$(\varphi - g)(x_\infty) \leq (\varphi - g)(x_p) < \inf\{(\varphi - g)(x); d(x, x_p) > \frac{1}{p}\} \leq (\varphi - g)(y_n)$$

for all n, which contradicts the fact that $\lim_n (\varphi - g)(y_n) = (\varphi - g)(x_\infty)$.

Examples of admissible classes

As we mentioned above, one can easily recover Ekeland's principle as well as the Theorem of Borwein-Preiss from Theorem 1.27. For that, we may consider the cones $\mathcal{A}_1$ (resp. $\mathcal{A}_2$) consisting of those functions f on (X, d) of the form

$$f(x) = -\sum_n \lambda_n d(x, x_n) \text{ (resp. } f(x) = -\sum_n \lambda_n d^2(x, x_n))$$

with $\lambda_n \geq 0$. We assume here, without loss of generality, that d is a bounded metric on X. Let us check that they are admissible cones.

Indeed, suppose F is a closed subset of $\tilde{X} = X \times \mathbb{R}$ such that $\tilde{h} = (h, -1) \in \tilde{\mathcal{A}}_1$ is bounded above on F. For small $\tau > 0$, consider a point $(x_0, s_0) \in F$ such that $h(x_0) - s_0 > \sup_F \tilde{h} - \tau^2$. Let $\tilde{k}$ be the functional in $\tilde{\mathcal{A}}_1$ defined by $\tilde{k} = (h - \tau d(\,, x_0), -1)$ and let S be the slice of F given by

$$S = \{(y, s) \in F; \tilde{k}(y, s) > \tilde{k}(x_0, s_0) - \tau^2\}.$$

It is easy to see that the d-diameter of S is less than τ, which means that $\tilde{X}$ is $\tilde{\mathcal{A}}_1$-uniformly dentable. A similar proof works for $\mathcal{A}_2$.

Note that Ekeland's theorem would then follow from Theorem 1.27 and the triangular inequality. Note also that we could have used the space $\mathcal{A}=\text{Lip}(X)$ of Lipschitz functions on X as an admissible space.

We shall now investigate the possibility of having classes of *bounded and smooth* functions as perturbation spaces. For simplicity, we shall only deal, in the sequel, with the case where X is a Banach space.

A function $b : X \to \mathbb{R}$ is said to be a *bump function* on X if it has a bounded and non-empty support.

Proposition 1.28: *Let $\mathcal{A}$ be a Banach space of continuous functions on a Banach space X satisfying the following properties:*
(a) *For each $g \in \mathcal{A}$, $\|g\|_{\mathcal{A}} \geq \|g\|_\infty = \sup\{|g(x)|; x \in X\}$.*
(b) *$\mathcal{A}$ is translation invariant, i.e. if $g \in \mathcal{A}$ and $x \in X$, then $\tau_x g : X \to \mathbb{R}$ given by $\tau_x g(t) = g(x + t)$ is in $\mathcal{A}$ and $\|\tau_x g\|_{\mathcal{A}} = \|g\|_{\mathcal{A}}$.*
(c) *$\mathcal{A}$ is dilation invariant, i.e. if $g \in \mathcal{A}$ and $\alpha \in \mathbb{R}$ then $g^\alpha : X \to \mathbb{R}$ given by $g^\alpha(t) = g(\alpha t)$ is in $\mathcal{A}$.*
(d) *There exists a bump function b in $\mathcal{A}$.*
   Then $\mathcal{A}$ is an admissible space.

Proof: According to (b) and (d), we can find a bump function b in $\mathcal{A}$ such that $b(0) \neq 0$. Using (c) and replacing $b(x)$ by $\alpha_1 b(\alpha_2 x)$ with suitable coefficients $\alpha_1, \alpha_2 \in \mathbb{R}$, we can assume that $b(0) > 0$, $\|b\|_{\mathcal{A}} < \varepsilon$ and $b(x) = 0$ whenever $\|x\| \geq \varepsilon$.

Let now $\tilde{g} = (g, -1) \in \tilde{\mathcal{A}}$ be bounded above on a closed subset F of $\tilde{X}$ and let $(x_0, s_0) \in F$ be such that $g(x_0) - s_0 > \sup_F \tilde{g} - b(0)$ and consider the function $h(x) = b(x - x_0)$ and $\tilde{k} = (g + h, -1)$.

By (b), $h \in \mathcal{A}$ and $\|h\|_{\mathcal{A}} = \|b\|_{\mathcal{A}} < \varepsilon$, which implies that $\|\tilde{g} - \tilde{k}\|_{\mathcal{A}} < \varepsilon$. On the other hand, consider the following slice of F,

$$S = \{(x, s) \in F; \tilde{k}(x, s) > \sup_F \tilde{k}\}.$$

It is non-empty since it contains (x_0, s_0). On the other hand, if $(x, s) \in F$ and $\|x - x_0\| \geq \varepsilon$, then $b(x - x_0) = 0$ and (x, s) cannot belong to S. It follows that the d-diameter of S is less than 2ε. Consequently, $\mathcal{A}$ is an admissible family.

Corollary 1.29: *Assume $\mathcal{A}$ is a Banach space of bounded continuous functions on X satisfying conditions (a)–(d) above. Then, for some constant $a > 0$, depending only on X and $\mathcal{A}$, the following holds:*

If $\varphi : X \to \mathbb{R} \cup \{+\infty\}$ is lower semi-continuous and bounded below with $D(\varphi) \neq \emptyset$ and if $y_0 \in X$ satisfies $\varphi(y_0) < \inf_X \varphi + a\varepsilon^2$ for some $\varepsilon \in (0, 1)$, then there exist $g \in \mathcal{A}$ and $x_0 \in X$ such that

(i) $\|x_0 - y_0\| \leq \varepsilon$.

(ii) $\|g\|_{\mathcal{A}} \leq \varepsilon$.

(iii) $\varphi + g$ attains its minimum at x_0.

Proof: We can clearly assume that there exists a bump function b in $\mathcal{A}$ with $b(0) = 1$, $0 \leq b \leq 1$ and such that the support of b is contained in the unit ball of X. Hypothesis (a) implies that $M := \|b\|_{\mathcal{A}} \geq \|b\|_\infty = 1$. Let $a = \frac{1}{4M}$ and suppose that ε and y_0 are given. Define the function

$$\tilde{\varphi}(x) = \varphi(x) - 2a\varepsilon^2 b\left(\frac{x - y_0}{\varepsilon}\right).$$

We have $\tilde{\varphi}(y_0) < \inf_X \varphi - a\varepsilon^2$ and $\tilde{\varphi}(y) \geq \inf_X \varphi$ whenever $\|y - y_0\| \geq \varepsilon$.

From Proposition 1.28 and Theorem 1.27, we can find $x_0 \in X$ and $k \in \mathcal{A}$ such that $\|k\|_{\mathcal{A}} \leq \min\{\varepsilon/2, a\varepsilon^2/2\}$ and $\tilde{\varphi} + k$ attains its minimum at x_0. The above conditions imply that $\|x_0 - y_0\| < \varepsilon$. Thus, the function $g \in \mathcal{A}$ defined by $g(x) = -2a\varepsilon^2 b\left(\frac{x - x_0}{\varepsilon}\right) + k(x)$ satisfies claims (i), (ii) and (iii) of the Corollary.

Remark 1.30: Again, we can recover Ekeland's variational principle on Banach spaces from the (easily verifiable) fact that the space $\mathcal{A}_1$ of all bounded Lipschitz functions on X equipped with the norm

$$\|f\|_{\mathcal{A}_1} = \sup\{|f(x)|; x \in X\} + \sup\left\{\frac{|f(x) - f(y)|}{\|x - y\|}; x \neq y\right\}$$

satisfies the conditions of Corollary 1.29 and hence it is an admissible space of perturbations. Note that, as an additional bonus, we get that the perturbation is also small in the uniform norm as well as in the Lipschitz norm.

To derive an analogue of the Borwein-Preiss Theorem, we can consider the space $\mathcal{A}_2$ of all bounded Lipschitz functions f on X that also verify the following second order condition

$$\|f\| = \sup\{\frac{|f(x+2h) - 2f(x+h) + f(x)|}{h^2}; x, h \in X\} < \infty.$$

The space $\mathcal{A}_2$ equipped with the norm

$$\|f\|_{\mathcal{A}_2} = \|f\|_{\mathcal{A}_1} + \|f\|$$

is also an admissible space of perturbations.

Clearly, the above norms will correspond to the C^1 and C^2-norms whenever the functions are differentiable. But since X is in general infinite dimensional, we have to deal with two types of difficulties: firstly, the appearance of various different and generally non-equivalent types of differentiability and secondly, the problem of *admissibility* of these spaces of differentiable functions which requires extra assumptions on the Banach spaces involved. We shall try to deal with some of these problems in the sequel.

1.8 Perturbed minimization and differentiability

Let again φ be a function defined on a Banach space X with values in $\mathbb{R} \cup \{+\infty\}$. In principle, we are only concerned with the concepts of *Fréchet* and *Gâteaux-differentiability*. However, since the arguments in both cases are similar, we shall avoid repetition, by working with the notion of differentiability associated with any bornology on X.

Recall that *a bornology* β is just a class of bounded subsets of X. It induces a topology τ_β on X^* which corresponds to the uniform convergence of linear functionals in X^* on the sets of β. Note that if β is the class of all bounded subsets (resp. all singletons) of X, then τ_β coincides with the norm (resp. weak*) topology on X^*.

Definition 1.31: Say that φ is β-*differentiable* at $x_0 \in D(f)$ with β-derivative $\varphi'(x_0) = p \in X^*$ if for any $A \in \beta$,

$$\lim_{t \to 0} t^{-1}[\varphi(x_0 + th) - \varphi(x_0) - \langle p, th \rangle] = 0$$

uniformly for $h \in A$.

When β is the class of all bounded subsets (resp. all singletons) of X, then β-differentiability coincides with the classical notion of *Fréchet-differentiability* (resp. *Gâteaux-differentiability*).

We shall denote by $C^1_\beta(X)$ the space of all real valued, bounded Lipschitz and β-differentiable functions g on X such that $g' : (X, \|\ \|) \to (X^*, \tau_\beta)$ is continuous, endowed with the norm

$$\|g\|_\beta = \sup\{|g(x)|; x \in X\} + \sup\{\|g'(x)\|; x \in X\} = \|g\|_\infty + \|g'\|_\infty.$$

We shall use the notation $C^1_F(X)$ (resp. $C^1_G(X)$) when we are dealing with Fréchet (resp. Gâteaux) differentiability. Analogously, we can define the spaces $C^2_F(X)$ (resp. $C^2_G(X)$) equipped with the C^2-norm.

Proposition 1.32: *Suppose X is a Banach space such that X (resp.*

X^*) is separable, then $C_G^1(X)$ (resp. $C_F^1(X)$) is an admissible space of perturbations.

If X is a Hilbert space, then $C_F^2(X)$ is an admissible space of perturbations.

Proof: If X (resp. X^*) is separable, then there exists an equivalent norm N on X which is Gâteaux (resp. Fréchet) differentiable on $X\backslash\{0\}$. Indeed, if $(x_n)_n$ is a dense sequence in the unit ball of X, we can associate the following equivalent *strictly convex* norm on X^*:

$$N^*(x^*) = \|x^*\| + \sum_n 2^{-n}|x^*(x_n)|.$$

Note that any $x \in X$ attains its maximum on the N^*-sphere at a unique point. We leave it to the reader to check that N^* is dual to a norm N on X that is Gâteaux-differentiable on $X \backslash \{0\}$ and whose derivative is (norm to weak*)-continuous from $X \backslash \{0\}$ to X^*.

In the case where X^* is separable, we let

$$N^*(x^*) = \|x^*\| + \sum_n 2^{-n}\text{dist}(x^*, E_n),$$

where $(E_n)_n$ is an increasing sequence of finite dimensional subspaces such that $X^* = \overline{\cup_n E_n}$. It is easy to see that N^* is *weak*-locally uniformly convex*: that is, a sequence $(x_n^*)_n$ on the N^*-unit sphere of X^* norm-converges to x^*, whenever it weak*-converges to x^* and $N^*(x^*) = 1$. The predual norm N on X is then Fréchet differentiable on $X \backslash \{0\}$ and its derivative is (norm to norm)-continuous from $X \backslash \{0\}$ to X^*. For more details, we refer the reader to the recent monograph [D-G-Z 2].

Consider now a continuously differentiable function $\ell : \mathbb{R} \to \mathbb{R}$, with support in $[1,3]$ and such that $\ell(2) \neq 0$. The function $b(x) = \ell(N(x))$ is a bump function in $C_G^1(X)$ (resp. $C_F^1(X)$) verifying condition (d) of Proposition 1.28.

The same reasoning holds when X is a Hilbert space, since the norm is then in $C^2(X)$.

We now apply the above results to the *subdifferentiability* of lower semi-continuous functions on Banach spaces.

Definition 1.33: If $\varphi : X \to \mathbb{R} \cup \{+\infty\}$ where X is a Banach space, we define the *β-subdifferential* of φ at a point $x_0 \in D(\varphi)$ by:

$$d_\beta^-\varphi(x_0) = \{p \in X^*; p = g'(x_0) \quad \text{for some } g \in C_\beta^1 \text{ such that}$$

$$\varphi - g \text{ attains its minimum at } x_0\}$$

We say that φ is *β-subdifferentiable* at x_0 if $d_\beta^-\varphi(x_0)$ is not empty.

We can introduce $C_\beta^+(x_0)$ and the notion of *β-superdifferentiability* in the same way. We can now prove the following:

Theorem 1.34: *Let X be a Banach space such that X (resp. X^*) is separable and let $\varphi : X \to \mathbb{R} \cup \{+\infty\}$ be lower semi-continuous, bounded below and such that $D(\varphi)$ is non-empty and open in X. Then, φ is Gâteaux-subdifferentiable (resp. Fréchet-subdifferentiable) on a dense subset of $D(\varphi)$.*

Proof: To deal with both cases at the same time, we can assume, in view of Proposition 1.32, that the space $C^1_\beta(X)$ is admissible, where β is the bornology that corresponds to Gâteaux (resp. Fréchet) differentiability.

Consider any $y_0 \in D(\varphi)$ and $\alpha > 0$. We are looking for a point of β-subdifferentiability x_0 of φ such that $x_0 \in B(y_0, \alpha)$, where $B(y_0, \alpha)$ denotes the ball in X of center y_0 and radius α. For $m > 0$, define

$$\psi(x) = \begin{cases} [b(m(x - y_0))]^{-2} & \text{if } b(m(x - y_0)) \neq 0, \\ +\infty & \text{otherwise} \end{cases}$$

where b is a β-differentiable bump function on X such that $b(0) \neq 0$. For m large enough, $\psi + \varphi$ is lower semi-continuous, bounded below and $D(\psi) \subset B(y_0, \alpha)$. Note that $D(\psi + \varphi) \neq \emptyset$ since it contains y_0. Applying Corollary 1.29, we can find $g \in C^1_\beta(X)$ such that $\varphi + \psi + g$ attains its minimum at some point $x_0 \in D(\varphi + \psi) \subset D(\psi) \subset B(y_0, \alpha)$. It follows that φ is β-subdifferentiable at x_0 and $p = -\psi'(x_0) - g'(x_0)$ belongs to $d^-_\beta \varphi(x_0)$.

We conclude this section with a proposition which will be crucial in the study of Hamilton-Jacobi equations.

Proposition 1.35: *Let X be a Banach space on which there is a bump function in $C^1_\beta(X)$. Let u, v be two bounded functions defined on X such that u is upper semi-continuous and v is lower semi-continuous. Then for each $\varepsilon > 0$, there exist $x, y \in X, p \in d^+_\beta u(x), q \in d^-_\beta v(y)$ such that*
 (i) $\|x - y\| < \varepsilon$ *and* $\|p - q\| < \varepsilon$,
 (ii) $v(z) - u(z) \geq v(x) - u(y) - \varepsilon$, *for all* $z \in X$.

Proof: Let $b \in C^1_\beta(X)$ be a bump function such that $b(0) = \max_X b = 1$ and with support in the unit ball. Let $\lambda > 4 \max(\|u\|_\infty, \|v\|_\infty) + \varepsilon$ and define the function $\omega : X \times X \to \mathbb{R}$ by

$$\omega(x, y) = v(y) - u(x) - \lambda b\left(\frac{x - y}{\varepsilon}\right)$$

Note that ω is lower semi-continuous and bounded below on $X \times X$. Moreover, the function B defined by $B(x, y) = b(x)b(y)$ is a bump function on $X \times X$ that belongs to $C^1_\beta(X \times X)$. This means that the latter is an admissible space of perturbations, and hence, by Corollary 1.29, there exist $(x, y) \in X \times X$ and $g \in C^1_\beta(X \times X)$ such that
 (a) $\|g\|_\infty < \varepsilon/2$ and $\|g'\|_\infty < \varepsilon/2$.
 (b) For every $(\hat{x}, \hat{y}) \in X \times X$,

$$v(\hat{y}) - u(\hat{x}) - \lambda b\left(\frac{\hat{x} - \hat{y}}{\varepsilon}\right) + g(\hat{x}, \hat{y}) \geq v(y) - u(x) - \lambda b\left(\frac{x - y}{\varepsilon}\right) + g(x, y).$$

If we fix $x = \hat{x}$, we see that v is β-subdifferentiable at y and that

$$q = \frac{\lambda}{\varepsilon} b' \left(\frac{x - y}{\varepsilon} \right) - g'_y(x, y) \in d^-_\beta v(y).$$

Similarly, u is β-superdifferentiable at x and

$$p = \frac{\lambda}{\varepsilon} b' \left(\frac{x - y}{\varepsilon} \right) + g'_x(x, y) \in d^+_\beta u(x).$$

Let us check Claims (i) and (ii). First, put $(\hat{x}, \hat{y}) = (z, z)$ in the inequality above, to obtain for every $z \in X$,

$$v(z) - u(z) \geq v(y) - u(x) + g(x, y) - g(z, z) + \lambda - \lambda b (\frac{x - y}{\varepsilon}).$$

Since $\|g\|_\infty < \varepsilon/2$ and $b(\frac{x-y}{\varepsilon}) \leq 1$, conclusion (ii) follows.

Moreover, if $\|x - y\| \geq \varepsilon$, the last term on the right-hand side above vanishes, and the resulting inequality contradicts our choice of λ. Thus $\|x - y\| < \varepsilon$. Finally, we use (a) to estimate

$$\|p - q\| = \|g'_x(x, y) + g'_y(x, y)\| < \varepsilon.$$

1.9 Infinite dimensional Hamilton-Jacobi equations

Let Ω be an open subset of a Banach space X. The general first-order Hamilton-Jacobi equation has the form

$$H(x, u, u') = 0, \qquad (1)$$

where $H : \Omega \times \mathbb{R} \times X^* \to \mathbb{R}$ is continuous.

Definition 1.36: Let Ω, X and H be as above.
 (i) A function $u : \Omega \to \mathbb{R}$ is a *classical β-solution* of (1) if u is a function in $C^1_\beta(\Omega)$ that verifies for every $x \in \Omega$,

$$H(x, u(x), u'(x)) = 0.$$

 (ii) A function $u : \Omega \to \mathbb{R}$ is *a viscosity β-subsolution* of (1) if u is upper semi-continuous and for every $x \in \Omega$ and $p \in d^+_\beta u(x)$

$$H(x, u(x), p) \leq 0.$$

(iii) A function $u : \Omega \to \mathbb{R}$ is a *viscosity β-supersolution* of (1) if u is lower semi-continuous and for every $x \in \Omega$ and $p \in d^-_\beta u(x)$

$$H(x, u(x), p) \geq 0.$$

Finally, u is called a *viscosity β-solution* of (1) if u is both a viscosity β-subsolution and viscosity β-supersolution of (1).

For simplicity, we shall only deal with Hamilton-Jacobi equations of the following type:

$$u + F(u') = f \quad \text{on} \quad X. \qquad (2)$$

Here is the main result of this section:

Theorem 1.37: *Let $f : X \to \mathbb{R}$ be a bounded and uniformly continuous function on a Banach space X and let $F : X^* \to \mathbb{R}$ be bounded.*

If X is separable and F norm-uniformly continuous and weak*- contin- uous (resp. X^* separable and F norm uniformly continuous) on X^*, then there exists a unique bounded viscosity Gâteaux-solution (resp. Fréchet-solution) of the equation $u + F(u') = f$ on X.

The uniqueness will follow immediately from the following

Lemma 1.38: *Let X be a Banach space with a bump function in $C_\beta^1(X)$. Let f, g be bounded and uniformly continuous functions on X and consider $F : X^* \to \mathbb{R}$ to be bounded and uniformly continuous. Assume that u is a bounded viscosity β-subsolution of $u + F(u') = f$ on X and that v is a bounded viscosity β-supersolution of $v + F(v') = g$ on X. Then,*

$$v - u \geq \inf_X (g - f).$$

Proof: Let u, v be as in the statement and $\varepsilon > 0$. Let x, y, p, q be given by Proposition 1.35. Since u is a viscosity β-subsolution of $u + F(u') = f$ on X and $p \in d_\beta^+ u(x)$, we have

$$u(x) + F(p) \leq f(x)$$

Similarly, since v is a viscosity β-supersolution of $v + F(v') = g$ on X and $q \in d_\beta^- v(y)$, we have

$$v(y) + F(q) \geq g(y).$$

So, for every $z \in X$,

$$\begin{aligned}
v(z) - u(z) &\geq v(y) - u(x) - \varepsilon \\
&\geq g(y) - f(x) + F(p) - F(q) - \varepsilon \\
&\geq \inf_X (g - f) + [f(y) - f(x)] + [F(p) - F(q)] - \varepsilon.
\end{aligned}$$

Moreover, $\|y - x\| \leq \varepsilon$, $\|p - q\| \leq \varepsilon$ and f and F are uniformly continuous. So passing to the limit when ε tends to zero, we get

$$v(z) - u(z) \geq \inf_X (g - f).$$

To prove the existence in Theorem 1.37, we shall establish a more general statement. First, recall that if Ω is an open subset of a Banach space X and if u is a function defined on Ω, the *upper semi-continuous envelope* u^* of u is defined by

$$u^* = \inf\{v; v \text{ is continuous on } \Omega \text{ and } v \geq u \text{ on } \Omega\}$$

The *lower semi-continuous envelope* u_* of u is defined similarly.

Theorem 1.37 will be a consequence of the following

Proposition 1.39: *Let Ω be an open subset of a Banach space X with a bump function in $C_\beta^1(X)$ and suppose $H : \Omega \times \mathbb{R} \times X^* \to \mathbb{R}$ is continuous when X is endowed with the norm topology and X^* is endowed with the τ_β-topology. Let u_0, v_0 be respectively, a viscosity β-subsolution and a* viscosity β-supersolution of $H(x, u, u') = 0$ on Ω, such that $u_0 \leq v_0$ on Ω. Then there is a function $u : X \to \mathbb{R}$ such that on Ω, $u_0 \leq u \leq v_0$, u^* is a viscosity β-subsolution and u_* is a viscosity β-supersolution of $H(x, u, u') = 0$.

Before establishing Proposition 1.39, we show how it implies the existence part in Theorem 1.37.

Let $u_0 \equiv -||f||_\infty - F(0)$ and $v_0 \equiv ||f||_\infty - F(0)$. It is easy to see that u_0 is a classical β-subsolution (hence u_0 is a viscosity β-subsolution) of $u + F(u') = f$ on X and v_0 is a classical β-supersolution of $u + F(u') = f$ on X. By Proposition 1.39, there exists u such that

$$-||f||_\infty - F(0) \leq u \leq ||f||_\infty - F(0),$$

u_* is a viscosity β-supersolution of $u + F(u') = f$ on X and u^* is a viscosity β-subsolution of $u + F(u') = f$ on X. Clearly $u_* \leq u^*$, while the comparison Lemma 1.38 yields that $u^* \leq u_*$. It follows that $u^* = u_*$ and the function $u = u_*$ is a viscosity β-solution of $u + F(u') = f$ on X.

The proof of Proposition 1.39 relies on the following

Lemma 1.40: Let Ω be an open subset of a Banach space X with a bump function in $C_\beta^1(X)$ and let $\mathcal{S}$ be a set of functions from Ω into $\mathbb{R}$ such that

$$u(x) = \sup\{v(x); v \in \mathcal{S}\} < +\infty \text{ for all } x \in \Omega.$$

Then, for every $x \in \Omega$ and every $p \in d_\beta^+ u^*(x)$, there exist sequences (v_n) in $\mathcal{S}$, (y_n) in Ω and (p_n) in X^* satisfying $p_n \in d_\beta^+ v_n^*(y_n)$ and
(i) $\lim_n v_n^*(y_n) = u^*(x)$.
(ii) $\lim_n y_n = x$ for the norm-topology.
(iii) $\lim_n p_n = p$ for the τ_β-topology.

Proof: Since $p \in d_\beta^+ u^*(x)$, we can find a lower semi-continuous function $\varphi : X \to \mathbb{R} \cup \{+\infty\}$ which is β-differentiable on $D(\varphi)$ and such that
 (a) $(u^* - \varphi)(x) = 0$ and $(u^* - \varphi)(y) \leq 0$ if $y \in \Omega$,
 (b) $p = \varphi'(x)$ and φ' is norm to τ_β continuous at x.
By definition of u^*, for every $n \geq 1$, there exist $v_n \in \mathcal{S}$ and $z_n \in X$ such that $||z_n - x|| \leq \frac{1}{n}$ and $(v_n - \varphi)(z_n) \geq -\frac{1}{n}$. Let us define $(v_n^* - \varphi)(y) = 0$ if $y \notin \Omega$. Note that $v_n^* - \varphi$ is then upper semi-continuous and bounded above by 0. According to Corollary 1.29, there exists $\psi_n \in C_\beta^1(X)$ such that
 (c) $\lim_n ||\psi_n||_\infty = 0$ and $\lim_n ||\psi_n'||_\infty = 0$.
 (d) $v_n^* - \varphi + \psi_n$ attains its supremum at some $y_n \in \Omega$.
 (e) $\lim_n ||y_n - z_n|| = 0$.
Using (e) and the fact that $||z_n - x|| < \frac{1}{n}$, we have $\lim_n ||y_n - x|| = 0$.

On the other hand, $p_n = \varphi'(y_n) - \psi'_n(y_n) \in d^+_\beta v^*_n(y_n)$ and

$$p_n - p = \varphi'(y_n) - \varphi'(x) - \psi'_n(y_n).$$

Thus (iii) follows from this last equality, the fact that $\lim_n ||\psi'_n||_\infty = 0$ and b). Finally,

$$0 \geq (v^*_n - \varphi)(y_n) \geq (v^*_n - \varphi)(z_n) - 2||\psi_n||_\infty \geq -\frac{1}{n} - 2||\psi_n||_\infty,$$

so that $\lim_n v^*_n(y_n) = \lim_n \varphi(y_n) = \varphi(x) = u^*(x)$ and the lemma is proved.

Proof of Proposition 1.39: Consider the non-empty class

$$S = \{\omega : X \to \mathbb{R}; u_0 \leq \omega \leq v_0 \quad \omega^* \text{ is a } \beta\text{-subsolution of (1) on } \Omega\}$$

and let $u = \sup\{\omega; \omega \in S\}$. We shall prove that u^* is the desired function.

Claim 1: u^* is a viscosity β-subsolution of $H(x, u, u') = 0$ on Ω.

Indeed, let $x \in X$ and $p \in d^+_\beta u^*(x)$. Let $(v_n), (y_n), (p_n)$ be as in Lemma 1.40. Since v^*_n is a viscosity β-subsolution of $H(x, u, u') = 0$ on Ω, we have $H(y_n, v^*_n(y_n), p_n) \leq 0$. So when n tends to infinity, using the continuity properties of H, we get $H(x, u^*(x), p) \leq 0$. Hence Claim 1 holds.

Claim 2: u_* is a viscosity β-supersolution of $H(x, u, u') = 0$ on Ω.

Indeed, if not, there exist $\varphi \in C^1_\beta(X)$ and $x_0 \in X$ such that φ' is norm to τ_β continuous at x_0 and

(a) $u_*(x_0) - \varphi(x_0) = 0$ while $u_*(x) - \varphi(x) \geq 0$ for all $x \in X$,

(b) $H(x_0, u_*(x_0), \varphi'(x_0)) < 0$.

We claim that $\varphi(x_0) < v_0(x_0)$; otherwise $\varphi(x) \leq v_0(x)$ and $\varphi(x_0) = u_*(x_0) = v_0(x_0)$ imply that $v_0 - \varphi$ attains its infimum at x_0. Since v_0 is a viscosity β-supersolution of $H(x, u, u')$ on Ω, we obtain that $H(x_0, v_0(x_0), \varphi'(x_0)) \geq 0$, which contradicts (b).

By continuity, there exists $\delta > 0$ and $b \in C^1_\beta(X)$ with support in the ball $B(x_0, \delta)$ such that $b(x_0) > 0$ and

$$H(x, \varphi(x) + b(x), \varphi'(x) + b'(x)) < 0 \text{ for all } x \in B(x_0, 2\delta),$$

while

$$\varphi(x) + b(x) \leq v_0(x) \text{ for all } x \in X.$$

Let

$$\omega(x) = \begin{cases} \max\{\varphi(x) + b(x), u(x)\} & \text{if } x \in B(x_0, 2\delta) \\ u(x) & \text{if } x \in X \backslash B(x_0, 2\delta) \end{cases}$$

Let $\Omega_1 = X \backslash \overline{B}(x_0, \delta)$ and $\Omega_2 = B(x_0, 2\delta)$. If $x \in B(x_0, 2\delta) \backslash \overline{B}(x_0, \delta)$, then $u(x) \geq \varphi(x) = \varphi(x) + b(x)$ so that $\omega(x) = u(x)$ on Ω_1 and ω^* is a viscosity β-subsolution on Ω_1. On the other hand, Claim 1 applied to the two functions whose supremum is w on Ω_2, yields that w^* is a viscosity β-subsolution on Ω_2. It is then easy to see that ω^* is a viscosity β-subsolution on $\Omega = \Omega_1 \cup \Omega_2$. Since $u_0 \le \omega \le v_0$, we obtain that $\omega \in \mathcal{S}$ and hence $\omega \le u$ on Ω. But $\omega(x) \ge \varphi(x) + b(x)$ on $B(x_0, \delta)$ yields $u(x) \ge \varphi(x) + b(x)$ so that $u_*(x) \ge \varphi(x) + b(x)$ on $B(x_0, \delta)$ and this contradicts the fact that $u_*(x_0) = \varphi(x_0)$. The proof of the Proposition is complete.

Notes and Comments: Ekeland's variational principle is the nonlinear version of the celebrated Bishop–Phelps theorem [Ph]. Its applications are numerous and documented in several surveys and monographs [A-E], [Ek 1], [Ph]. We have chosen to include some of its applications that will be needed later on in this monograph. We learned Corollary 1.5.(ii) from Brezis and Nirenberg [B-N 3]. Corollaries 1.10 and 1.11 are standard but we formulated Lemma 1.9 in order to derive them simultaneously. As mentioned above, Theorem 1.12 is due to G. Tarantello [T 1] and will be continued in Chapter 8.

Theorem 1.18 is due to Borwein and Preiss [B-P] and its applications have been so far limited to the study of differentiability properties in Banach spaces [Ph]. Actually, a variant of this theorem was already established and used by Preiss [P] in his remarkable proof of the *almost everywhere* differentiability of Lipschitz functions defined on Hilbert spaces. Our analysis of the Hartree-Fock equation follows closely P. L. Lions [L].

The idea of identifying classes of functions that are suitable *perturbation spaces* for a minimization principle goes back to the series of lectures given by B. Maurey at the Banff conference on Banach space theory (1988), which in turn are based on the work of Ghoussoub and Maurey [G-M 1,2,3] and Ghoussoub, Maurey and Schachermayer [G-M-S 1,2]. (See the discussion in the next chapter.) The elegant approach (via bump functions) to the smooth variational principle is due to Deville, Godefroy and Zizler [D-G-Z 1]. The Perron-type method was first used in the theory of Hamilton-Jacobi equations by Ishii [I], while the idea of using perturbed minimization principles in this context is due to Crandall and Lions. For more details, we refer the reader to a series of their papers [C-L 1,2,3,4].

2

LINEAR AND PLURISUBHARMONIC

PERTURBED MINIMIZATION PRINCIPLES

In this chapter, we present two results of Ghoussoub and Maurey: the first is concerned with linear perturbations and is applicable in reflexive Banach spaces. It will be used in Theorems 2.12 and 2.13 below to get generic minimization results in the case of critical exponents and non-zero data. The second one deals with the possibility of finding plurisubharmonic perturbations and can be applied to spaces as "bad" as L^1. The proofs essentially boil down to showing that these new classes of functions are, in the terminology of Chapter 1, *admissible cones of perturbations*. However, the methods here are slightly more involved than the ones used earlier. Moreover, the minima for the perturbed functionals are not necessarily close to any given minimizing sequence of the original function, and the perturbations we are seeking here, can never be bounded on the whole Banach space.

Actually, our main goal for this chapter, is to introduce the reader to new and different methods for proving perturbed variational principles and especially, to the martingales techniques used in Theorem 2.17.

2.1 A minimization principle with linear perturbations

We shall first consider a situation where the perturbations can be taken to be linear. First, we recall the following

Definition 2.1: Let C be a subset of a Banach space X. A real-valued function φ is said to *strongly expose C from below at a point $x \in C$* if:
  (i) $\varphi(x) = \inf \varphi(C)$;
  (ii) every minimizing sequence $(x_n)_n$ for φ in C norm-converges to x.

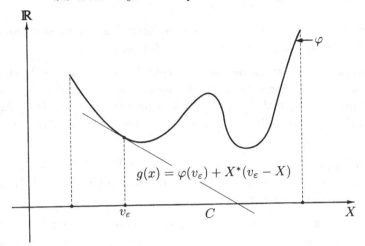

$$g(x) = \varphi(v_\varepsilon) + X^*(v_\varepsilon - X)$$

Fig. 2.1.

We define in a similar fashion the functions that expose C from above.

With an eye on the applications, we shall first state the *linear perturbed principle* in the special case of reflexive spaces.

Theorem 2.2 (Fig 2.1): *Let $\varphi : X \to (-\infty, \infty]$ be a lower semi-continuous function on a separable reflexive Banach space X.*

(i) *If φ is bounded below on a closed bounded subset C of X, then the set $\{x^* \in X^*; \varphi + x^* \text{ strongly exposes } C \text{ from below}\}$ is a dense G_δ in X^*.*

(ii) *If φ is such that $\varphi + x^*$ is bounded below for any $x^* \in X^*$, then the set $\{x^* \in X^*; \varphi + x^* \text{ strongly exposes } X \text{ from below}\}$ is a dense G_δ in X^*.*

We shall deduce the above theorem from a more general result. We need the following concept.

2.2 Weak*-H_δ subsets

Definition 2.3: Let C be a subset of a dual Banach space Y^*. We say that C is *a strict w^*-H_δ set* in a subset D of Y^* if $D \backslash C = \bigcup_n K_n$ where each K_n is w^*-compact and convex in Y^* and $\text{dist}(K_n, C) > 0$.

The relevance of this definition to Theorem 2.2 comes from the following simple but crucial observation.

Lemma 2.4: *If X is a separable reflexive Banach space, then any norm closed subset C of X is a strict w-H_δ in X.*

In particular, if $\varphi : X \to (-\infty, +\infty]$ is norm lower semi-continuous, then the epigraph of φ is a strict $w\text{-}H_\delta$ subset of $X \times \mathbb{R}$.

Proof: Since X is separable, we may consider a sequence $(x_n)_n$ that is dense in $X \backslash C$. For each n, let K_n be the ball centered at x_n and with radius $r_n = \frac{1}{4}\mathrm{dist}(x_n, C)$. Since X is reflexive, each K_n is weakly compact and convex. It is also easy to see that $X \backslash C = \cup_n K_n$ with $\mathrm{dist}(K_n, C) > r_n$.

In the sequel, if C is a subset of a dual space Y^*, we shall denote by $\overline{C}^*$ (resp. $\overline{\mathrm{conv}}^*(C)$) its closure (resp. the closure of its convex hull) in Y^* for the weak*-topology.

Theorem 2.5: *Let C be a norm separable subset of a dual Banach space Y^* that is a strict $w^*\text{-}H_\delta$ set in $D = \overline{\mathrm{conv}}^*(C)$. Assume that V is a norm-open subset of Y such that $\sup y(C) < \infty$ for every $y \in V$. Then the following hold:*

(i) *The set of functionals in V that strongly expose D at a point in C is a dense G_δ in V.*

(ii) *If $\varphi : C \to (-\infty, +\infty]$ is a norm lower semi-continuous such that $\varphi + y$ is bounded below for each $y \in Y$, then the set*

$$\{y \in Y; \varphi + y \quad \text{strongly exposes } X \text{ from below}\}$$

is a dense G_δ in Y.

For the proofs, we shall need the following notions and notation. For each subset L of D we denote by $F(L, D)$ the set of functions in Y that achieve their maximum on D at a point in L, i.e.

$$F(L, D) = \{y \in Y; \exists \ell \in L, y(\ell) \geq y(d) \quad \text{for every} \quad d \in D\}.$$

We shall use the easy fact that $F(L, D)$ is norm-closed whenever L is w^*-compact.

Recall that *a slice of C by a function ℓ* is a set of the form

$$S(C, \ell, \alpha) = \{x \in C; \ell(x) \geq \sup \ell(C) - \alpha\}$$

where α is a strictly positive real number (Fig 2.2).

In the sequel, we shall denote by $B_Y(y, \delta)$ the closed ball in the Banach space Y that is centered in y and with radius δ. The unit ball will be denoted by B_Y.

Lemma 2.6: *Let D be a subset of a dual space Y^* and K a w^*-compact convex subset of D. Suppose $y \in Y$ and $\alpha > 0$ are such that $B_Y(y, \alpha) \subseteq F(K, D)$, then for any $\varepsilon > 0$ we have $S(D, y, \varepsilon) \subset K + \frac{\varepsilon}{\alpha} B_{Y^*}$.*

In particular, we have that $\mathrm{dist}(K, C) = 0$ for every subset C of D such that $D \subseteq \overline{\mathrm{conv}}^*(C)$.

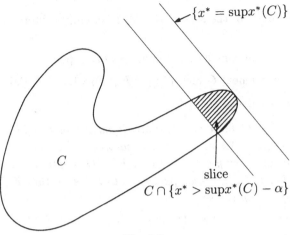

Fig. 2.2.

Proof: Suppose x^* is not in $K + \frac{\varepsilon}{\alpha} B_{Y^*}$; there exists then z in Y, $\|z\| \le 1$ and $\sup z(K) < z(x^*) - \frac{\varepsilon}{\alpha}$. We claim that $x^* \notin S(D, y, \varepsilon)$. Indeed, if not then $y(x^*) > \max y(D) - \varepsilon$ and $y + \alpha z \in B_Y(y, \alpha) \subseteq F(K, D)$, hence there exists $k_0 \in K$ such that $(y + \alpha z)(k_0) \ge (y + \alpha z)(x^*)$. But

$$(y + \alpha z)(x^*) > (y(k_0) - \varepsilon) + \alpha \left(z(k_0) + \frac{\varepsilon}{\alpha} \right) = y(k_0) + \alpha z(k_0).$$

A contradiction. The second assertion holds since $S(D, y, \varepsilon)$ must intersect C for every $\varepsilon > 0$.

Lemma 2.7: *Assume that V is a norm-open subset of Y such that $\sup y(C) < \infty$ for every $y \in V$. Then every $y \in V$ attains its maximum on $D = \overline{\mathrm{conv}}^*(C)$.*

Proof: Note first that $\sup y(C) = \sup y(D)$. Note also that the slice $S = D \cap \{y^*; y(y^*) > \lambda\}$ is bounded for every $y \in V$ and $\lambda \in R$; indeed since V is open, we can find $\varepsilon > 0$ such that

$$\sup(y + \varepsilon z)(S) \le \sup(y + \varepsilon z)(D) < \infty \quad \text{for every } z \in B_Y$$

and this yields immediately that $\sup z(S) < \infty$ for every $z \in Y$. Since S is bounded, we deduce that every $y \in V$ attains its maximum on D.

Lemma 2.8: *Let C be a subset of a dual Banach space Y^* that is a strict w^*-H_δ set in $D = \overline{\mathrm{conv}}^*(C)$. Assume V is a norm-open subset of Y such that $\sup y(C) < \infty$ for every $y \in V$. Then the set of functionals in V that attain their maximum on D at a point of C contains a dense G_δ in Y.*

Proof: Write $D \backslash C = \cup_n K_n$ where each K_n is w^*-compact convex with $\mathrm{dist}(K_n, C) > 0$. From Lemma 2.7, every $y \in V$ attains its maximum on D. By Lemma 2.6, the sets $F(K_n, D)$ have empty interiors. Hence the set

$$\{y \in V; y \text{ attains its maximum on } D \text{ at a point of } C\}$$

contains the complement G in V of $\bigcup_n F(K_n, D)$ which, by Baire's category theorem, is a dense G_δ in V.

Lemma 2.9: *Let C be a norm separable subset of a dual space Y^* and $D = \overline{\mathrm{conv}}^*(C)$. Assume V is a norm-open subset of Y such that $\sup y(C) < \infty$ for every $y \in V$, and that $F(C, D)$ contains a dense G_δ-subset of V. Let K be a w^*-compact subset of D such that $K \cap C = \emptyset$. Then for every $\varepsilon > 0$, the set*

$$O(K, \varepsilon) = \{y \in V; \exists \tau > 0, \overline{S}^*(D, y, \tau) \cap K = \emptyset, \ \mathrm{diam}(\overline{S}^*(D, y, \tau)) \le \varepsilon\}$$

is open and dense in V.

Proof: Since $S(D, y, \tau)$ is bounded for every $y \in V$, it follows easily that $O(K, \varepsilon)$ is open. To show that they are norm-dense in V, let Ω be any non-empty open subset of V. Use the norm separability of C to find a countable family of w^*-compact convex subsets (C_k) of D such that
(a) $C \subseteq \bigcup_k C_k$.
(b) $\mathrm{dist}(C_k, K) > 0$ for each k.
(c) $\mathrm{diam}(C_k) \le \varepsilon/2$ for each k.

Note that $\Omega \supset \bigcup_k F(C_k, D) \cap \Omega \supset F(C, D) \cap \Omega$ and the latter contains a dense G_δ set in Ω, hence by Baire's category theorem, there exists k_0 such that $F(C_{k_0}, D) \cap \Omega$ has a non-empty interior. If y is in such an interior, we get from Lemma 2.6 that for each $\eta > 0$, there exists $\tau > 0$ such that $S(D, y, \tau) \subseteq C_{k_0} + \eta B_{Y^*}$. It is now enough to choose $\eta < \inf\{\varepsilon/2, \mathrm{dist}(C_{k_0}, K)\}$ to obtain that $y \in \Omega \cap O(K, \varepsilon)$.

Proof of Theorem 2.5: (i) Suppose $D \backslash C = \bigcup_n K_n$, where each K_n is w^*-compact convex with $\mathrm{dist}(K_n, C) > 0$. Lemma 2.8 gives that $F(C, D)$ contains a dense G_δ in V. Lemma 2.9 then yields that for each N, the set $O_N = O\left(K_0 \cup K_1 \cup \cdots \cup K_N, \frac{1}{N}\right)$ is a dense open set in V. Note that by the Baire category theorem, the set $O = \bigcap_N O_N$ is a dense G_δ in V consisting of the functionals in V that strongly expose D from above at a point of C.

(ii) Now we turn to the problem of minimizing non-linear functions on strict w^*-H_δ sets. Let φ be a $(-\infty, +\infty]$-valued function on C. Note that $C \times \mathbb{R}$ is a strict w^*-H_δ in $D \times \mathbb{R}$ and since φ is lower semicontinuous, its epigraph $\mathrm{Epi}(\varphi)$ is a strict w^*-H_δ in $D \times \mathbb{R}$, therefore also in $\overline{\mathrm{conv}}^*(\mathrm{Epi}(\varphi))$. The hypothesis implies that every function $(y, \alpha) \in$

$Y \times R$ has a finite supremum on $\text{Epi}(\varphi)$ when $\alpha < 0$, hence part (i) applies to $\text{Epi}(\varphi)$, with $V = \{(y, \alpha); y \in Y, \alpha < 0\}$. It is thus possible to find for each $\varepsilon > 0$ a functional (y, α) in V such that $\|(y, \alpha) - (0, -1)\| \leq \varepsilon$ and which strongly exposes $\text{Epi}(\varphi)$ at a point (x_0^*, λ_0). If $0 < \varepsilon < 1/2$, then $-3/2 < \alpha < -1/2$ and for each $(x^*, \lambda) \in \text{Epi}(\varphi)$ such that $x^* \neq x_0^*$, we have $y(x_0^*) + \alpha\lambda_0 > y(x^*) + \alpha\lambda$. Consequently, $\frac{1}{\alpha} y(x_0^*) + \varphi(x_0^*) < \frac{1}{\alpha} y(x^*) + \varphi(x^*)$ for each $x^* \neq x_0^*$ in C. Note that if $z = \frac{y}{\alpha}$ then $\|z\| \leq 2\varepsilon$ and $\varphi + z$ attains its minimum at x_0^* in C.

If (x_n^*) is a minimizing sequence for $\varphi + z$ on C then $(x_n^*, \varphi(x_n^*))$ is a maximizing sequence for the functional (y, α) that strongly exposes $\text{Epi}(\varphi)$ at (x_0^*, λ_0), hence (x_0^*) norm-converges to x_0^*. This completes the proof of Theorem 2.5.

The following corollary is now immediate.

Corollary 2.10: *Let C be a bounded subset of a dual Banach space Y^* such that C is a strict w^*-H_δ in $D = \overline{\text{conv}}^*(C)$. If C is norm separable, then for any concave, bounded below and lower semi-continuous function φ and any $\varepsilon > 0$, there exist y in Y, $\|y\| \leq \varepsilon$ and x_0^* in C such that*

(i) $(\varphi + y)(x^*) > (\varphi + y)(x_0^*)$ *for all x^* in $C, x^* \neq x_0^*$.*
(ii) *Every minimizing sequence for $\varphi + y$ in C, norm-converges to x_0^*.*
(iii) x_0^* *is an extreme point of C.*

Remark 2.11: The hypothesis imposed on the set C above are optimal for obtaining a variational principle with linear perturbations. Indeed, in [G-M 1], it is proved that for a bounded convex separable subset C of a dual space Y^*, the following conditions are equivalent
(1) C is a strict w^*-H_δ in $D = \overline{C}^*$.
(2) For every bounded below convex and lower semi-continuous function $\varphi : C \to \mathbb{R} \cup \{+\infty\}$ with $\varphi \not\equiv +\infty$, the set

$$F(\varphi, C) = \{y \in Y; \varphi + y \text{ attains its minimum in } C\}$$

contains a dense G_δ in Y.

2.3 Generic results for non-homogeneous elliptic problems

We start with the following minimization problem with critical exponent and non-zero data.

Let Ω be a smooth domain in $\mathbb{R}^N$ with $N \geq 3$. Let $f \in H^{-1}(\Omega)$ and consider the following minimization problem

$$C_f = \min\{\int_\Omega (\frac{1}{2}|\nabla v|^2 - fv)\, dx; v \in H_0^1(\Omega), \|v\|_{2^*} = 1\}$$

where $2^* = \frac{2N}{N-2}$ is the critical exponent for the Sobolev embedding.

It is well known that C_0 (i.e., when $f = 0$) is not achieved, while if $f \neq 0$, Brezis and Nirenberg [B-N 2] have shown recently, with a fairly technical argument, that C_f is achieved.

It is quite surprising that with such a general result as the linear perturbation principle, one can get the following generic result.

Theorem 2.12: Let $\varphi_f(v) = \int_\Omega (\frac{1}{2}|\nabla v|^2 - fv)\,dx$. Then the set of $f \in H^{-1}(\Omega)$ such that φ_f attains its minimum uniquely on the unit sphere of $L^{2^*}(\Omega)$ contains a dense G_δ in $H^{-1}(\Omega)$.

Moreover, all the minimizing sequences for such $\varphi_f's$, necessarily norm converge to the unique minimum.

Of course, one does not get the Brezis-Nirenberg result for all $f \neq 0$, but one gains the generic uniqueness result. It may be interesting if this can be used to simplify the proof in [B-N 2]. Another advantage of Theorem 2.2, is that one can deal with more general situations as we shall show in the following.

Let again Ω be a smooth domain in $\mathbb{R}^N$ with $N \geq 3$ and consider the equation

$$-\Delta u = f(x,u) \quad \text{in } \Omega \quad \text{and} \quad u = 0 \text{ on } \partial\Omega.$$

where f is a Caratheodory function satisfying

$$|f(x,s)| \leq C|s|^{p-1} + b(x) \tag{1}$$

where $C > 0$, $1 \leq p \leq \frac{2N}{N-2}$ and $b \in L^{p'}$ ($\frac{1}{p} + \frac{1}{p'} = 1$).

Again, the functional φ defined on $H_0^1(\Omega)$ by

$$\varphi(u) = \int_\Omega (\frac{1}{2}|\nabla u|^2 - F(x,u))\,dx$$

where $F(x,s) = \int_0^s f(x,t)\,dt$, is C^1 and its critical points are weak solutions for the above equation (Appendix B).

Let λ_1 be the first eigenvalue of $-\Delta$ in $H_0^1(\Omega)$ with homogeneous boundary value conditions. Recall that λ_1 is given by the Rayleigh-Ritz quotient

$$\lambda_1 = \inf\{\frac{\|u\|_{H_0^1}^2}{\|u\|_2^2}; u \in H_0^1(\Omega), u \neq 0\}. \tag{2}$$

Theorem 2.13: Suppose that f is a Caratheodory function satisfying (1) and the following growth condition

$$F(x,s) \leq cs^2/2 + a(x)s \tag{3}$$

where $a \in L^{q'}(\Omega)$ for some $1 \leq q \leq \frac{2N}{N-2}$ and $c < \lambda_1$. Then the following hold:

(i) The set of functions $h \in H^{-1}(\Omega)$ such that the corresponding functional

$$\varphi_h(u) := \int_\Omega (\frac{1}{2}|\nabla u|^2 - F(x, u) + hu)\, dx$$

attains its minimum uniquely on H_0^1, contains a dense G_δ subset of $H^{-1}(\Omega)$.

(ii) In particular, the set of functions $h \in H^{-1}(\Omega)$ such that

$$-\Delta u = f(x, u) + h(x) \quad \text{in } \Omega \quad \text{and} \quad u = 0 \quad \text{on } \partial\Omega.$$

has a solution, contains a dense G_δ subset of $H^{-1}(\Omega)$.

(iii) Moreover, if $f(x, s)$ is non-decreasing in s, then the set of functions $h \in H^{-1}(\Omega)$ such that

$$-\Delta u = f(x, u) + h(x) \quad \text{in } \Omega \quad \text{and} \quad u = 0 \quad \text{on } \partial\Omega$$

has a unique solution, contains a dense G_δ subset of $H^{-1}(\Omega)$.

Proof: In view of Theorem 2.2, we need only to check that $\varphi + x_h^*$ is bounded below for all $h \in (H_0^1)^*$ where x_h^* is the linear functional on H_0^1 associated to h. Note that by the Sobolev embedding and (3), we have for all $u \in H_0^1$ that

$$(\varphi + x_h^*)(u) \geq \int_\Omega \frac{1}{2}(|\nabla u|^2 - c|u|^2)\, dx - \|a\|_{q'}\|u\|_q - \|h\|_{H^{-1}}\|u\|_{H_0^1}$$

$$\geq \frac{1}{2}(1 - \frac{c}{\lambda_1})\int_\Omega |\nabla u|^2\, dx - K\|u\|_{H_0^1}$$

Since $1 - \frac{c}{\lambda_1} > 0$, we get that $\varphi + x_h^*$ is bounded below and in fact $(\varphi + x_h^*)(u) \to \infty$ as $\|u\| \to \infty$.

Remark 2.14: Condition (3) above implies that $\varphi + x_h^*$ is coercive, so that if $p < \frac{2N}{N-2}$, $\varphi + x_h^*$ always attains its minimum, since φ is then weakly lower semicontinuous. The only gain in this case is that one can perturb φ so that the resulting functional has a unique minimum. However, if $p = \frac{2N}{N-2}$, φ is only norm lower semi-continuous and by perturbing, we get existence and uniqueness.

2.4 Analytic martingales and optimization in L^p $(0 < p \leq 1)$

We shall now present a result due to Ghoussoub and Maurey. It is relevant when we are dealing with non-reflexive spaces like the space L^1. Actually, we shall work in the setting of quasi-Banach spaces to include the Lebesgue spaces L^p and the Hardy spaces H^p for $0 < p < 1$. Note that in this case, linear functionals are not available and the techniques à la Hahn-Banach that we have used so far, cannot play any role. Instead, we shall use *martingale methods*. Unfortunately, we cannot present here in details all the ramifications of this rich theory. We shall only recall the basic concepts needed for the proof of the main theorem, hoping to give the reader an enticing flavor of this beautiful theory.

We first recall the relevant concepts. Let X be a vector space. A map $x \to \|x\|$ from X into $\mathbb{R}^+$ is called a *quasi-norm* if

(i) $\|x\| > 0$ when $x \neq 0$.

(ii) $\|\alpha x\| = |\alpha| \, \|x\|$, for $\alpha \in \mathbb{C}, x \in X$.

(iii) $\|x_1 + x_2\| \leq C(\|x_1\| + \|x_2\|)$ for all $x_1, x_2 \in X$. Here C is a constant larger or equal to one.

We call $\| \ \|$ a *p-norm* for $0 < p \leq 1$, if in addition it is p-subadditive, that is

(iv) $\|x_1 + x_2\|^p \leq \|x_1\|^p + \|x_2\|^p$ for $x_1, x_2 \in X$.

The Aoki-Rolewicz theorem asserts that every quasi-norm is equivalent to a p-norm for some p $(0 < p \leq 1)$. A complete quasi-normed vector space X will be called a *quasi-Banach space*. If the quasi-norm on X is also p-subadditive, we say that X is a *p-Banach space*. For the basics about non-locally convex vector spaces, we refer the reader to the book [K-P-R].

Let $\Delta = \{z \in \mathbb{C}, |z| < 1\}$ be the open unit disc. Denote by $\partial\Delta$ or $\mathbb{T}$ the unit circle $\{z \in \mathbb{C}; |z| = 1\}$ and by $\bar{\Delta} = \Delta \cup \partial\Delta$.

Definitions 2.15: Let $g : X \to [-\infty, +\infty)$ be an upper semi-continuous function.

(i) g is said to be *plurisubharmonic* on X if for every $x, y \in X$ we have

$$g(x) \leq \int_0^{2\pi} g(x + e^{i\theta}y)\frac{d\theta}{2\pi}.$$

We denote by $PSH(X)$ the class of all such functions.

(ii) $LIP_p(X)$ will denote the set of functions g on X satisfying for some $K > 0$, $|g(x) - g(y)| \leq K\|x - y\|^p$ for all x, y in X.

We shall write $PSH_p(X)$ for $PSH(X) \cap LIP_p(X)$ and we shall equip it with the following norm

$$\|g\|_p = \max\{|g(0)|, \sup\{|g(x) - g(y)|/\|x - y\|^p; x, y \in X, x \neq y\}\}$$

We now introduce the notion of *Analytic martingale*. In the sequel, λ will denote the normalized Lebesgue measure on the torus $\mathbb{T}$. Also λ^n (resp. $\lambda^{\mathbb{N}}$) will denote the product Lebesgue measure on $\mathbb{T}^n$ (resp. $\mathbb{T}^{\mathbb{N}}$).

Definitions 2.16: Let X be a quasi-Banach space. A sequence of X-valued random variables $(F_n)_{n=0}^\infty$ defined on $\mathbb{T}^{\mathbb{N}}$ and such that $F_0 = x_0$ is called *an analytic martingale* if for $n \in \mathbb{N}$ and $(\theta_1, \theta_2, \ldots, \theta_{n-1}) \in \mathbb{T}^{n-1}$, we have

$$F_n(\theta_1, \theta_2, \ldots, \theta_n) - F_{n-1}(\theta_1, \theta_2, \ldots, \theta_{n-1}) = f_n(\theta_1, \theta_2, \ldots, \theta_{n-1})e^{2\pi i \theta_n}$$

where the coefficients $f_k \colon [0, 2\pi]^{k-1} \to X$ are X-valued random variables.

Here is the *Plurisubharmonic perturbed minimization principle*.

Theorem 2.17: *Let X be a p-Banach space for some p $(0 < p \leq 1)$. The following assertions are equivalent:*

(a) *All L^p-bounded X-valued analytic martingales converge a.e.*

(b) *For any closed bounded subset C of X and every bounded below, lower semi-continuous function φ on C, the set*

$$\{g \in PSH_p(X); \varphi - g \text{ strongly exposes } C \text{ from below } \}$$

is a dense G_δ in $PSH_p(X)$.

Before proving Theorem 2.17, we mention the following:

Corollary 2.18: *Suppose X is one of the following classical spaces: Lebesgue spaces L^p, Hardy spaces H^p or the Schatten spaces C^p where $0 < p \leq 1$. Then, for any closed bounded subset C of X and every bounded below, lower semi-continuous function φ on C, the set*

$$\{g \in PSH_p(X); \varphi - g \text{ strongly exposes } C \text{ from below}\}$$

is a dense G_δ in $PSH_p(X)$.

Remark 2.19: Note that when $1 < p < \infty$, the above mentioned spaces are all reflexive Banach spaces and therefore one can apply the stronger *linear perturbation principle*.

2.5 A variational characterization of the plurisubharmonic envelope of a function

For the proof of Theorem 2.17, we need the following characterizations for the *plurisubharmonic envelope* of a given function.

Proposition 2.20 : *Let $g \colon X \to \mathbb{R} \cup \{-\infty\}$ be an upper semi-continuous function on a quasi-normed complex vector space X. Set $g_0 = g$ and for each $n > 0$, define inductively on X the functions*

$$g_{n+1}(x) := \inf \left\{ \int_0^{2\pi} g_n(x + e^{i\theta}v)\frac{d\theta}{2\pi}; v \in X \right\}.$$

(i) *The sequence $(g_n)_n$ then decreases pointwise on X to the largest function $\hat{g}$ in $PSH(X)$ that is dominated by g.*

(ii) *If g is a function in $LIP_p(X)$ such that $\hat{g}$ is not identically $-\infty$, then $\hat{g}$ belongs to $PSH_p(X)$ and $\|\hat{g}\|_p \leq \|g\|_p$.*

(iii) For each $n, g_n(x) = \inf E[g(F_n)]$ where the infimum is taken over all analytic martingales $(F_k)_{k=0}^n$ with $F_0 = x$ and whose coefficients are finitely valued in X.

Proof: (i) We start by showing that $\hat{g}$ is upper semi-continuous. Inductively, assume that g_{n-1} is upper semi-continuous and let (x_k) be a sequence that converges to $x_0 \in X$. Let $y_0 \in X$ and note that the upper semi-continuous function g_{n-1} is bounded above on the compact set $\cup_k \{x_k + \bar{\Delta} y_0\}$, hence we obtain from Fatou's lemma that

$$\int_0^{2\pi} g_{n-1}(x_0 + e^{i\theta} y_0)\frac{d\theta}{2\pi} \geq \limsup_{k\to\infty} \int_0^{2\pi} g_{n-1}(x_k + e^{i\theta} y_0)\frac{d\theta}{2\pi}$$
$$\geq \limsup_{k\to\infty} g_n(x_k)$$

and therefore that $g_n(x_0) \geq \limsup_{k\to\infty} g_n(x_k)$. Hence each g_n and consequently $\hat{g}$ is upper semi-continuous.

Suppose now $h \in PSH(X)$ is such that $h \leq g = g_0$ on X. Inductively, assume that $h \leq g_{n-1}$ on U. For x_0 and $y_0 \in X$, we have

$$\int_0^{2\pi} g_{n-1}(x_0 + e^{i\theta} y_0)\frac{d\theta}{2\pi} \geq \int_0^{2\pi} h(x_0 + e^{i\theta} y_0)\frac{d\theta}{2\pi} \geq h(x_0)$$

whence $g_n(x_0) \geq h(x_0)$. This clearly implies that $\hat{g} \geq h$ on X.

Finally to show the mean value inequality for $\hat{g}$, take again x_0 and $y_0 \in X$, we get from the Beppo-Levi's monotone convergence theorem that

$$\hat{g}(x_0) = \lim_{n\to\infty} g_n(x_0) \leq \lim_{n\to\infty} \int_0^{2\pi} g_{n-1}(x_0 + e^{i\theta} y_0)\frac{d\theta}{2\pi}$$
$$= \int_0^{2\pi} \hat{g}(x_0 + e^{i\theta} y_0)\frac{d\theta}{2\pi}$$

and the proof of (i) is complete.

To prove (ii), suppose that $g \in PSH_p(X)$ and let x, y be two points in X. For each $\varepsilon > 0$, there is a v in X so that $y + \bar{\Delta} v \subset X$ and

$$g_1(x) - g_1(y) \leq g_1(x) - \int_0^{2\pi} g(y + e^{i\theta} v)\frac{d\theta}{2\pi} + \varepsilon.$$

It follows that

$$g_1(x) - g_1(y) \leq \int_0^{2\pi} [g(x + e^{i\theta} v) - g(y + e^{i\theta} v)]\frac{d\theta}{2\pi} + \varepsilon \leq K\|x - y\|^p + \varepsilon,$$

where K is the p-Lipschitz constant of g. An easy induction implies that $\hat{g} \in PSH_p(X)$.

The proof of (iii) is again by induction on n. For $n = 0$, we take $F_0 = x$ so that $g_0(x) = g(x) = E[g(X_0)]$. Assume the formula true for

$n-1$. Fix $x \in X$ and $\varepsilon > 0$. Choose $v \in X$ so that

$$g_n(x) \leq \int_0^{2\pi} g_{n-1}(x + e^{i\theta}v)\frac{d\theta}{2\pi} + \varepsilon/2.$$

Let F_1 have the uniform distribution on the circle $C = x + \bar{\Delta}v$. To get the formula for n, it is enough to use the Yankov-Von Neumann selection theorem to choose measurably for each θ, an analytic martingale $(F_k)_{k=1}^n$ with $F_1(\theta) = x + e^{i\theta}v$ and $g_{n-1}(x+e^{i\theta}v) \leq E[g(F_n)]+\varepsilon/2$. The analytic martingale $(F_k)_{k=0}^n$ with $F_0 = x$ will clearly satisfy $g_n(x) \leq E[g(F_n)]+\varepsilon$.

The details of the proof as well as the fact that the coefficients can be chosen to be "step functions" are left for the interested reader. In the sequel, we shall denote by $PSH_p^1(X)$ the class of all functions g in $PSH_p(X)$ such that $\|g\|_p \leq 1$.

2.6 Proof of the plurisubharmonically perturbed minimization principle

Proof of theorem 2.17: We shall prove that assertions (a) and (b) of that theorem are actually equivalent to the following:

(c) For every map F from a set K into X such that $F(K)$ is separable and any real-valued function φ on K satisfying for some $\alpha \in \mathbb{R}$, that $\varphi(t) \leq \alpha - \|F(t)\|^p$ for all $t \in K$, there exist for every $\varepsilon > 0$, a $\tau > 0$ and $g \in PSH_p^1(X)$ such that

(i) $\rho = \sup\{\varphi(t) + g(F(t)); t \in K\} < \infty$ and

ii) $\operatorname{diam}\{F(t); t \in K, \varphi(t) + g(F(t)) > \rho - \tau\} \leq \varepsilon$.

To prove that (a) implies (c), we define for each $t \in K$, the following function on X,

$$\varepsilon_t(y) = \inf\{\varphi(t)-\varphi(u)+\|y-F(u)\|^p; u \in K \text{ and } \|F(u)-F(t)\|^p > \varepsilon^p/2\}.$$

If Δ is a countable subset of K such that $F(\Delta)$ is dense in $F(K)$, it is clear that the above infimum can be restricted to the elements of Δ. Note also that the function ε_t is in $LIP_p(X)$ and is bounded below by the constant $\varphi(t) - \alpha$. It then follows from Proposition 2.20 that $\hat{\varepsilon}_t$ is also finite and belongs to $PSH_p^1(X)$. To establish the above claim, it is enough to prove that:

$$\text{there exists } t \in \Delta \text{ such that } \hat{\varepsilon}_t(F(t)) > 0. \tag{1}$$

Indeed, in this case the set $A = \{s \in K; \varphi(s) + \hat{\varepsilon}_t(F(s)) > \varphi(t)\}$ is non-empty since it contains t.

Moreover, if $s \in K$ is such that $\|F(s)-F(t)\|^p > \varepsilon^p/2$, then by taking $u = s$ to get an upper bound for $\varepsilon_t(F(s))$ we obtain

$$\hat{\varepsilon}_t(F(s)) \leq \varepsilon_t(F(s)) \leq \varphi(t) - \varphi(s).$$

This means that $s \notin A$ and consequently $\operatorname{diam} F(A) \leq \varepsilon$.

Back to the claim and let us assume it is not true: that is $\hat{\varepsilon}_t(F(t)) \leq 0$ for all $t \in K$. Let $(\tau_n)_n$ be a sequence of positive reals so that $\tau = \sum_{n=1}^{\infty} \tau_n < \infty$. We shall construct two sequences of random variables $(T_n)_{n \geq 0}$ and $(U_n)_{n \geq 1}$ on $\mathbf{T}^{\mathbf{N}}$ such that for each $n \in \mathbf{N}$,

(+) T_n is Δ-valued and $\|F(T_{n+1}) - F(T_n)\|^p > \varepsilon^p/2$.

(++) U_n is the k_n−th variable of an X-valued analytic martingale starting at 0 and

$$\mathbf{E}[\varphi(T_n) - \varphi(T_{n+1}) + \|F(T_n) + U_{n+1} - F(T_{n+1})\|^p] < \tau_{n+1}.$$

Start with any t_0 in Δ and set $T_0 = t_0$. Suppose T_j and U_j have been constructed for $j \leq n$. Since for every $\omega \in \mathbf{T}^n$ we have that $\hat{\varepsilon}_{T_n(\omega)}(F(T_n(\omega))) \leq 0$, we can use Proposition 2.20 to find U_{n+1} that is the k_{n+1}−th variable of an X-valued analytic martingale starting at 0 such that

$$\mathbf{E}[\varepsilon_{T_n(\omega)}(F(T_n(\omega)) + U_{n+1})] < \tau_{n+1}.$$

Use now the definition of $\varepsilon_{T_n(\omega)}$ to find a Δ-valued random variable T_{n+1} such that (+) and (++) hold.

Note that (++) gives that

$$\mathbf{E}[\varphi(T_0) - \varphi(T_{n+1})] \leq \tau$$

and hence that

$$\mathbf{E}[\|F(T_{n+1})\|^p] \leq \alpha - \mathbf{E}[\varphi(T_{n+1})] \leq \alpha + \tau - \varphi(t_0). \qquad (2)$$

On the other hand, we have

$$\mathbf{E}[\sum_{n=0}^{\infty} \|F(T_n) - F(T_{n+1}) + U_{n+1}\|^p] < \tau - \varphi(T_0) + \alpha. \qquad (3)$$

Hence

$$\mathbf{E}[\|F(T_{n+1}) - F(T_0) - M_{n+1}\|^p] \leq \tau - \varphi(t_0) + \alpha \qquad (4)$$

where $(M_n)_n$ is the subsequence of an analytic martingale defined by $M_0 = 0$ and $M_n = U_1 + ... + U_n$ for $n > 0$. Note that $(M_n)_n$ is clearly L^p-bounded and hence must converge a.e. by the hypothesis. But (3) implies that the sequence $(F(T_n))_n$ also converges a.e. This clearly contradicts (+) and the claim in (c) is therefore established.

To prove that (c) implies (b), let C be a closed bounded subset of X and let $\varphi : C \to \mathbf{R}$ be a bounded above, upper semi-continuous function. Apply the above claim to φ/ε, $K = C$ and $F(x) = x$ to obtain for each $\varepsilon > 0$ a $g \in PSH_p(X)$ with $\|g\|_p \leq \varepsilon$ and $\tau > 0$ such that the slice

$$S(C, \varphi + g, \tau) = \{x \in C; \ (\varphi + g)(x) > \sup_C(\varphi + g) - \tau\}$$

has a diameter less than ε. This means that $PSH_p(X)$ is an admissible cone of perturbations on C. The result now follows from Theorem 1.27.

We now show how (c) implies the convergence of analytic martingales. For simplicity, we shall assume that the quasi-norm is a plurisubharmonic function on the space X. This is trivially the case when X is a Banach space, since $\| \; \|$ is then convex. In general, it can be shown that assertion (a) actually implies that one can equip X with an equivalent p-norm which is also plurisubharmonic on X. For details we refer to [G-M 2].

We shall use the following property of analytic martingales that can be verified by using elementary facts in complex variables: for every $g \in PSH_p(X)$, the sequence $(g(M_n))_n$ is a real-valued submartingale: that is $\int_A g(M_n)\, dP \leq \int_A g(M_{n+1})\, dP$ for every n and any $A \in \Sigma_n$, where Σ_n denotes the σ-field generated by the first n coordinates in $\mathbf{T}^\mathbf{N}$ and $P = \lambda^\mathbf{N}$. This is actually the definition of a *PSH-martingale*. See Appendix E.

Suppose now that f is a measurable function from a probability space $(\Omega, \mathcal{F}, P)$ into a complete metric space (Z, d). We shall say that a point $\omega \in \Omega$ is *regular for* $(f, \mathcal{F})$ if for every $\varepsilon > 0$, there exists $B \in \mathcal{F}$ with $P(B) > 0$ such that for every $\omega' \in B$ we have $d(f(\omega), f(\omega')) \leq \varepsilon$. It is easy to see that if $(f_n)_n$ is a countable sequence of random variables and if (Z, d) is separable, then there is $\Omega' \subset \Omega$ with $P(\Omega') = 1$ such that every $\omega \in \Omega'$ is regular for each f_n.

Let now $(M_n)_n$ be an X-valued and L^p-bounded analytic martingale. By a standard exhaustion argument, it is enough to prove the following:

Claim: For every measurable set $A \subset \mathbf{T}^\mathbf{N}$ with $P(A) > 0$ and any $\varepsilon > 0$, there exists a measurable set $A' \subset A$ with $P(A') > 0$ such that for all $\omega \in A'$, we have $\limsup_{m,n} \| M_n(\omega) - M_m(\omega) \| < \varepsilon$.

Since $\| \; \|^{p/2}$ is a plurisubharmonic function on X, we get from the above observations that $(\| M_n \|^{p/2})_n$ is an L^2-bounded real-valued submartingale. It follows from Doob's inequality that $\sup_n \| M_n \|^p \in L^1$. The real submartingale convergence theorem gives the L^1 as well as the almost sure convergence of $(\| M_n \|^p)_n$ to a random variable that we denote by Z. (See Appendix E).

Fix $A \subset \mathbf{T}^\mathbf{N}$ and $\varepsilon > 0$ and let $\lambda > 0$ be such that $A_\lambda = A \cap \{Z \leq \lambda\}$ has non-zero measure. Set $D_\lambda = \mathbf{T}^\mathbf{N} \backslash A_\lambda$, $h = -1_{D_\lambda}(Z + \lambda + 1)$ and let $h_n = \mathbf{E}[h; \Sigma_n]$ be the conditional expectation of h with respect to Σ_n. By the above remark we can find a measurable set $\Omega \subset \mathbf{T}^\mathbf{N}$ of full measure such that for all $\omega \in \Omega$ we have $h(\omega) = \lim_n h_n(\omega)$, $Z(\omega) = \lim_n \| M_n(\omega) \|^p$ and ω is regular for the sequence $\{(h_n, M_n), \Sigma_n; n \geq 0\}$.

We want to apply assertion (c) to the set $K = \mathbf{N} \times \Omega$ and the functions

$F(n,\omega) = M_n(\omega)$ and $\varphi(n,\omega) = h_n(\omega)$. For that let us check that we have the right hypothesis. First φ is clearly bounded above by 0. On the other hand, we have for all $(n,\omega) \in K$,

$$\varphi(n,\omega)) \leq \lambda - \|F(n,\omega)\|^p. \tag{5}$$

Indeed, if $(n,\omega) \in K$ and since ω is regular, there exists for every $\varepsilon > 0$, a set $C \in \Sigma_n$ with $P(C) > 0$ such that

$$\|M_n(\omega)\|^p \leq \frac{1}{P(C)} \int_C \|M_n\|^p \, dP + \varepsilon$$

$$\leq \frac{1}{P(C)} \int_C Z \, dP + \varepsilon$$

$$\leq \frac{1}{P(C)} \int_C (\lambda + 1_{D_\lambda}(Z + \lambda + 1)) \, dP + \varepsilon$$

$$= \lambda - \frac{1}{P(C)} \int_C h_n \, dP \leq \lambda - h_n(\omega) + 2\varepsilon.$$

Apply now assertion (c) to obtain $g \in PSH_p(X)$ with $\|g\|_p \leq 1$ such that

$$\rho = \sup\{h_n(\omega) + g(M_n(\omega)); (n,\omega) \in K\} < \infty$$

and a τ $(0 < \tau < 1)$ such that $\operatorname{diam}(F(K_0)) \leq \varepsilon$, where

$$K_0 = \{(n,\omega) \in K; h_n(\omega) + g(M_n(\omega) > \rho - \tau\}$$

The real submartingale $g(M_n)_n$ converges a.e. and in L^1 to a random variable ψ. Let

$$A' = \{\omega \in \Omega'; (h + \psi)(\omega) > \rho - \tau\}$$

where Ω' is the subset of Ω on which $g(M_n)_n$ converges to ψ. It is clear that if $\omega \in A'$ then $(n,\omega) \in K_0$ for n large enough which implies that

$$\limsup_{n,m} \|M_n(\omega) - M_m(\omega)\| < \varepsilon. \tag{6}$$

So it remains to prove that $A' \subset A$ while having a non-zero measure. First we show that

$$\psi \leq Z + \lambda + \rho \quad \text{on the set } \Omega'. \tag{7}$$

Indeed, for $\omega_0 \in A_\lambda \cap \Omega'$ we have $\lim_m h_m(\omega_0) = 0$, hence

$$\psi(\omega_0) = \lim_m (h_m(\omega_0) + g(M_n(\omega_0))) \leq \rho. \tag{8}$$

It follows that for any $\omega \in \Omega'$,

$$\psi(\omega) - \psi(\omega_0) \leq \lim_n \|M_n(\omega) - M_n(\omega_0)\|^p \leq Z(\omega) + Z(\omega_0) \leq Z(\omega) + \lambda$$

and therefore (7) is verified. But this implies that $A' \subset A_\lambda$ since if $\omega \in \Omega' \backslash A_\lambda$ we have

$$(h + \psi)(\omega) = (\psi - 1_{D_\lambda}(Z + \lambda + 1))(\omega) = \psi(\omega) - Z(\omega) - \lambda - 1 < \rho - \tau.$$

Hence $\omega \notin A'$. To show that the latter has a non zero measure, pick $(n_1, \omega_1) \in K_0$. Since ω_1 is regular for $\{(h_{n_1}, M_{n_1}), \Sigma_{n_1}\}$ there exists $C \in \Sigma_{n_1}$ with $P(C) > 0$ and $h_{n_1}(\omega) + g(M_{n_1}(\omega)) > \rho - \tau$ for all $\omega \in C$. By the submartingale property, we get

$$\rho - \tau < \frac{1}{P(C)} \int_C (h_{n_1} + g(M_{n_1})) \, dP \leq \frac{1}{P(C)} \int_C (h + \psi) \, dP.$$

This clearly implies that $P(A') > 0$ and assertion a) of the theorem is proved.

It remains to show that (c) is only a formal strengthening of (b). Indeed, if K, φ, F and ε are as in the hypothesis of Claim (c), we consider the function ψ defined on $C = F(K)$ by

$$\psi(x) = \sup\{\varphi(t); \ t \in K \text{ and } F(t) = x\}.$$

It is clearly bounded above on C. Let $\tilde{\psi}$ be its *upper semi-continuous regularization* on the closure $\bar{C}$, i.e., for every $x \in \overline{C}$,

$$\tilde{\psi}(x) = \limsup_{y \in C, y \to x} \psi(y).$$

Apply the plurisubharmonic optimization principle (b) to find a function $g \in PSH_p(X)$ with $\|g\|_p \leq 1$ such that the set

$$C_0 := \{x \in \overline{C} : \ \tilde{\psi}(x) + g(x) > 0\}$$

is non-empty and $\text{diam}(C_0) < \varepsilon$. It is clear that the set $C_1 := \{x \in C : \psi(x) + g(x) > 0\}$ is also non-empty and is contained in C_0. Finally, it is easy to verify that the image under F of

$$K_0 := \{t \in K : \varphi(t) + g(F(t)) > 0\}$$

is non-empty and is contained in C_0. It follows that $\text{diam}(F(K_0)) \leq \varepsilon$.

Proof of Corollary 2.18: We shall only give a (sketch of) proof when X is the space L^1. For the rest, we refer to [K 1] and [K 2].

If μ is a Radon probability measure on X, we say that a vector $x \in X$ is a *Jensen barycenter* for μ, if for every $g \in PSH_1(X)$, we have

$$g(x) \leq \int g \, d\mu. \tag{9}$$

The norm of L^1 is *uniformly plurisubharmonic* in the following sense.

There exists an increasing and positive continuous function θ on $[0, \infty)$ with $\theta(0) = 0$ and $\theta(t) > 0$ for $t > 0$ such that for any tight Radon probability measure μ on X with a Jensen barycenter $x_0 \in X$, $\|x_0\| = 1$, we have

$$\int \|x\| \, d\mu(x) \geq 1 + \int \theta(\|x - x_0\|) \, d\mu(x). \tag{10}$$

For the space L^1, the modulus $\theta(t) = \frac{1}{16}\frac{t^2}{1+t}$ will do the job.

Clearly (10) implies that for any Radon probability measure μ with a Jensen barycenter $x_0 \in X$, $x_0 \neq 0$, we have

$$\int \|x\| d\mu(x) \geq \|x_0\| + \|x_0\| \int \theta(\frac{\|x - x_0\|}{\|x_0\|}) \, d\mu(x). \qquad (11)$$

If now $(F_n)_n$ is an X-valued analytic martingale, $F_0 = x_0$ is a Jensen barycenter for the probability distribution of each F_n. If we denote by $\mathbf{E}$ the integral (or expectation) with respect to Lebesgue measure on the infinite torus, it follows that

$$\mathbf{E}[\|F_n\|] \geq \|x_0\| + \|x_0\| \, \mathbf{E}[\theta(\frac{\|F_n - x_0\|}{\|x_0\|})]. \qquad (12)$$

The *conditional* version of (12) is that

$$\mathbf{E}[\|F_n\|; \Sigma_m] \geq \|F_m\| + \|F_m\| \, \mathbf{E}[\theta(\frac{\|F_n - F_m\|}{\|F_m\|}); \Sigma_m]. \qquad (13)$$

Integrating this, we get

$$\mathbf{E}[\|F_n\|] \geq \mathbf{E}[\|F_m\|] + \mathbf{E}[\|F_m\|\theta(\frac{\|F_n - F_m\|}{\|F_m\|})]. \qquad (14)$$

Since $(\|F_n\|)_n$ is an L^1-bounded real-valued submartingale, we get that it is convergent almost surely and also that $\mathbf{E}[\|F_n\|]$ is increasing and convergent (Appendix E). This implies that

$$\lim_{m,n \to \infty} \mathbf{E}[\|F_m\|\theta(\frac{\|F_n - F_m\|}{\|F_m\|})] = 0.$$

Thus $\|F_m\|\theta(\frac{\|F_n - F_m\|}{\|F_m\|})$ converges to 0 in probability. Since $(\|F_n\|)_n$ converges almost surely, we get that $\theta(\frac{\|F_n - F_m\|}{\|F_m\|})$ converges to zero in probability on the set where $(\|F_n\|)_n$ does not converge to zero. It follows that $\|F_n - F_m\|$ also converges to zero in probability. Since $(F_n)_n$ is a martingale, one can easily show that this is enough for its almost sure convergence.

Notes and Comments: Linear perturbation results belong to the theory of *Banach spaces with the Radon-Nikodym property* and results of that sort were established, with various degrees of generality, by Ekeland-Lebourg [Ek 1], Phelps [P], Bourgain [B] and Stegall [St]. Theorem 2.5 was established by Ghoussoub-Maurey. We refer to [G-M 1] for various versions and applications of that theorem. We note that Theorem 2.5 can also be proved via martingale theory. In any case, it turns out that bounded separable w^*-H_δ sets are exactly those sets which "trap" martingales. See [G-M 1] for details.

The first minimization result with plurisubharmonic perturbations was established in Banach spaces by Ghoussoub-Lindenstrauss-Maurey [G-L-M]. However, the perturbations were only proved to be C_0-small. The version included here was established in [G-M 2]. Other perturbed variational principles with different types of perturbing functions can be found in [G-M 1,3] and [G-M-S 1,2]. The concept of a *uniformly plurisubharmonic* quasi-norm as well as many other related results are due to Edgar ([E 1,2]) and Kalton ([K 1,2].

3

THE CLASSICAL MIN-MAX PRINCIPLE

In this chapter, we state and prove the min-max theorem under the standard boundary condition (see $(F0)$ below). Besides the fact that we give here *a C^1-proof*, what is relevant for us is that later, we shall be able to establish the more refined versions of the min-max principle by reducing their proofs to the one used in this chapter for the classical version of the theorem. For these reductions, we will only need to perturb appropriately the functional under study and to change the class of sets over which we are min-maxing. This will be done in the next chapter.

3.1 Homotopy-stable families

The following notion plays a central role throughout this monograph.

Definition 3.1: Let B be a closed subset of X. We shall say that a class $\mathcal{F}$ of compact subsets of X is a *homotopy-stable family with boundary B* provided
(a) every set in $\mathcal{F}$ contains B.
(b) for any set A in $\mathcal{F}$ and any $\eta \in C([0,1] \times X; X)$ satisfying $\eta(t,x) = x$ for all (t,x) in $(\{0\} \times X) \cup ([0,1] \times B)$ we have that $\eta(\{1\} \times A) \in \mathcal{F}$.

The above definition as well as all statements proved below are still valid if the boundary B is empty, provided we follow the usual convention of defining $\sup(\emptyset) = -\infty$. In this case, we will just say that $\mathcal{F}$ is a homotopy-stable family.

The following theorem covers many of the standard results like *the mountain pass* and *the saddle point* theorems as well as the various variational principles of *Ljusternik-Schnirelmann* type.

Theorem 3.2: *Let φ be a C^1-functional on a complete connected C^1-Finsler manifold X (without boundary) and consider a homotopy-stable family $\mathcal{F}$ of compact subsets of X with a closed boundary B. Set $c = c(\varphi, \mathcal{F}) = \inf\limits_{A \in \mathcal{F}} \max\limits_{x \in A} \varphi(x)$ and suppose that*

(F0) $$\sup \varphi(B) < c.$$

Then, for any sequence of sets $(A_n)_n$ in $\mathcal{F}$ such that $\limsup\limits_{n}\limits_{A_n} \varphi = c$, there exists a sequence $(x_n)_n$ in X such that

(i) $\lim\limits_{n} \varphi(x_n) = c$

(ii) $\lim\limits_{n} \|d\varphi(x_n)\| = 0$

(iii) $\lim\limits_{n} \mathrm{dist}(x_n, A_n) = 0$.

Moreover, if $d\varphi$ is uniformly continuous, then x_n can be chosen to be in A_n for each n.

The following weakening of the Palais-Smale condition turns out to be relevant for our study.

Definition 3.3: *Say that φ verifies $(PS)_c$ along a sequence $(A_n)_n$ in $\mathcal{F}$ if every sequence $(x_n)_n$ that verifies (i), (ii) and (iii) above has a convergent subsequence.*

Definition 3.4: *Say that a sequence $(A_n)_n$ in $\mathcal{F}$ is min-maxing for φ if $\limsup\limits_{n}\limits_{A_n} \varphi = c(\varphi, \mathcal{F})$.*

Note that in order to obtain $x_n \in A_n$ in the above theorem, we do not need the full strength of the uniform continuity of the derivative of φ. We only need the following

Definition 3.5: *Say that $d\varphi$ is uniformly continuous on almost critical sequences at the level c if whenever $(y_n)_n$ and $(x_n)_n$ are such that $\mathrm{dist}(x_n, y_n) \to 0$ and if $\varphi(x_n) \to c$ and $d\varphi(x_n) \to 0$, then $d\varphi(y_n) \to 0$.*

It is easy to see that if φ verifies $(PS)_c$, then $d\varphi$ is necessarily uniformly continuous on almost critical sequences at the level c.

In the sequel, we shall denote by A_∞ the set

$$A_\infty = \{x \in X; \underline{\lim}_n \, \mathrm{dist}(x, A_n) = 0\}.$$

The following is now immediate.

Corollary 3.6: *Under the hypothesis of Theorem 3.2 and assuming that φ verifies $(PS)_c$ along a min-maxing sequence $(A_n)_n$, then there* exists a sequence $(x_n)_n$ in X with $x_n \in A_n$ that converges to a point in the set $K_c \cap A_\infty$.

Before proceeding with the proofs we shall recall the various properties of Finsler manifolds that we will need in the sequel.

Let X be a C^1-Banach manifold with $T(X)$ as tangent bundle and $T_x(X)$ as tangent space at the point x. Recall that a *Finsler structure* on $T(X)$ is a continuous function $\| \ \| : T(X) \to [0, +\infty)$ such that

(a) for each $x \in X$, the restriction $\| \ \|_x$ of $\| \ \|$ to $T_x(X)$ is a norm on the latter.

(b) for each $x_0 \in X$ and $k > 1$, there is a trivializing neighborhood U of x_0 such that

$$\frac{1}{k} \| \ \|_x \leq \| \ \|_{x_0} \leq k \| \ \|_x \text{ for all } x \in U.$$

If $\sigma : [a,b] \to X$ is a C^1-path, the length of σ is defined by $L(\sigma) = \int_a^b \|\dot{\sigma}(t)\| \, dt$. The distance $\rho(x,y)$ between two points x and y in the same connected component of X is defined as the infimum of $L(\sigma)$ over all σ joining x and y. ρ is then a metric on each component of X (called the *Finsler metric*) and it is consistent with the topology of X.

If $\varphi \in C^1(X, \mathbb{R})$, the differential of φ at x—denoted by $d\varphi(x)$—is an element of the cotangent space of X at x which is the dual space $T_x(X)^*$. The latter equipped with the dual norm induces a dual Finsler structure on the cotangent bundle $T(X)^*$.

Suppose now that x is not a critical point for φ, there exists then a vector $v(x)$ in $T_x(X)$ such that

(c) $\|v(x)\| < 2\|d\varphi(x)\|$ and $\langle v(x), d\varphi(x) \rangle > \|d\varphi(x)\|^2$.

Moreover, by using a suitable partition of unity, one can construct the *pseudo-gradient vector field* $v(x)$ continuously on the set $R = \{x \in X : d\varphi(x) \neq 0\}$ of regular points of φ.

Indeed, since $d\varphi$ is continuous, we can insure that for each $x \in R$, there exists a vector $v(x)$ in $T_x(X)$ and a neighborhood $W(x)$ of x such that for every $w \in W(x)$,

$$\|v(x)\| < 2\|d\varphi(w)\| \text{and}   \langle v(x), d\varphi(w) \rangle > \|d\varphi(w)\|^2.$$

Since a Finsler manifold is necessarily paracompact, there exists a locally finite refinement $(W_i)_{i \in I}$ of the cover $\{W(x); x \in R\}$. Choose now a continuous partition of unity $(\ell_i)_{i \in I}$ subordinate to $(W_i)_{i \in I}$: that is, a family of continuous functions $\ell_i : X \to [0,1]$ that are supported on W_i such that $\sum_{i \in I} \ell_i = 1$ on R. Finally, we let $v(x) = \sum_{i \in I} \ell_i(x)v(x_i)$. It is easy to verify that v is the required vector field. For more details about pseudogradients on Finsler manifolds we refer to Palais [Pa 1].

3.2 The basic deformation lemma for C^1-manifolds

The following lemma will be frequently used throughout this monograph.

Lemma 3.7: *Let φ be a C^1-functional on a complete connected C^1-Finsler manifold X and let B and C be two closed and disjoint subsets of X. Suppose that C is compact and that $\|d\varphi(x)\| > 2\varepsilon > 0$ for every $x \in C$. Then, for each $k > 1$, there exist a positive continuous function g on X and a deformation α in $C([0,1] \times X; X)$ such that for some $t_0 > 0$, the following holds for every $t \in [0, t_0)$,*

(i) $\alpha(t, x) = x$ for every $x \in B$.

(ii) $\rho(\alpha(t, x), x)) \leq kt$ for every $x \in X$.

(iii) $\varphi(\alpha(t, x)) - \varphi(x) \leq -\varepsilon g(x)t$ for every $x \in X$.

(iv) $g(x) = 1$ for all $x \in C$.

Proof: Fix $k > 1$ and use (a), (b) and (c) above to find for each $x_i \in C$, a chart $f_i : U_i \to T_{x_i}(X)$ and $v_i \in T_{x_i}(X)$ such that

$$\frac{1}{k} \| \; \|_x \leq \| \; \|_{x_i} \leq k \| \; \|_x \quad \text{for all } x \in U_i \tag{1}$$

and

$$\langle ((\varphi \circ f_i^{-1})'(y), \frac{v_i}{\|v_i\|} \rangle \geq \frac{1}{2} \|d\varphi(x_i)\| \quad \text{for all } y \in f_i(U_i). \tag{2}$$

Let $V_i \subset U_i$ be an open neighborhood of x_i such that for some $\delta_i > 0$,

$$B(V_i, \delta_i) \subset U_i \quad \text{and} \quad B(f_i(V_i), \delta_i) \subset f_i(U_i) \tag{3}$$

where $B(W, \delta)$ denotes the δ-neighborhood of the set W in the appropriate metric.

Since C is compact, we can find a finite covering $(V_i)_{i=1}^m$ to which we can associate a continuous partition of unity $(\chi_i)_{i=1}^m$. Let $\ell : X \to [0,1]$ be a continuous function such that

$$\ell(x) = \begin{cases} 1 & \text{if } x \in C \\ 0 & \text{if } x \in (X \backslash \cup_{i=1}^m V_i) \bigcup B. \end{cases}$$

Let $\delta_0 = \min\{\delta_1, ...\delta_m\}$ and $t_0 = \frac{\delta_0}{1+k^2}$. Starting with $\alpha_0(t, x) = x$, we define by induction on j $(1 \leq j \leq m)$, the functions

$$\alpha_j(t, x) = \begin{cases} f_j^{-1}(f_j(\alpha_{j-1}(t, x)) - t\ell(x)\chi_j(x)\frac{v_j}{\|v_j\|}) & \text{if } \alpha_{j-1}(t, x) \in U_j \\ \alpha_{j-1}(t, x) & \text{otherwise.} \end{cases}$$

We now prove that if $t \in (0, t_0)$, we have for each $1 \leq j \leq m$,

$$\alpha_j(t, x) \quad \text{is well defined and is continuous} \tag{4}$$

(i.e., $f_j(\alpha_{j-1}(t, x)) - t\ell(x)\chi_j(x)\frac{v_j}{\|v_j\|} \in f_j(U_j)$ if $\chi_j(x) \neq 0$ and $\alpha_{j-1}(t, x) \in U_j$.)

$$\rho(\alpha_{j-1}(t, x), \alpha_j(t, x)) \leq k\ell(x)\chi_j(x)t \tag{5}$$

and

$$\varphi(\alpha_j(t,x)) - \varphi(\alpha_{j-1}(t,x)) \le -\varepsilon\ell(x)\chi_j(x)t. \tag{6}$$

Indeed, for $j = 1$, (4) follows immediately from (3). To show (5), let x be any point in U_1 and consider the path $\sigma_1(s) = \alpha_1(s,x), (0 \le s \le t)$ joining x to $\alpha_1(t,x)$. We have

$$\rho(x,\alpha_1(t,x)) \le \int_0^t \|\sigma_1'(s)\| ds \le k \int_0^t \|\frac{d}{ds} f_1(\sigma_1(s))\|_{x_1} ds = k\ell(x)\chi_1(x)t.$$

To show (6), let $x \in U_1$ and use the mean value theorem to find $\theta \in (0,1)$ such that

$$\varphi(\alpha_1(t,x)) - \varphi(x)$$
$$= \varphi \circ f_1^{-1}(f_1(x) - t\ell(x)\chi_1(x)\frac{v_1}{\|v_1\|}) - \varphi \circ f_1^{-1}(f_1(x))$$
$$= t\ell(x)\chi_1(x)\langle(\varphi \circ f_1^{-1})'(f_1(x) - \theta t\ell(x)\chi_1(x)\frac{v_1}{\|v_1\|}, \frac{v_1}{\|v_1\|}\rangle.$$

It follows from (2), that if $x \in U_1$ then

$$\varphi(\alpha_1(t,x)) - \varphi(x) \le -\varepsilon\ell(x)\chi_1(x)t$$

The same trivially holds when $x \notin U_1$ since then $\chi_1(x) = 0$.

Suppose now (4), (5) and (6) verified up to $j-1$ and let us prove them for $\alpha_j(t,x)$. First note that (5) and the triangular inequality give

$$\rho(x,\alpha_{j-1}(t,x)) \le k t\ell(x)\sum_{k=1}^{j-1}\chi_k(x) \le kt. \tag{7}$$

Since $t < \delta_0/(1+k^2)$ we get that $\rho(x,\alpha_{j-1}(t,x)) \le \delta_0/2$. But this implies that for any $x \in \mathrm{supp}(\chi_j)$ with $\alpha_{j-1}(t,x) \in U_j$ we have

$$\rho(x,\alpha_{j-1}(t,x)) = \inf\{L(\sigma); \sigma \text{ joining } x \text{ to } \alpha_{j-1}(t,x) \text{ and } \sigma \subset U_j\} \tag{8}$$

because if such a path σ leaves U_j we would have that $L(\sigma) \ge \delta_0$ in view of the fact that $\rho(x, X\backslash U_j) \ge \delta_0$. But if $\sigma \subset U_j$ joins x to $\alpha_{j-1}(t,x)$ we have

$$L(\sigma) = \int_a^b \|\sigma'(s)\| ds \ge \frac{1}{k}\int_a^b \|\frac{d}{ds}f_j(\sigma(s))\|_{x_j} ds$$
$$\ge \frac{1}{k}\|\int_a^b \frac{d}{ds}f_j(\sigma(s)) ds\|_{x_j}$$
$$= \frac{1}{k}\|f_j(\alpha_{j-1}(t,x)) - f_j(x)\|_{x_j}.$$

It follows that

$$\|f_j(\alpha_{j-1}(t,x)) - f_j(x)\|_{x_j} \le k\rho(x,\alpha_{j-1}(t,x)) \le k^2 t.$$

Hence

$$\|f_j(\alpha_{j-1}(t,x)) - t\ell(x)\chi_j(x)\frac{v_j}{\|v_j\|} - f_j(x)\|_{x_j} \le k^2 t + t < \delta_0.$$

This implies that $f_j(\alpha_{j-1}(t,x)) - t\ell(x)\chi_j(x)\frac{v_j}{\|v_j\|} \in f_j(U_j)$ whenever $x \in \text{supp}(\chi_j)$ and $\alpha_{j-1}(t,x) \in U_j$ which clearly proves (4).

For (5) it is enough to consider the path

$$\sigma_j(s) = f_j^{-1}(f_j(\alpha_{j-1}(t,x)) - s\ell(x)\chi_j(x)\frac{v_j}{\|v_j\|})$$

for $0 \leq s \leq t$ and to note that

$$\rho(\alpha_{j-1}(t,x),\alpha_j(t,x)) \leq L(\sigma_j) \leq \int_0^t \|\sigma_j'(s)\|\, ds \leq k\ell(x)\chi_j(x)t.$$

Assertion (6) is also shown as in the case $j=1$ by using the mean value theorem between $\alpha_{j-1}(t,x)$ and $\alpha_j(t,x)$. The induction is complete.

Set now $\alpha(t,x) = \alpha_m(t,x)$ and $g(x) = \ell(x)\sum_{j=1}^m \chi_j(x)$. They clearly satisfy the properties claimed in the lemma.

Proof of Theorem 3.2: Let A be a set in $\mathcal{F}$ such that

$$c \leq \sup \varphi(A) < c + \varepsilon^2.$$

Consider the subspace $\mathcal{L}$ of $C([0,1] \times X; X)$ consisting of all continuous deformations η such that

$$\eta(t,x) = x \quad \text{for all } (t,x) \text{ in } K_0 = (\{0\} \times X) \cup ([0,1] \times B)$$

and $\sup\{\rho(\eta(t,x),x); t \in [0,1], x \in X\} < +\infty$.

The space $\mathcal{L}$ is a complete metric space once equipped with the following metric

$$\delta(\eta,\eta') = \sup\{\rho(\eta(t,x),\eta'(t,x)); (t,x) \in [0,1] \times X\}.$$

Define a function $I : \mathcal{L} \to \mathbb{R}$ by $I(\eta) = \sup\{\varphi(\eta(1,x)); x \in A\}$. Let $\bar{\eta}$ be the identity in $\mathcal{L}$, that is $\bar{\eta}(t,x) = x$ for all (t,x) in $[0,1] \times X$ and note that

$$I(\bar{\eta}) = \sup\{\varphi(x); x \in A\} < c + \varepsilon^2 \leq \inf\{I(\eta); \eta \in \mathcal{L}\} + \varepsilon^2.$$

Apply now Ekeland's theorem to get η_0 in $\mathcal{L}$ such that

$$I(\eta_0) \leq I(\bar{\eta}) \tag{1}$$

$$\delta(\eta_0,\bar{\eta}) \leq \varepsilon \tag{2}$$

$$I(\eta) \geq I(\eta_0) - \varepsilon\delta(\eta,\eta_0) \quad \text{for all } \eta \text{ in } \mathcal{L}. \tag{3}$$

Let $C = \{x \in \eta_0(\{1\} \times A); \varphi(x) = I(\eta_0)\}$ and note that hypothesis (F0) implies that $C \cap B = \emptyset$.

It is enough to prove the following:

Claim: There exists $x_\varepsilon \in C$ such that $\|d\varphi(x_\varepsilon)\| \leq 4\varepsilon$.

Indeed, in view of (1) and (2), any point verifying the claim necessarily satisfies $c \leq \varphi(x_\varepsilon) < c + \varepsilon^2$ and $\text{dist}(x_\varepsilon, A) \leq \varepsilon$.

Suppose now that the above claim false. Fix $1 < k < 2$ and apply

Lemma 3.7 above to the sets C and B to get $\alpha(t, x)$ satisfying the conclusion of that Lemma with a suitable function g and a time $t_0 > 0$. For $0 < \lambda < t_0$, we consider the function $\eta_\lambda(t, x) = \alpha(t\lambda, \eta_0(t, x))$. It belongs to $\mathcal{L}$ since it is clearly continuous on $[0, 1] \times X$ and since for all $(t, x) \in (\{0\} \times X) \cup ([0, 1] \times B)$, we have

$$\eta_\lambda(t, x) = \alpha(t\lambda, \eta_0(t, x)) = \alpha(t\lambda, x) = x$$

Since $\delta(\eta_\lambda, \eta_0) < kt\lambda \leq k\lambda$, we get from (3) that $I(\eta_\lambda) \geq I(\eta_0) - \varepsilon k\lambda$. Since A is compact, let $x_\lambda \in A$ be such that $\varphi(\eta_\lambda(1, x_\lambda)) = I(\eta_\lambda)$. We have that

$$\varphi(\eta_\lambda(1, x_\lambda)) - \varphi(\eta_0(1, x)) \geq -\varepsilon k\lambda \quad \text{for every } x \in A. \tag{4}$$

On the other hand, by (iii) of lemma 3.7, we have for all x_λ

$$\begin{aligned}
\varphi(\eta_\lambda(1, x_\lambda)) - \varphi(\eta_0(1, x_\lambda)) &= \varphi(\alpha(\lambda, \eta_0(1, x_\lambda))) - \varphi(\eta_0(1, x_\lambda)) \\
&\leq -2\varepsilon\lambda g(\eta_0(1, x_\lambda)).\}
\end{aligned} \tag{5}$$

Combining (4) and (5) we get

$$-\varepsilon k \leq -2\varepsilon g(\eta_0(1, x_\lambda)). \tag{6}$$

If now x_0 is any cluster point of (x_λ) when $\lambda \to 0$, we have from (4) that $\eta_0(1, x_0) \in C$ and hence $g(\eta_0(1, x_0)) = 1$. Since $k < 2$, this clearly contradicts (6) and therefore the initial claim was true. The proof of the theorem is complete.

3.3 The standard (PS) condition and a subcritical elliptic boundary value problem

We first present a simple situation where $(PS)_{F,c}$ holds for any real c and any bounded set F.

Proposition 3.8: Let φ be a C^1-functional on a Banach space X such that φ' can be decomposed into $L + K$ where L is a linear isomorphism and K is a compact map from X into X^*. Then φ verifies $(PS)_{F,c}$ for any real c and any bounded set F.

Proof: If $(x_n)_n$ is bounded, then by the compactness of K, we can assume, modulo passing to a subsequence, that $(K(x_n))_n$ is convergent. If now $\varphi'(x_n) = L(x_n) + K(x_n) \to 0$, it follows that $(L(x_n))_n$ and therefore $(x_n)_n$ are also convergent.

Remark 3.9: It follows from Proposition 3.8 and Corollary 1.5 that such a functional φ verifies $(PS)_c$ for any $c \in \mathbb{R}$ if and only if it is coercive.

We now present a typical example where the Palais-Smale condition holds. It deals with elliptic equations having *subcritical* non-linearities.

A subcritical semilinear elliptic boundary value problem

Let Ω be smoothly bounded domain in $\mathbb{R}^N$, $N \geq 3$, and suppose that $g : \Omega \times \mathbb{R} \to \mathbb{R}$ is a Caratheodory function with primitive $G(\cdot, u) = \int_0^u g(\cdot, v)dv$. We consider the following equation

$$
\begin{cases}
-\Delta u = g(x, u) & x \in \Omega, \\
u = 0 & x \in \partial\Omega.
\end{cases}
\tag{1}
$$

Theorem 3.10: *Suppose g is a Caratheodory function that verifies the following conditions*

$$
\lim_{s \to 0} \frac{g(x, s)}{s} < \lambda_1,
\tag{2}
$$

$$
|g(x, s)| \leq C(1 + |s|^{p-1}) \text{ for almost all } x \in \Omega \text{ and all } s,
\tag{3}
$$

$$
0 < qG(x, s) \leq g(x, s)s \text{ for almost all } x \in \Omega \text{ and } |s| \geq R_0,
\tag{4}
$$

where $2 < p < 2^* = \frac{2N}{N-2}$, $q > 2$ and R_0 is some positive constant.
Then, problem (1) admits a non-trivial weak solution in $H_0^1(\Omega)$.

Proof: Define

$$
\varphi(u) = \frac{1}{2} \int_\Omega |\nabla u|^2 dx - \int_\Omega G(x, u)\, dx.
$$

By appendix B, we know that condition (3) implies that φ is Fréchet differentiable on $X = H_0^1(\Omega)$ and that the assertion of the theorem is equivalent to finding a non-trivial critical point for φ. The proof consists of applying the classical Mountain pass lemma which is a particular case of Theorem 3.2. We first prove the following:

Lemma 3.11: *If g is a Caratheodory function that satisfies condition (3) above, then φ verifies $(PS)_{F,c}$ for any $c \in \mathbb{R}$ and any bounded set F in $X = H_0^1$.*
If in addition, g verifies condition (4), then φ verifies $(PS)_c$ for any c.

Proof of Lemma 3.11: Let $L : X \to X^*$ be the duality map. That is for u and $v \in X$,

$$
Lu(v) = \int_\Omega \nabla u \cdot \nabla v\, dx.
$$

Note that

$$
\varphi'(u) = Lu - \mathcal{G}'(u)
\tag{5}
$$

where $\mathcal{G}(u) = \int_\Omega G(x, u)dx$, and the operator $\mathcal{G}' = N_g$ is compact from X into X^*, in view of the compactness of the Sobolev imbedding (Appendix B). Proposition 3.8 then applies to give that φ verifies $(PS)_{F,c}$ for any $c \in \mathbb{R}$ and any bounded set F in $X = H_0^1$.
Suppose now that g verifies (4) and that $|\varphi(u_n)| \leq M$ for all n. It

remains to show that $(u_n)_n$ is bounded in X. To do that, let $T(x,s) = q^{-1}g(x,s)s - G(x,s)$ which is non-negative for almost all x and for large s. We have

$$
\begin{aligned}
M + q^{-1}\|u_n\| &\geq \varphi(u_n) - q^{-1}\langle \varphi'(u_n), u_n\rangle \\
&= \left(\frac{1}{2} - \frac{1}{q}\right)\|u_n\|^2 + \int_\Omega T(x,u_n)dx \\
&\geq \left(\frac{1}{2} - \frac{1}{q}\right)\|u_n\|^2 \\
&\quad + \int_{\{|u_n|\geq R_0\}} T(x,u_n)dx + \int_{\{|u_n|<R_0\}} T(x,u_n)dx.
\end{aligned}
\tag{6}
$$

Note that the first term on the last line in (6) is non-negative while the second term is bounded by a constant independent of n. Since $(2^{-1} - q^{-1}) > 0$, we can deduce easily from (6) that (u_n) is bounded in X. The proof of the lemma is complete.

To set up the appropriate min-max principle, we need the following:

Claim (i): There is $\rho > 0$ and $a > 0$ so that $\inf \varphi(S_\rho(0)) \geq a$.

By (2), we get for some $0 < \mu < \lambda_1$, a $\delta > 0$ such that

$$|s| \leq \delta \text{ implies } |G(x,s)| \leq \tfrac{1}{2}\mu|s|^2 \text{ for all } x \in \Omega.$$

By (3), there is a constant $M(\delta) > 0$ such that

$$|s| \geq \delta \text{ implies } |G(x,s)| \leq M|s|^p \text{ for all } x \in \Omega.$$

Combining these two estimates, we get for all u and all $x \in \Omega$,

$$|G(x,s)| \leq \frac{1}{2}\mu|s|^2 + M|s|^p.$$

From the Poincaré and the Sobolev inequalities, we get

$$\varphi(u) = \frac{1}{2}\|u\|^2 - \mathcal{G}(u) \geq \frac{1}{2}(1 - \frac{\mu}{\lambda_1})\|u\|^2 - M\|u\|^p.$$

Since $p > 2$, this implies Claim (i).

Claim (ii): There exists $e \in X \setminus B_\rho(0)$ such that $\varphi(e) < 0$.

Indeed, condition (4) implies that for two positive constants C, D, we have

$$G(x,s) \geq C\|s\|^q - D \quad \text{for all } x \in \Omega \text{ and } s \in \mathbb{R}.$$

It follows that for a fixed $u \in H_0^1$, we have for any $R > 0$, that

$$\varphi(Ru) = \frac{R^2}{2}\int_\Omega |\nabla u|^2 dx - \int_\Omega G(x,Ru)dx \leq \frac{R^2}{2}\|u\|^2 - R^q C\|u\|_q^q + D|\Omega|,$$

where $|\Omega|$ denotes the measure of the domain Ω. This implies Claim (ii).

Consider now $\mathcal{F}_0^e$ to be the class of all continuous paths joining 0 to e

in X. It is clear that $\mathcal{F}_0^e$ is homotopy-stable with boundary $B = \{0, e\}$. Moreover,

$$\max\{\varphi(0), \varphi(e)\} \leq 0 < a \leq \inf \varphi(B_\rho(0)) \leq c(\varphi, \mathcal{F}_0^e).$$

Hence Theorem 3.2 applies and we get a nontrivial critical point at level $c > 0$ since by Lemma 3.11, φ verifies the $(PS)_c$ condition.

3.4 The (PS) condition along a min-maxing sequence and a resonance problem

We now present an example where the Palais-Smale condition is not satisfied at all levels. However it will be satisfied along an appropriate min-maxing sequence.

An asymptotically linear boundary value problem with strong resonance

Let $\Omega \subset \mathbf{R}^N$ be a bounded regular domain, and let λ_s be an eigenvalue of the problem

$$-\Delta u = \lambda u \quad \text{in} \quad \Omega$$

$$u = 0 \quad \text{on} \quad \partial\Omega.$$

We consider the following resonance problem

$$-\Delta u - \lambda_s u = g(u) + h \quad \text{in} \quad \Omega, \tag{1}$$

$$u = 0 \quad \text{on} \quad \partial\Omega,$$

where g is a continuous function that verifies one of the following two conditions

$$\lim_{|s| \to \infty} g(s) = 0 \quad \text{and} \quad \int_{-\infty}^{+\infty} G(s)\, ds = 0 \tag{2}$$

or, in case g oscillates at infinity, we will assume that

$$g \text{ has mean zero and is } T\text{-periodic for some } T > 0. \tag{3}$$

We shall prove the following

Theorem 3.12: *Let $h \in H^{-1}(\Omega)$ be orthogonal, in the duality between $H_0^1(\Omega)$ and $H^{-1}(\Omega)$, to the eigenspace corresponding to λ_s and assume that g verifies one of the conditions (2) or (3). Then, problem (1) has at least one (weak) solution.*

To apply the variational approach, we need to consider on the space $X := H_0^1(\Omega)$ the functional $\varphi \in C^1(X)$ defined as

$$\varphi(u) := J(u) + \mathcal{G}(u)$$

where

$$J(u) := \int_\Omega (|\nabla u|^2 - \lambda_s |u|^2)\, dx - \int_\Omega hu\, dx \tag{4}$$

and

$$\mathcal{G}(u) := \int_\Omega G(u)\, dx$$

G being an antiderivative of g.

Suppose that the multiplicity of λ_s is j. We will write W for the linear span of $\{e_{k+1}, ..., e_n\}$ where $j = n - k$ and $e_{k+1}, ..., e_n$ are the eigenfunctions associated to the eigenvalue λ_s. We also consider a solution $\bar{u}$ in X to

$$-\triangle u - \lambda_s u = h$$

which exists because $h \in W^\perp$ and we can assume $\bar{u} \in W^\perp$.

We shall need the following Riemann-Lebesgue type lemma whose proof is left to the reader (See also [S 2]). We recall that N_g denotes the Nemitskii operator associated with g. See Appendix B for the relevant properties of such an operator.

Lemma 3.13: *Let g be a continuous function satisfying either condition (2) or (3). If $\ell, \ell_n \in C^1(\bar{\Omega}), \ell = \lim_n \ell_n$ in $C^1(\bar{\Omega})$ and $\nabla \ell \neq 0$ almost everywhere, and if U is a precompact subset of $H_0^1(\Omega)$, then*
(a) $\lim_n N_g(u + n\ell_n) = 0$ *weakly in* $H^{-1}(\Omega)$,
(b) $\lim_n \mathcal{G}(u + n\ell_n) = 0$,
both (a) and (b) holding uniformly for $u \in U$.

The following Lemma shows the type of restricted compactness we have to deal with.

Lemma 3.14: *Under the hypothesis of Theorem 3.12, we have*
(i) *The functional φ verifies $(PS)_{W^\perp,c}$ for all $c \in \mathbb{R}$.*
(ii) *φ verifies $(PS)_c$ for all $c \neq J(\bar{u})$.*

Proof: (i) Let P denote the projection on $W^\perp$ and suppose $(u_n)_n \in X$ is such that $\varphi'(u_n) \to 0$. That is

$$-\triangle u_n - \lambda_s u_n = g(u_n) + h + \rho_n \tag{5}$$

with $\rho_n \to 0$ in H^{-1}.

Write $u_n = v_n + w_n$ with $w_n = (I - P)(u_n) \in W$ and $v_n = P(u_n) \in W^\perp$. It is enough to show that v_n is convergent in X, since the hypothesis implies that $(w_n)_n$ is eventually bounded in the finite dimensional space W and hence it has a converging subsequence.

To do that, project equation (5) by P to get

$$-\triangle v_n - \lambda_s v_n = Pg(u_n) + h + \varepsilon_n \tag{6}$$

with $\varepsilon_n \to 0$ in H^{-1}.
Since λ_s is not an eigenvalue for $-\triangle$ on $W^\perp$ and $(v_n)_n$ is contained in

$W^\perp$, the latter sequence is clearly bounded. Moreover, since $\varepsilon_n \in W^\perp$, we let z_n be solutions of

$$-\Delta z_n - \lambda_s z_n = \varepsilon_n. \tag{7}$$

By (6) and (7), we see that $v_n - z_n - P\bar{u}$ is actually bounded in $W^{2,p}(\Omega)$ for all p and therefore it can be supposed to be converging. Since also $(z_n)_n$ is going to zero, we can deduce that $(v_n)_n$ itself is converging.

(ii) It will be enough to prove that if $\|w_n\| \to \infty$, then $\varphi(u_n) \to J(\bar{u})$. For infinitely many $n \in N$, there exists k_n such that $\lim_n n^{-1}\|w_{k_n}\| = 1$. For these n's we define $\ell_n = n^{-1}w_{k_n}$ and by passing to a further subsequence, we can assume that $\ell_n \to \ell$. Moreover, since ℓ is a normalized eigenfunction, we have that $\nabla \ell \neq 0$ almost everywhere. The above lemma now implies that $N_g(v_n + n\ell_n) \to 0$ weakly (note that we can assume that $\ell_n = \ell$ for those n for which ℓ_n has not yet been defined). By passing to the weak limit in (5), one gets $-\Delta v - \lambda_s v = h$, i.e. $v = \bar{u}$. If we pass to the limit in (4) we get from part (b) of the lemma that $\varphi(u_n) = J(v_n) + \mathcal{G}(u_n) \to J(v) = J(\bar{u})$. The proof of the lemma is complete.

To implement the min-max principle, set $E_k := \text{span}\{e_1, ..., e_k\}$ and for any bounded neighborhood U of $\bar{u}$ in $\bar{u}+E_k$, consider the class $\mathcal{H}_k(U)$ of all continuous $\eta : X \to X$ such that $\eta(u) = u$ for all $u \in \partial U$ (where ∂ denotes the relative boundary of a set in the space spanned by itself). We then consider the following classes of sets

$$\mathcal{A}_k(U) : = \{A \subset E; \exists \eta \in \mathcal{H}_k : A = \eta(\bar{U})\},$$

$$\bar{\mathcal{A}}_k(U) : = \{A \subset E; A \text{ compact and } \forall \eta \in \mathcal{H}_k : \eta(A) \cap (\bar{u} + E_k^\perp) \neq \emptyset\}.$$

Lemma 3.15: *The classes $\mathcal{A}_k(U)$ and $\bar{\mathcal{A}}_k(U)$ are two homotopy-stable families with boundary ∂U. They also satisfy the following properties:*
(i) $\mathcal{A}_k(U) \subset \bar{\mathcal{A}}_k(U)$.
(ii) If V is a neighborhood of $\bar{u}$ in $\bar{u} + E_n$ $(1 \leq k < n)$ such that $V \cap \{e_{k+1}, ..., e_n\}^\perp = U$, then for every $A \in \mathcal{A}_n(V)$, we have

$$A \cap \{e_{k+1}, ..., e_n\}^\perp \in \bar{\mathcal{A}}_k(U).$$

Proof: We shall assume without loss of generality that $\bar{u} = 0$, that $U = B_k := B_k(0, \bar{r})$ is the open ball in E_k with radius $\bar{r}$ fixed and centered at 0, while $V = B_n := B_n(0, \bar{r})$ is the open ball in E_n with the same radius. Denote by P_k the orthogonal projection of X on E_k and by Q the orthogonal projection of X on $W = \text{Span}\{e_{k+1}, ..., e_n\}$.

To prove (i), take any $\eta \in \mathcal{H}_k(U)$ and note that $P_k \circ \eta : \bar{U} \to E_k$ is continuous and is equal to the identity on the boundary of U. Brouwer's fixed point theorem gives then a point $x \in \bar{U}$ such that $P_k \circ \eta(x) = 0$

(Appendix D). But this means that $E_k^\perp \cap \eta(\bar{U}) \neq \emptyset$ and our claim is proved.

For (ii), suppose $\eta' \in \mathcal{H}_k$ is given. We need to show that

$$\eta'(A \cap W^\perp) \cap E_k^\perp \neq \emptyset. \qquad (8)$$

Let $\eta \in \mathcal{H}_n$ such that $\eta(\bar{B}_n) = A$. We must prove the existence of a zero for the map $P_k \circ \eta'$ in the set of zeros for the map Q; in other words, we are looking for the zeroes of the product map

$$G := P_k \circ \eta' \circ \eta \times Q \circ \eta : B_n \to E_k \times W$$

$$u \to (P_k \circ \eta' \circ \eta(u), Q \circ \eta(u)).$$

Consider the following change of variable ζ that sends the cylinder $C = B_k \times B$, where B is the ball of W centered at zero and with radius r, into the ball B_n: Note that $u \in C$ means $\|v\| < r, \|w\| < r$ while $u \in B_n$ means $\|v\|^2 + \|w\|^2 < r^2$, if $u = v + w, v \in E_k, w \in W$, so we may define $\zeta : \bar{C} \to \bar{B}_n$ as follows: for $u = v + w \in \bar{C}$, let $\zeta(u) = v + r^{-1}w\sqrt{(r^2 - \|v\|^2)}$; this map is clearly a homeomorphism from C onto B_n.

Observe that $\partial C = (\partial B_k \times \bar{B}) \cup (B_k \times \partial B)$ where

$$\partial B_k \times \bar{B} = \{u \in \bar{C}; \|v\| = r\},$$

$$B_k \times \partial B = \{u \in \bar{C}; \|w\| = r, \|v\| < r\}.$$

We also have

$$\text{for all } v + w \in \partial B_k \times \bar{B}, \quad \zeta(v + w) = v \in \partial B_k,$$

$$\text{for all } v + w \in B_k \times \partial B, \quad \zeta(v + w) \in \partial B_n.$$

Our problem is therefore equivalent to proving the existence in C of zeroes of the map $\bar{G} := G \circ \zeta$ and we obtain this by proving that $\deg(\bar{G}, C, 0) \neq 0$ (Appendix D). First of all, it is clear that its degree is well defined because no zeroes of $\bar{G}$ are on ∂C. In fact, as $\eta_{|\partial B_k} = \text{id}$ and $\eta'_{|\partial B_k} = \text{id}$, one has

$$\forall v + w \in \partial B_k \times \bar{B} \qquad P_k \circ \eta' \circ \eta \circ \zeta(u) = v \neq 0,$$

$$\forall u \in B_k \times \partial B \qquad Q \circ \eta \circ \zeta(u) = \frac{\sqrt{r^2 - \|v\|^2}}{r} w \neq 0. \qquad (9)$$

Furthermore, the homotopy $H(t, u) := (1 - t)\bar{G} + tI$ is admissible, where $I = P_k \times P$. In fact, by (9),

$$\forall u \in \partial B_k \times \bar{B} \qquad \pi_1 \circ \bar{G}(u) = v = P_k(u),$$

$$\forall u \in B_k \times \partial B \qquad \pi_2 \circ \bar{G}(u) = \frac{\sqrt{r^2 - \|v\|^2}}{r} w = \frac{\sqrt{r^2 - \|v\|^2}}{r} P(u),$$

where π_1, π_2 are the usual projections on the first and second components respectively. That means $\deg(\bar{G}, C, 0) = 1$ and the proof is complete.

Proof of Theorem 3.12: Since J is concave on $\bar{u} + E_k$ and $J'(\bar{u}) = 0$ we get that

$$J(\bar{u}) = \max_{\bar{u}+E_k} J. \tag{10}$$

On the other hand, since φ is bounded below on $\bar{u} + E_k^\perp$ and since it goes quadratically to $-\infty$ on E_k, we have for some $\bar{r} > 0$,

$$\sup_{\partial B_k(\bar{u},\bar{r})} \varphi < \inf_{\bar{u}+E_k^\perp} \varphi. \tag{11}$$

Actually, we can strengthen (11) to

$$\sup_{\partial B_k(\bar{u},\bar{r})+W} \varphi < \inf_{\bar{u}+E_k^\perp} \varphi \tag{12}$$

because the component belonging to W gives only a bounded contribution in the evaluation of φ. Let

$$c_k := \inf_{A\in\mathcal{A}_k} \sup_A \varphi \quad \text{and} \quad \bar{c}_k := \inf_{A\in\bar{\mathcal{A}}_k} \sup_A \varphi$$

where $\mathcal{A}_k$ and $\bar{\mathcal{A}}_k$ are the classes corresponding to $U = B_k(\bar{u},\bar{r})$. We shall distinguish the following three cases:

Case (1): $c_k \neq J(\bar{u})$ or $\bar{c}_k \neq J(\bar{u})$.

We then have that at least one of them is a critical level for φ. Indeed, by (11) and Lemma 3.15.(i), we have

$$\sup_{\partial B_k(\bar{u},\bar{r})} \varphi < \inf_{\bar{u}+E_k^\perp} \varphi \leq \bar{c}_k \leq c_k.$$

Hence the classical min-max Theorem 3.2 coupled with Lemma 3.14.(ii) yield our claim.

It remains to prove that φ has some critical level in the case where $c_k = \bar{c}_k = J(\bar{u})$. To try another level, consider for each $\alpha \in \mathbb{N}$, the ball $B(0,\alpha)$ in E centred at zero with radius α and let $C(\alpha) = B_k(\bar{u},\bar{r})+B(0,\alpha)\cap W$. We shall consider the class $\mathcal{A}_n(C(\alpha))$ and its corresponding min-max value $c_n(\alpha) = c(\varphi, \mathcal{A}_n(C(\alpha)))$. Also define

$$t(\alpha) = \sup_{B_k+\partial B(0,\alpha)\cap W} \varphi \quad \text{and} \quad a(\alpha) = \sup_{\partial C(\alpha)} \varphi.$$

It is easily seen that

$$\lim_{\alpha\to\infty} t(\alpha) = J(\bar{u}). \tag{13}$$

Indeed, one can first show that for $v \in B_k$, and any $w_\alpha \in \partial B(0,\alpha)\cap W$, $\varphi(v+w_\alpha)$ converges to $J(v)$ and then use (10) and Lemma 3.13 to deduce the uniform convergence to $J(\bar{u})$. But this implies the following:

Claim : If $\bar{c}_k \leq J(\bar{u})$ then for large enough α, we have that $t(\alpha) = a(\alpha)$.

Indeed, by (12) and the fact that $\inf_{\bar{u}+E_k^\perp} \varphi \leq \bar{c}_k$, one has

$$\sup_{\partial B_k + W} \varphi < \bar{c}_k \leq J(\bar{u})$$

so that if α is large, (13) will give $a(\alpha) = t(\alpha)$.

Case (2): Assume $\bar{c}_k = c_k = J(\bar{u})$ and there exists an α such that $c_n(\alpha) > a(\alpha)$.

Here again we can apply Theorem 3.2 and Lemma 3.14.(ii) to deduce that $c_n(\alpha)$ is a critical level since the following boundary condition holds

$$J(\bar{u}) = \bar{c}_k = \inf_{A \in \mathcal{A}_k^*} \sup_A \varphi \leq \sup_{\partial C(\alpha)} \varphi = a(\alpha) < c_n(\alpha).$$

Case (3): Assume $\bar{c}_k = c_k = J(\bar{u})$ and for all α, $c_n(\alpha) \leq a(\alpha)$.

In this case, we can find for all $\alpha \in \mathbb{N}$, an $A_\alpha \in \mathcal{A}_n(C(\alpha))$ such that

$$\sup_{A_\alpha} \varphi < a(\alpha) + \frac{1}{\alpha}.$$

By Lemma 3.10.(ii), we have that $A_\alpha \cap W^\perp \in \bar{\mathcal{A}}_k$ and therefore,

$$\bar{c}_k \leq \lim_\alpha \sup_{A_\alpha \cap W^\perp} \varphi \leq \lim_\alpha t(\alpha) = J(\bar{u}) = \bar{c}_k,$$

Hence $A_\alpha \cap W^\perp \in \bar{\mathcal{A}}_k$ is a min-maxing sequence along which the (PS) condition is satisfied by Lemma 3.12.(i). Theorem 3.2 applies and we get a critical point for φ at the level $\bar{c}_k$.

Notes and Comments: The min-max theorem under the standard boundary condition is well known. The only novelty here is that the manifold is only supposed to be C^1-smooth while one can see it proved in the literature for C^{2-}- Finsler manifolds. The reason is that we use Ekeland's variational principle as opposed to the deformation lemma which requires more smoothness since it involves solving a Cauchy problem on the manifold. The construction is based on ideas of Szulkin [Sz]. The form of the basic deformation lemma given here is new and seems to be sufficient for proving all the results that require *deformation techniques*.

Theorem 3.10 is due to Ambrosetti-Rabinowitz [A-R]. Theorem 3.12 is due to Lupo-Solimini [Lu-S] and is a refinement of a result of Solimini [S 2]. Another approach that uses the existence of critical points on certain dual sets will be given in Chapter 4.

4

A STRONG FORM OF

THE MIN-MAX PRINCIPLE

We shall now state, prove and motivate a strong form of the min-max principle. This theorem will be relevant for most of the results in this monograph. In this chapter, we shall start illustrating its usefulness by giving some of its immediate applications in relaxing boundary conditions, studying the structure of the critical set as well as to obtaining multiplicity results. More elaborate examples will follow in the next chapters.

4.1 Duality and the location of critical points

We have seen how the classical min-max theorem identifies a potential *critical level*. We shall now give a refined min-max principle that will, in addition, help us locate *critical points* on that level.

Theorem 4.1: *Let φ be a C^1-functional on a complete connected C^1-Finsler manifold X (without boundary) and consider a homotopy stable family $\mathcal{F}$ of compact subsets of X with a closed boundary B. Assume $c = c(\varphi, \mathcal{F}) = \inf\limits_{A \in \mathcal{F}} \max\limits_{x \in A} \varphi(x)$ is finite and let F be a closed subset of X satisfying*

(F1) $\qquad\qquad F \cap B = \emptyset$ and $F \cap A \neq \emptyset$ for all A in $\mathcal{F}$

and

(F2) $\qquad\qquad \inf \varphi(F) \geq c.$

Then, for any sequence of sets $(A_n)_n$ in $\mathcal{F}$ such that $\limsup\limits_{n} \varphi = c$, there exists a sequence $(x_n)_n$ in X such that

(i) $\lim_n \varphi(x_n) = c$

(ii) $\lim_n \|d\varphi(x_n)\| = 0$

(iii) $\lim_n \operatorname{dist}(x_n, F) = 0$

(iv) $\lim_n \operatorname{dist}(x_n, A_n) = 0.$

Moreover, if $d\varphi$ is uniformly continuous on almost critical sequences at the level c, then $(x_n)_n$ can be chosen in such a way that $x_n \in F$ (or $x_n \in A_n$) for each $n \in \mathbb{N}$.

Definition 4.2: Say that φ verifies $(PS)_{F,c}$ along a sequence $(A_n)_n$ in $\mathcal{F}$ if any sequence $(x_n)_n$ in X satisfying $\lim_n \varphi(x_n) = c$, $\lim_n d\varphi(x_n) = 0$, $\lim_n \operatorname{dist}(x_n, F) = 0$ and $\lim_n \operatorname{dist}(x_n, A_n) = 0$, has a convergent subsequence.

One can also introduce the *uniform continuity of $d\varphi$ on almost critical sequences at the level c around the set F*. Such a property is clearly satisfied whenever φ verifies $(PS)_{F,c}$. The following is now immediate.

Corollary 4.3: Under the hypothesis of Theorem 4.1 and assuming that φ verifies $(PS)_{F,c}$ along a min-maxing sequence $(A_n)_n$, there exists a sequence $(x_n)_n$ with $x_n \in F$ (or $x_n \in A_n$) that converges to a point in the set $F \cap K_c \cap A_\infty$.

Remark 4.4: To see that the above theorem is stronger than the classical min-max principle (Theorem 3.2), notice that under the standard boundary condition $(F0)$ (i.e., $\sup \varphi(B) < c$), the set $F = \{\varphi \geq c\}$ verifies conditions $(F1)$ and $(F2)$ of Theorem 4.1.

We shall actually prove the following quantitative version of Theorem 4.1 that will be needed in Chapters 8 and 11.

Theorem 4.5: Let X, φ, B and $\mathcal{F}$ be as in Theorem 4.1. Let F be a closed set satisfying $(F1)$ and

$(F2)_\delta$ $\qquad\qquad\qquad \inf \varphi(F) \geq c - \delta.$

Suppose $0 < \delta < \max\left\{\frac{1}{32}\operatorname{dist}^2(B, F); \frac{1}{8}[\inf \varphi(F) - \sup \varphi(B)]\right\}$, then for any A in $\mathcal{F}$ satisfying $\max \varphi(A) \leq c + \delta$, there exists $x_\delta \in X$ such that

(i) $c - \delta \leq \varphi(x_\delta) \leq c + 9\delta.$

(ii) $\|d\varphi(x_\delta)\| \leq 18\sqrt{\delta}.$

(iii) $\operatorname{dist}(x_\delta, F) \leq 5\sqrt{\delta}.$

(iv) $\operatorname{dist}(x_\delta, A) \leq 3\sqrt{\delta}.$

In view of Theorem 4.5, the following notions are in order.

Definition 4.6: Say that *a family of closed sets $\mathcal{F}^*$ is dual to $\mathcal{F}$* if every

$F \in \mathcal{F}^*$ verifies condition (F1) with respect to $\mathcal{F}$; that is if

$$F \cap B = \emptyset \quad \text{and} \quad F \cap A \neq \emptyset \quad \text{for every} \quad A \in \mathcal{F}.$$

Note that for such a dual family, we readily have

$$c^* := \sup_{F \in \mathcal{F}^*} \inf_{x \in F} \varphi(x) \leq \inf_{A \in \mathcal{F}} \max_{x \in A} \varphi(x) =: c.$$

Definition 4.7: A sequence $(F_n)_n$ in $\mathcal{F}^*$ is said to be *a suitable max-mining sequence* for φ in $\mathcal{F}^*$ if

$$\liminf_{n} \varphi = \sup_{F \in \mathcal{F}^*} \inf_{x \in F} \varphi(x) =: c^*$$

and

$$\lim_{n} \text{dist}(F_n, B) > 0.$$

Note that when $\sup \varphi(B) < c^*$, then any max-mining sequence in $\mathcal{F}^*$ is necessarily a *suitable* one.

If now $c = c^*$, we shall say that φ verifies $(PS)_c$ *along a min-maxing sequence* $(A_n)_n$ *in* $\mathcal{F}$ *and a max-mining sequence* $(F_n)_n$ *in* $\mathcal{F}^*$ if any sequence $(x_n)_n$ in X satisfying $\lim_n \varphi(x_n) = c$, $\lim_n \|d\varphi(x_n)\| = 0$, $\lim_n \text{dist}(x_n, F_n) = 0$ and $\lim_n \text{dist}(x_n, A_n) = 0$, has a convergent subsequence.

Theorem 4.5 yields now the following

Corollary 4.8: *Let* X, φ *and* $\mathcal{F}$ *be as in Theorem 4.1 and consider a family of sets* $\mathcal{F}^*$ *that is dual to* $\mathcal{F}$. *Assume that*

$$\sup_{F \in \mathcal{F}^*} \inf_{x \in F} \varphi(x) = \inf_{A \in \mathcal{F}} \max_{x \in A} \varphi(x) = c.$$

If φ *verifies* $(PS)_c$ *along a min-maxing sequence* $(A_n)_n$ *in* $\mathcal{F}$ *and a suitable max-mining sequence* $(F_n)_n$ *in* $\mathcal{F}^*$, *then there exists a sequence* $(x_n)_n$ *with* $x_n \in F_n$ *(or* $x_n \in A_n$*) that converges to a point in* $A_\infty \cap F_\infty \cap K_c \neq \emptyset$.

Roughly speaking, the above corollary implies that, whereas the min-max procedure on $\mathcal{F}$ determines the critical level of φ, the max-min procedure on $\mathcal{F}^*$ locates the critical points on that level.

Remark 4.9: We can also define the *dual of* $\mathcal{F}$ *at the level c* to be

$$\mathcal{F}_c^* = \{F \subseteq \{\varphi \geq c\};\ F \text{ closed and } F \cap A \neq \emptyset \text{ for all } A \text{ in } \mathcal{F}\}$$

A reformulation of Theorem 4.1 when φ verifies $(PS)_c$ is that K_c belongs to the *second dual* $\mathcal{F}^{**}$ of $\mathcal{F}$ at the level c.

Proof of Theorem 4.5: Let $\delta = \varepsilon^2/8$. The hypothesis implies that

$$0 < \varepsilon < \max \left\{ \frac{1}{2} \text{dist}(B, F);\ \sqrt{[\inf \varphi(F) - \sup \varphi(B)]^+} \right\}$$

where $\alpha^+ = \alpha \vee 0$ and that $\inf \varphi(F) \geq c - \varepsilon^2/8$.

We shall prove the existence of $x_\varepsilon \in X$ such that

(i) $c - \varepsilon^2/8 \leq \varphi(x_\varepsilon) \leq c + 9\varepsilon^2/8$.

(ii) $\mathrm{dist}(x_\varepsilon, F) \leq 3\varepsilon/2$.

(iii) $\|d\varphi(x_\varepsilon)\| \leq 6\varepsilon$.

(iv) $\mathrm{dist}(x_\varepsilon, A) \leq \varepsilon/2$.

This will clearly imply the claim of the theorem.

Let $F_\varepsilon = \{x \in X; \mathrm{dist}(x, F) < \varepsilon\}$ and consider the subspace $\mathcal{L}$ of $C([0,1] \times X; X)$ consisting of all deformations η such that

$$\eta(t, x) = x \text{ for all } (t, x) \in K_0 = (\{0\} \times X) \cup ([0,1] \times (A \setminus F_\varepsilon) \cup B)$$

and $\sup\{\rho(\eta(t, x), x); t \in [0,1], x \in X\} < +\infty$.

Since $(\{0\} \times X) \cup ([0,1] \times B) \subset K_0$, we get that $\eta(\{1\} \times A) \in \mathcal{F}$ for all η in $\mathcal{L}$.

As in Chapter 3, the space $\mathcal{L}$ equipped with the uniform metric δ is a complete metric space.

Set now $\psi(x) = \max\{0, \varepsilon^2 - \varepsilon\, \mathrm{dist}(x, F)\}$ and define a lower semi-continuous function $I : \mathcal{L} \to \mathbb{R}$ by

$$I(\eta) = \sup\{(\varphi + \psi)(\eta(1, x)); x \in A\})$$

Let $d = \inf\{I(\eta); \eta \in \mathcal{L}\}$. Since $\eta(\{1\} \times A) \in \mathcal{F}$ for all $\eta \in \mathcal{L}$ and since $\psi = \varepsilon^2$ on F we get from conditions (F1) and (F2)$_\delta$ that

$$I(\eta) \geq \sup\{(\varphi + \psi)(x); x \in \eta(\{1\} \times A) \cap F\} \geq c - \varepsilon^2/8 + \varepsilon^2.$$

Hence

$$d \geq c + 7\varepsilon^2/8. \tag{1}$$

Consider again the identity element $\bar{\eta}$ in $\mathcal{L}$ and note that

$$d \leq I(\bar{\eta}) = \sup\{(\varphi + \psi)(x); x \in A\} < c + \varepsilon^2/8 + \varepsilon^2 = c + 9\varepsilon^2/8. \tag{2}$$

Combine (1) and (2) to get that $\bar{\eta}$ verifies

$$I(\bar{\eta}) < c + 9\varepsilon^2/8 \leq d + \varepsilon^2/4 = \inf\{I(\eta); \eta \in \mathcal{L}\} + \varepsilon^2/4. \tag{3}$$

Apply Ekeland's theorem to get η_0 in $\mathcal{L}$ such that

$$I(\eta_0) \leq I(\bar{\eta}) \tag{4}$$

$$\delta(\eta_0, \bar{\eta}) \leq \varepsilon/2 \tag{5}$$

$$I(\eta) \geq I(\eta_0) - (\varepsilon/2)\delta(\eta, \eta_0) \quad \text{for all } \eta \text{ in } \mathcal{L}. \tag{6}$$

Let $C = \{x \in \eta_0(\{1\} \times A); (\varphi + \psi)(x) = I(\eta_0)\}$ and $B' = (A \setminus F_\varepsilon) \cup B$. We need to show that

$$C \cap B' = \emptyset. \tag{7}$$

For that, we shall prove that

$$\sup(\varphi + \psi)((A \setminus F_\varepsilon) \cup B) \leq d - 3\varepsilon^2/4. \tag{8}$$

Indeed, since $\psi = 0$ outside F_ε we get from (1) that
$$\sup(\varphi + \psi)(A \setminus F_\varepsilon) \le \sup \varphi(A) < c + \varepsilon^2/8 \le d - 3\varepsilon^2/4.$$
We now distinguish two cases:

— either $0 < \varepsilon < \frac{1}{2} \operatorname{dist}(B, F)$ which means that $B \subset A \setminus F_\varepsilon$ and we are done;

— or $0 < \varepsilon < \sqrt{[\inf \varphi(F) - \sup \varphi(B)]^+}$ which means that the latter is strictly positive and $\sup \varphi(B) \le \inf \varphi(F) - \varepsilon^2 \le c - \varepsilon^2$ in view of hypothesis (F1). Hence $\sup(\varphi + \psi)(B) \le c \le d - 7\varepsilon^2/8$ by (1).

In both cases, (8) is verified.

We shall now prove the following:

Claim: There exists $x_\varepsilon \in C$ such that $\|d\varphi(x_\varepsilon)\| \le 6\varepsilon$.

Before proving it, let us show how it implies Theorem 4.5. First note that since $x_\varepsilon \in C$ we have by (3) and (4) that $d \le (\varphi + \psi)(x_\varepsilon) \le c + 9\varepsilon^2/8$. Since $0 \le \psi \le \varepsilon^2$, we get from (1) that $c - \varepsilon^2/8 \le \varphi(x_\varepsilon) \le c + 9\varepsilon^2/8$ which is assertion (i). For (ii) write $x_\varepsilon = \eta_0(1, x)$ where, in view of (7), x is necessarily in F_ε. Hence $\operatorname{dist}(x, F) \le \varepsilon$. On the other hand, by (5) we have $\rho(x_\varepsilon, x) = \rho(\eta_0(1, x), x) \le \delta(\eta_0, \bar\eta) \le \varepsilon/2$. Hence $\operatorname{dist}(x_\varepsilon, F) \le 3\varepsilon/2$. Note finally that (iv) is satisfied since $x \in A$.

Back to the above claim. Suppose it is false. Fix $1 < k < 2$ and apply Lemma 3.7 to the sets C and B' to get $\alpha(t, x)$ satisfying the conclusion of that lemma with a suitable function g and a time $t_0 > 0$.

For $0 < \lambda < t_0$, consider the function $\eta_\lambda(t, x) = \alpha(t\lambda, \eta_0(t, x))$. It belongs to $\mathcal{L}$ since it is clearly continuous on $[0, 1] \times X$ and since for all $(t, x) \in (\{0\} \times X) \cup ([0, 1] \times B')$, we have $\eta_\lambda(t, x) = \alpha(t\lambda, \eta_0(t, x)) = \alpha(t\lambda, x) = x$.

Since $\delta(\eta_\lambda, \eta_0) < kt\lambda \le k\lambda$, we get from (6) that $I(\eta_\lambda) \ge I(\eta_0) - \varepsilon k\lambda$. Since A is compact, let $x_\lambda \in A$ be such that $(\varphi + \psi)(\eta_\lambda(1, x_\lambda)) = I(\eta_\lambda)$. We have
$$(\varphi + \psi)(\eta_\lambda(1, x_\lambda)) - (\varphi + \psi)(\eta_0(1, x)) \ge -\varepsilon k\lambda/2 \quad \text{for every } x \in A. \tag{9}$$
Since the Lipschitz constant of ψ is less than ε we get
$$\varphi(\eta_\lambda(1, x_\lambda)) - \varphi(\eta_0(1, x_\lambda)) \ge -3\varepsilon\lambda k/2. \tag{10}$$
On the other hand, by (iii) of lemma 3.7, we have for each x_λ
$$\begin{aligned}\varphi(\eta_\lambda(1, x_\lambda)) - \varphi(\eta_0(1, x_\lambda)) &= \varphi(\alpha(\lambda, \eta_0(1, x_\lambda)) - \varphi(\eta_0(1, x_\lambda)) \\ &\le -3\varepsilon\lambda g(\eta_0(1, x_\lambda)).\end{aligned} \tag{11}$$
Combining (10) and (11) we get
$$-3\varepsilon k/2 \le -3\varepsilon g(\eta_0(1, x_\lambda)). \tag{12}$$
If now x_0 is any cluster point of (x_λ) when $\lambda \to 0$, we have from (9)

that $\eta_0(1, x_0) \in C$ and hence $g(\eta_0(1, x_0)) = 1$. Since $k < 2$, this clearly contradicts (12) and therefore the initial claim was true. The proof of the theorem is complete.

Remark 4.10: A careful consideration of the proof above shows that all that is needed is the following, less restrictive, hypothesis on the class $\mathcal{F}$: For each $A \in \mathcal{F}$, $\eta(1, A) \in \mathcal{F}$ whenever η is an isotopy in $C([0, 1] \times X, X)$ such that

— $\eta(t, x) = x$ for (t, x) in $(\{0\} \times X) \cup ([0, 1] \times B)$.
— $\varphi(\eta(t, x)) \leq \varphi(x)$ for all x in X.

The duality condition (F1) can then be weakened to read: $B \cap F = \emptyset$ and there is $\varepsilon > 0$ such that $A \cap F \neq \emptyset$ for all A in

$$\mathcal{F}_{c+\varepsilon} = \{A \in \mathcal{F}; \sup \varphi(A) \leq c + \varepsilon\}.$$

In other words, F needs only to be dual to the class $\mathcal{F}_{c+\varepsilon}$ for some $\varepsilon > 0$.

On the other hand, if X is a Hilbert space and if P is a closed convex subset of X such that $(I - \varphi')(P) \subset P$, then the deformation $\alpha(t, x)$ in Lemma 3.7 can be constructed in such a way that it preserves P for any t. (See Hofer [H 1]). It follows from the above proof that Theorem 4.1 is still valid if the family $\mathcal{F}$ is stable under the more restrictive class consisting of those homotopies that preserve the set P, along which φ is decreasing, and which of course fix the boundary B.

Remark 4.11: Suppose $\mathcal{F} = \bigcup_\alpha \mathcal{F}_\alpha$ where each $\mathcal{F}_\alpha$ is homotopy stable with boundary B_α. Since $c(\varphi, \mathcal{F}) = \inf_\alpha c(\varphi, \mathcal{F}_\alpha)$, Theorem 4.1 holds then for $\mathcal{F}$ provided (F1) is replaced by

$$F \cap B_\alpha = \emptyset \text{ for all } \alpha \text{ and } F \cap A \neq \emptyset \text{ for all } A \in \mathcal{F}$$

A relevant situation covered by the above is when $\mathcal{F}$ itself is considered to be homotopy stable but the boundary B is *not supposed* to be contained in all the sets of $\mathcal{F}$. In this case, one can write $\mathcal{F} = \bigcup_{A \in \mathcal{F}} \mathcal{F}_A$ where $\mathcal{F}_A$ consists of all the homotopic images of A that preserves $A \cap B$. The latter can be considered to be the boundary of $\mathcal{F}_A$. Theorem 4.1 then holds with (F1) replaced by

$$B \cap A \cap F = \emptyset \text{ and } A \cap F \neq \emptyset \text{ for all } A \text{ in } \mathcal{F}$$

We shall now give some of the immediate applications of Theorems 4.1 and 4.5. More elaborate examples will be given in the next chapters.

4.2 The (PS) condition around a dual set and a relaxed boundary condition

We have seen, in Theorem 3.12, a functional φ which does not satisfy the standard Palais-Smale condition at all levels. In order to deal with the difficulty this might create, we had to use the fact that φ verifies (PS) along a carefully chosen min-maxing sequence of sets. We shall now show that the same example can be handled via Theorem 4.1, since in the very case where we had difficulty applying the classical min-max theorem, it turns out that there exists a dual set satisfying the assumption of the new min-max principle, and around which φ verifies the (PS) condition.

With the same notation as in Theorem 3.12, we recall that the corresponding functional φ verifies $(PS)_{c,W^\perp}$ for all $c \in \mathbb{R}$, while it verifies $(PS)_c$ for all $c \neq J(\bar{u})$. The classical min-max theorem can be applied in the first two cases: that is when c_k or $\bar{c}_k$ is different from $J(\bar{u})$ or when $\bar{c}_k = c_k = J(\bar{u})$ and there exists an α such that $c_n(\alpha) > a(\alpha)$.

The remaining and most interesting case for us, is when $\bar{c}_k = c_k = J(\bar{u})$ and for all α, $c_n(\alpha) \leq a(\alpha)$. In this case, we consider the generalized family $\mathcal{F}_n = \cup_\alpha A_n(C(\alpha))$ and we note that

$$c_n = c(\varphi, \mathcal{F}_n) = \inf_\alpha c_n(\alpha) = \lim_\alpha a(\alpha) = \lim_\alpha t(\alpha) = J(\bar{u}) = \bar{c}_k.$$

On the other hand, we get from Lemma 3.15.(ii) that for every $A \in \mathcal{F}_n$, $A \cap W^\perp \in \bar{A}_k$ and therefore $A \cap W^\perp \cap \{\varphi \geq \bar{c}_k\} \neq \emptyset$. It follows that the set $F = W^\perp \cap \{\varphi \geq c_n\}$ is dual to the family $\mathcal{F}_n$ while verifying $(F2)$. Since φ verifies the (PS) condition around F, Corollary 4.3 applies to yield the result.

We now turn to the problem of relaxing the boundary condition. Note that, in general, the number $c = c(\varphi, \mathcal{F})$ is unknown and hypothesis (F2) is not easily verifiable in applications. Actually, one usually establishes first the duality between F and $\mathcal{F}$ which readily implies that $\inf \varphi(F) \leq c$ and then verifies that $\sup \varphi(B) < \inf \varphi(F)$, which then ensures that the standard boundary condition $(F0)$ holds. Theorem 4.1 will allow us to treat the case where the above inequality is not strict.

Corollary 4.12: *Under the hypothesis of Theorem 4.1, replace (F2) by*

(F3) $$\sup \varphi(B) \leq \inf \varphi(F).$$

If φ verifies $(PS)_c$, then c is a critical value for φ.

Proof: Indeed, hypothesis (F1) always implies that $\inf \varphi(F) \leq c$. So we distinguish the two cases:
(a) Either $\sup \varphi(B) < c$, which means that the set $F' = \{\varphi \geq c\}$ verifies (F1) and (F2).
(b) or $\sup \varphi(B) = \inf \varphi(F) = c$, which means that the set F itself verifies (F1) and (F2).

In both cases, Theorem 4.1 applies and, in the second case, a critical point can be found on F.

We now give a simple application of Corollary 4.12. More elaborate examples will be given in Chapter 5.

4.3 The forced pendulum

Let $G \in C^1(\mathbb{R}, \mathbb{R})$ be a 2π-periodic function and let $f \in C(\mathbb{R}, \mathbb{R})$ be a T-periodic function. We consider the problem of existence of T-periodic solutions of the equation

$$\ddot{u} + G'(u) = f(t). \tag{1}$$

In particular, if $G(u) = -\cos(u)$, we obtain the forced pendulum equation

$$\ddot{u} + \sin(u) = f(t).$$

Let H be the Hilbert space of absolutely continuous T-periodic functions on $\mathbb{R}$ such that $\int_0^T |\dot{u}|^2 \, dt < \infty$ with inner product

$$(u, v) = \int_0^T (\dot{u}\dot{v} + uv) \, dt.$$

The solutions of (1) correspond to the critical points of the action functional

$$\varphi(u) = \int_0^T (\frac{1}{2}\dot{u}^2 - G(u) + fu) \, dt$$

We shall prove the following:

Theorem 4.13: *If $\int_0^T f(t) \, dt = 0$, then equation (1) has at least two T-periodic solutions which do not differ by a multiple of 2π.*

Proof: For $u \in H$, write $v = u - \frac{1}{T}\int_0^T u(t) \, dt$ and set $w = \frac{1}{T}\int_0^T u(t) \, dt$ so that

$$\varphi(u) = \int_0^T (\frac{1}{2}\dot{v}^2 - G(v + w) + fv) \, dt.$$

This clearly implies that φ is 2π-periodic: i.e., $\varphi(u + 2\pi) = \varphi(u)$.

Claim (i): φ is bounded below on H.

Indeed, let $\alpha = \max_{\mathbb{R}} G$. By the Cauchy-Schwarz and Wirtinger inequalities, we obtain

$$\varphi(u) \geq \int_0^T \frac{1}{2}\dot{v}^2 \, dt - \alpha T - \left(\int_0^T f^2 \, dt\right)^{1/2} \left(\int_0^T v^2 \, dt\right)^{1/2}$$

$$\geq \int_0^T \frac{1}{2}\dot{v}^2 \, dt - \alpha T - \frac{T}{2\pi} \left(\int_0^T f^2 \, dt\right)^{1/2} \left(\int_0^T \dot{v}^2 \, dt\right)^{1/2}. \tag{1}$$

This proves the first claim.

Note that the periodicity of φ implies that the set of critical points is either empty or unbounded, which means that φ cannot satisfy the Palais-Smale condition in the classical sense. However, if we consider, for every $N \in \mathbb{N}$ the set

$$F_N = \{u \in H; |\frac{1}{T} \int_0^T u(t)\, dt| \in [0, 2N\pi]\},$$

then we have:

Claim (ii): φ verifies $(PS)_{F_N, c}$ for any $N \in \mathbb{N}$ and any $c \in \mathbb{R}$.

Indeed, suppose $\varphi'(u_k) \to 0$ and $\varphi(u_k) \to c$. From (1) and Wirtinger's inequality, it follows that the sequence $(v_k)_k$ is bounded in H. On the other hand, since $(u_k)_k$ approachs F_N, we can assume that $(w_k)_k$ and hence $(u_k)_k$ is bounded in H. In view of the compact embedding of H into L^∞, and modulo passing to a subsequence, we can assume that

$$u_k \to u \quad \text{uniformly on } [0, T] \quad \text{and} \quad \dot{u}_k \to \dot{u} \quad \text{weakly in } L^2. \quad (2)$$

On the other hand, we have

$$\langle \varphi'(u_n) - \varphi'(u_m), u_n - u_m \rangle = \int_0^T (\dot{u}_n - \dot{u}_m)^2\, dt$$

$$- \int_0^T \langle V'(u_n) - V'(u_m), u_n - u_m \rangle\, dt$$

$$\geq \|\dot{u}_n - \dot{u}_m\|_2^2 - K\|u_n - u_m\|_\infty$$

where K is some positive constant. Consequently $\|\dot{u}_n - \dot{u}_m\|_2 \to 0$ as $n, m \to +\infty$, and therefore $(u_k)_k$ is norm convergent in H.

To prove Theorem 4.13, note first that Claims (i) and (ii) coupled with the periodicity of φ gives that φ has a global minimum $u \in H$ such that $\varphi(u + 2\pi) = \varphi(u) = \inf_H \varphi$.

Consider now $\mathcal{F}_u^v$ to be the class of all continuous paths joining a point u to $v = u + 2\pi$ in X. It is clear that $\mathcal{F}_u^v$ is homotopy-stable with boundary $B = \{u, v\}$ and that any sphere $F = S_\rho(u)$ centered at u is dual to $\mathcal{F}_u^v$ provided $\rho < \|u - v\|$. Condition (F3) above is trivially satisfied. Let $c = c(\varphi, \mathcal{F}_u^v)$. We distinguish two cases:

(i) Either $c = \varphi(u)$ which means that every sphere S_ρ centered at u and separating u from v, verifies conditions $(F1)$ and $(F2)$ and we obtain a solution on S_ρ since necessarily $S_\rho \subset F_N$ for some N. Such a solution does not differ by a multiple of 2π from u.

(ii) Or $c > \varphi(u)$ and hence there is an almost critical sequence $(u_n)_n$ such that $\varphi(u_n) \to c$. By the periodicity of φ, we can assume that $u_n \in F_1$ and hence we get from Claim (ii), that a critical point exists at a level $c \neq \varphi(u)$ which means that again, we find a solution that does not differ by a multiple of 2π from u.

4.4 The structure of the critical set and the existence of good paths

We now show how Theorem 4.1 gives information about the type of critical points it generates. The idea is simple: The largest dual set verifying (F2) is the set $\{\varphi \geq c\}$. The trick is to find smaller subsets F of $\{\varphi \geq c\}$ that are still dual to $\mathcal{F}$, since the smaller F is, the more information we have about the critical points it contains. Here is the simplest example.

Under the classical condition $\sup \varphi(B) < c$ and if we assume, that all the sets in the class $\mathcal{F}$ are pathwise connected, then the boundary F of the set $\{\varphi \geq c\}$ is also dual to $\mathcal{F}$. Hence there exist critical points located F at level c. But these points cannot be local minima, hence we have proved the following

Corollary 4.14: *Let φ be as in Theorem 4.1 and suppose that $\mathcal{F}$ is a homotopy-stable class with a non-empty boundary B and whose members are all pathwise connected sets. If φ verifies $(PS)_c$ and the boundary condition (F0), then it has a critical point on the level c which is not a local minimum.*

Many other, more elaborate, choices for the dual set F will be given in the next chapters.

We now denote by

$$G_c = \{x \in X; \varphi(x) < c\} \quad \text{and} \quad L_c = \{x \in X; \varphi(x) \geq c\}$$

Theorem 4.15: *Under the hypothesis of Corollary 4.3. For any $\delta > 0$ and $\varepsilon > 0$, the following hold:*

(1) There exists $A \in \mathcal{F}$ such that

$$\sup \varphi(A) < c + \varepsilon \quad \text{and} \quad A \subset (X \backslash F) \cup (F \cap K_c \cap A_\infty)^\delta.$$

(2) If $\sup \varphi(B) < c$, there exists $A \in \mathcal{F}$ such that

$$\sup \varphi(A) < c + \varepsilon \quad \text{and} \quad A \subset G_c \cup (K_c \cap A_\infty)^\delta.$$

Proof: (1) If not then the set $F' = F \backslash (F \cap K_c \cap A_\infty)^\delta$ would be dual to $\mathcal{F}$. By Corollary 4.3, this would imply that $F' \cap K_c \cap A_\infty \neq \emptyset$ which is absurd.

(2) follows from the same reasoning applied with $F = L_c$.

Theorem 4.16: *Under the hypothesis of Theorem 4.1 and assuming that $d\varphi$ is uniformly continuous on almost critical sequences at the level c and around the set F, we have the following:*

(i) For every $\varepsilon > 0$, there exist $\delta > 0$ and $A \in \mathcal{F}$ such that
$$\sup \varphi(A) < c + \delta \text{ and } \|d\varphi(x)\| < \varepsilon \text{ whenever } x \in A \cap F^\delta \cap L_{c-\delta}.$$

(ii) If $\sup \varphi(B) < c$, then for every $\varepsilon > 0$, there exist $\delta > 0$ and $A \in \mathcal{F}$ such that
$$\sup \varphi(A) < c + \delta \text{ and } \|d\varphi(x)\| < \varepsilon \text{ whenever } x \in A \cap L_{c-\delta}.$$

Proof: If (i) does not hold, there exists $\varepsilon > 0$ such that for every $\delta > 0$, the set
$$F' = F^\delta \cap L_{c-\delta} \cap \{x; \|d\varphi(x)\| \geq \varepsilon\}$$
verifies conditions $(F1)$ and $(F2)_\delta$ of Theorem 4.5. Hence, there exists x_δ that verifies the conclusions of that Theorem. By letting $\delta \to 0$, we construct an almost critical sequence (x_n) whose distance to F and to the set $\{x; \|d\varphi(x)\| \geq \varepsilon\}$ goes to zero. This contradicts the uniform continuity of $d\varphi$.

4.5 Multiplicity results of Ljusternik-Schnirelmann type

One tool to measure the "size" of a set is the so-called *Ljusternik-Schnirelmann category*. For a subset A of a topological space X, it is defined as *the least integer k such that A can be covered by k closed sets that are contractible in X*. When X is a Finsler manifold, the category Cat_X verifies the following properties:

(C1) $\text{Cat}_X(A) = 0$ if and only if $A = \emptyset$.

(C2) If A_1 and A_2 are closed in X and $\eta : [0, 1] \times A_1 \to Y$ is a continuous deformation of A_1 (i.e. $\eta(0, x) = x$, $\forall x \in A_1$) with $\eta(1, A_1) \subset A_2$ then $\text{Cat}_X(A_1) \leq \text{Cat}_X(A_2)$.

(C3) For any closed set K, there exists a closed neighborhood $K^\delta = \{x; \text{dist}(x, K) \leq \delta\}$ of K so that $\text{Cat}_X(K^\delta) = \text{Cat}_X(K)$.

(C4) $\text{Cat}_X(A_1 \cup A_2) \leq \text{Cat}_X(A_1) + \text{Cat}_X(A_2)$ for all closed subsets A_1, A_2.

Indeed, properties (C1) and (C4) are obvious. To prove (C2), we can clearly assume that $A_2 = \eta(1, A_1)$ and that $\text{Cat}_X(A_2) = k < \infty$. Let $(C_i)_{i=1}^k$ be closed contractible sets in X that cover A_2 and let $(f_i)_{i=1}^k$ be their associated deformations. The sets $D_i = \eta_1^{-1}(C_i)$ $(1 \leq i \leq k)$ clearly cover A_1. So it remains to prove that each one of them is contractible to a point. But it is easy to see that the maps $f_i * \eta : [0, 1] \times D_i \to X$ defined by
$$f_i * \eta(t, x) = \begin{cases} \eta(2t, x) & \text{if } 1 \leq t \leq 1/2 \\ f_i(2t - 1, \eta(1, x)) & \text{if } 1/2 \leq t \leq 1 \end{cases}$$
do just that.

To prove (C3), again we may assume $\text{Cat}_X(K) = k < \infty$ and let $(C_i)_{i=1}^k$ be closed contractible sets in X that cover K and let $(f_i)_{i=1}^k$ be their associated deformations that contract them to the points $(a_i)_{i=1}^k$. It suffices to prove that each C_i has a closed neighborhood U_i that is contractible in X. To do that, fix i and consider the set $E_i = ([0,1] \times C_i) \cup (\{0\} \times X) \cup (\{1\} \times X)$ which is closed in $[0,1] \times X$. On E_i, consider the following continuous function $h_i : E_i \to X$:

$$h(t,x) = \begin{cases} f_i(t,x) & \text{if } (t,x) \in [0,1] \times C_i \\ x & \text{if } (t,x) \in \{0\} \times X \\ a_i & \text{if } (t,x) \in \{1\} \times X \end{cases}$$

Since the Finsler manifold X is an *absolute neighborhood extensor* (See [Du]), h_i has a continuous extension $\tilde{f}_i$ defined on a closed neighborhood N_i of E_i. By the compactness of $[0,1]$, we can find a closed neighborhood U_i of C_i such that $[0,1] \times U_i \subset N_i$. It is now clear that $\tilde{f}_i$ contracts U_i to the same point a_i.

We can now give a simple proof of a celebrated theorem of Ljusternik-Schnirelmann.

Corollary 4.17: *Let φ be a bounded below C^1-functional satisfying (PS) on a C^1-Finsler manifold X. Assume that $\text{Cat}_X(X) = N$ and define for $1 \leq n \leq N$, the families*

$$\mathcal{F}_n = \{A; \ A \text{ compact in } X \text{ and } \text{Cat}_X(A) \geq n\}.$$

Let $c_n = c(\varphi, \mathcal{F}_n)$ and assume $c_j = c_{j+p}$ for $1 \leq j \leq j+p \leq N$, then for every min-maxing sequence $(A_n)_n$ in $\mathcal{F}_{j+p}$ we have

$$\text{Cat}_X(K_{c_j} \cap A_\infty) \geq p+1.$$

In particular, φ has at least N distinct critical points.

Proof: Suppose $1 \leq j \leq j+p \leq N$ are such that $c_j = c_{j+p}$. If $\text{Cat}_X(K_{c_j} \cap A_\infty) \leq p$, use (C3) to find a neighborhood U of $K_{c_j} \cap A_\infty$ such that $\text{Cat}_X(\overline{U}) \leq p$. By (C4) we have $A \setminus U \in \mathcal{F}_j$ for every A in $\mathcal{F}_{j+p}$. Hence $(A \setminus U) \cap \{\varphi \geq c_j\} \neq \emptyset$ for such an A. It follows that

$$F' := \{\varphi \geq c_{j+p}\} \setminus U = \{\varphi \geq c_j\} \setminus U$$

is dual to $\mathcal{F}_{j+p}$. By Theorem 4.1, the set $K_{c_j} \cap A_\infty \setminus U = K_{c_{j+p}} \cap A_\infty \cap F' \neq \emptyset$ which is clearly a contradiction.

Remark 4.18: The N-dimensional torus $\mathbb{T}^N$ has a category of $N+1$. So the above corollary implies the well known result stating that a bounded below C^1-functional on $\mathbb{T}^N$ has at least $N+1$ distinct critical points.

More generally, we have the following:

Corollary 4.19: *Let G be a discrete subgroup of a Banach space X*

such that the dimension of the linear subspace it spans in X is finite and equal to N. Assume φ is a G-invariant C^1-functional that verifies (PS) on X. If φ is bounded below, then it has at least $N + 1$ distinct critical orbits.

Proof: If G is a discrete subgroup of the Banach space X that generates a finite dimensional subspace V of dimension N, then X is isomorphic to $\mathbb{R}^N \times W$ where W is the linear complement of V. Note also that G is isomorphic to $\mathbb{Z}^N$ and $\pi(X) = X/G$, where $\pi : X \to X/G$ is the quotient map, is a C^1-Finsler manifold isomorphic to $\mathbb{T}^N \times W$. If we consider the subset $A = [0,1]^N \times \{0\}$ of X (identified with $\mathbb{R}^N \times W$), we obtain that

$$\text{Cat}_{\pi(X)}(\pi(A)) = \text{Cat}_{\mathbb{T}^N \times W}(\mathbb{T}^N \times \{0\}) = \text{Cat}_{\mathbb{T}^N}(\mathbb{T}^N) = N + 1.$$

It follows that the classes

$$\mathcal{F}_j = \{A; \ A \text{ compact in } X \text{ and } \text{Cat}_{\pi(X)}(\pi(A)) \geq j\}$$

are not empty for any $1 \leq j \leq N + 1$. So if we assume that φ is a C^1, G-invariant functional on X, it defines unambiguously a C^1-functional $\tilde{f}$ on the C^1-Finsler manifold X/G. Moreover, to every critical point $\pi(x)$ for $\tilde{\varphi}$ corresponds a critical orbit $\pi^{-1}(\pi(x))$ for φ. The proof now follows immediately from Corollary 4.17.

Remark 4.20: The above corollary yields another approach for the problem of the forced pendulum. Indeed, the 2π-periodicity of φ, implies that it is G-invariant under the action of the linear group $G = \{2k\pi; k \in \mathbb{Z}\}$. The type of (PS) condition satisfied by φ is equivalent to saying that $\tilde{\varphi}$ verifies the (PS) condition on the quotient manifold $\pi(X) = X/G$ which is what is really needed for Corollary 4.19 to hold. Since the linear span of G in H^1 is one dimensional, it follows that φ has at least two distinct orbits and hence the equation has two T-periodic solutions which do not differ by a multiple of 2π.

We also have the following new type of multiplicity results.

Corollary 4.21: *Under the hypothesis of Corollary 4.3, we have*

$$\text{Cat}_X(K_c \cap F \cap A_\infty) \geq \inf\{\text{Cat}_X(A \cap F); A \in \mathcal{F}\}.$$

Remark 4.22: With the above assumptions on φ and $\mathcal{F}$,
(i) If $\sup \varphi(B) < c$, we can apply Corollary 4.21 to $F = \{\varphi \geq c\}$ and we obtain

$$\text{Cat}_X(K_c \cap A_\infty) \geq \inf\{\text{Cat}_X(A \cap \{\varphi \geq c\}); A \in \mathcal{F}\}.$$

Proof of Corollary 4.21: Let $n = \inf\{\text{Cat}_X(A \cap F); A \in \mathcal{F}\}$. By property (C3), there exists a neighborhood U of $K_c \cap F \cap A_\infty$ such that

$\text{Cat}_X(\overline{U}) = \text{Cat}_X(K_c \cap F \cap A_\infty)$. Let $F' = X \setminus U$. By property (C4), we have for any A in $\mathcal{F}$

$$\text{Cat}_X(A \cap F) \leq \text{Cat}_X(A \cap F \setminus U) + \text{Cat}_X(\overline{U})$$
$$= \text{Cat}_X(A \cap F \cap F') + \text{Cat}_X(K_c \cap F \cap A_\infty).$$

It follows that if $\text{Cat}_X(K_c \cap F \cap A_\infty) \leq n-1$, then $\text{Cat}_X(A \cap F \cap F') \geq 1$ and in particular $A \cap F \cap F' \neq \emptyset$ for all A in $\mathcal{F}$ by property (C1). On the other hand $\inf \varphi(F \cap F') \geq \inf \varphi(F) \geq c$. Hence Corollary 4.3 applies to the dual set $F \cap F'$ and we get that $K_c \cap F \cap F' \cap A_\infty \neq \emptyset$ which is clearly a contradiction.

We also have the following:

Corollary 4.23: *Let X, φ, $\mathcal{F}$ and $\mathcal{F}^*$ be as in Corollary 4.8. Then*

$$\text{Cat}_X(A_\infty \cap F_\infty \cap K_c) \geq \inf\{\text{Cat}_X(A \cap F); A \in \mathcal{F}, F \in \mathcal{F}^*\}.$$

Proof: Indeed, if not, we proceed as above to find a neighborhood U of $K_c \cap F_\infty \cap A_\infty$ such that $\text{Cat}_X(\overline{U}) = \text{Cat}_X(K_c \cap F_\infty \cap A_\infty)$. Let $F' = X \setminus U$. The same computation as in Corollary 4.21 shows that for any A in $\mathcal{F}$ and any $F \in \mathcal{F}^*$, $\text{Cat}_X(A \cap F \cap F') \geq 1$. In particular $A \cap F \cap F' \neq \emptyset$ for all A in $\mathcal{F}$ and $F \in \mathcal{F}^*$. In other words, the family

$$\mathcal{F}_1^* = \{F \cap F'; F \in \mathcal{F}^*\}$$

is also dual to $\mathcal{F}$. Note now that $\lim_n \inf \varphi(F_n \cap F') = c$ which implies by Corollary 4.8 that $K_c \cap F_\infty \cap F' \cap A_\infty \neq \emptyset$. A contradiction.

4.6 Functions with a common critical level and multiplicity of non-radial solutions on the annulus

Here is another immediate and useful application of theorem 4.1.

Corollary 4.24: *Let φ be a C^1-functional on a complete connected C^1-Finsler manifold X and consider a homotopy stable family $\mathcal{F}$ of compact subsets of X with a closed boundary B.*

Let ψ be another continuous functional such that $\psi \leq \varphi$ on X while $c = c(\varphi, \mathcal{F}) = c(\psi, \mathcal{F})$. Suppose $\sup \psi(B) < c$ and that φ verifies $(PS)_c$, then there exists $x \in X$, such that $\varphi'(x) = 0$, $\varphi(x) = c$ and $\psi(x) = c$.

Proof: It is enough to realize that the set $F = \{\psi \geq c\}$ is then dual to $\mathcal{F}$ while $\inf \varphi(F) = c$. Theorem 4.1 then applies to give the claim.

Multiplicity of non-radial solutions for elliptic equations on the annulus.

For $a > 0$, we let $\Omega_a = \{a < |x| < a + 1\} \subset \mathbb{R}^3$ be the corresponding

3-dimensional annulus. We consider the equation

$$\begin{cases} -\Delta u = f(u) & \text{if } x \in \Omega_a \\ \quad u > 0 & \text{if } x \in \Omega_a \\ \quad u = 0 & \text{if } x \in \partial\Omega_a. \end{cases} \tag{1}_a$$

where $f : \mathbb{R} \to \mathbb{R}$ is a continuous function with $f(0) = 0$. Let $F(s) = \int_0^s f(t)\, dt$. We shall show the following:

Theorem 4.25: *Suppose that f verifies the following conditions for some p, $1 < p < 5$, $\sigma > 0$ and $K > 0$:*

$$0 \le f(u) \le Ku^p, \tag{2}$$

$$F(u) \le (\frac{1}{2} - \sigma)f(u)u \tag{3}$$

for all $u > 0$. Then, for any number N and for a sufficiently large, problem $(1)_a$ admits N distinct non-radial solutions.

Proof: Let $X^a = H_0^1(\Omega_a)$ and consider the C^1- functional $\varphi^a : X^a \to \mathbb{R}$ defined by

$$\varphi^a(v) = \int_{\Omega_a} (\frac{1}{2}|\nabla v|^2 - F(v^+))\, dx. \tag{4}$$

Note that, as in Theorem 3.10, and since $2^* = 6$, the conditions on f guarantee that φ^a verifies $(PS)_c$ for every c, and that each nonzero critical point $u \in X^a$ of φ^a becomes a classical solution for $(1)_a$.

We let $\sum_a$ denote the solution set of $(1)_a$ and we write

$$\sum_a^0 = \{u \in \sum_a; u = u(|x|)\}$$

for the set of radial solutions. We shall also consider the set

$$\sum_a^1 = \{u \in \sum_a; u = u(r, z)\},$$

where $r = (x_1^2 + x_2^2)^{1/2}$ and $z = x_3$ for $x = (x_1, x_2, x_3) \in \Omega_a$. It is clear that

$$\sum_a \supset \sum_a^1 \supset \sum_a^0.$$

We write $(x_1, x_2) = (r\cos\theta, r\sin\theta)$ for $x = (x_1, x_2, x_3) \in \Omega_a \subset \mathbb{R}^3$. For $\ell = 1, 2, ...$ denote by T_ℓ the rotation by $\frac{2\pi}{\ell}$ in the θ-direction, in the (x_1, x_2)-plane,

$$T_\ell(x_1, x_2) = (r\cos(\theta + \frac{2\pi}{\ell}), r\sin(\theta + \frac{2\pi}{\ell}))$$

and consider the following groups $G_\ell = \{T_\ell^j \oplus \{id\}; 0 \le j \le \ell-1\} \subset O(3)$ that are acting on Ω_a. We write X_ℓ^a for the subspace of X^a consisting of the G_ℓ-invariant functions.

Introduce the transformation $S_\ell : \Omega_a \to \Omega_a$ by

$$S_\ell x = (r\cos\ell\theta, r\sin\ell\theta, z).$$

It induces a bijective mapping $S_\ell^* : X_\ell^a \to X^a$. For $w \in X^a$ we define $\varphi_\ell^a(w) = \varphi(S_\ell^{*-1}w)$.

We shall see that, as in Theorem 3.10, the hypothesis will allow us to set up the Mountain pass theorem for φ_ℓ^a, for each $a > 0$ and each integer ℓ. If $e \in X^a$, we shall write

$$\Gamma^a(e) = \{\gamma \in C([0,1]; X^a); \gamma(0) = 0, \gamma(1) = e\},$$

and $c_\ell(a) := c(\varphi_\ell^a; \Gamma^a(e)) = \inf_{\gamma \in \Gamma(e)} \max_{0 \le t \le 1} \varphi_\ell^a(\gamma(t))$.

We first establish

Claim 1: The following properties hold for every $a > 0$,
 (i) $\varphi_1^a \le \varphi_2^a \le ... \le \varphi_\ell^a \le ...$
 (ii) If $e \in X^a$ is such that $\varphi_\ell^a(e) < 0$ and $c_\ell(a) > 0$ for $1 \le \ell \le N$, then each $c_\ell(a)$ is a critical value for φ_ℓ^a and

$$c_1(a) \le c_2(a) \le ... \le c_N(a).$$

(iii) If $c_\ell(a) = c_{\ell-1}(a)$, then $c_\ell(a) \in \varphi^a(\sum_a^1)$.

Proof of Claim 1: Note that

$$\varphi_\ell^a(w) = \int_0^{2\pi} d\theta \int_{\Omega_+^a} \{\frac{1}{2}(w_r^2 + \ell^2 r^{-2} w_\theta^2 + |\nabla_z w|^2 - F(w^+))\} r \, dr \, dz \quad (5)$$

where $\Omega_+^a = \{(r, z); r > 0, a^2 < r^2 + z^2 < (a+1)^2\} \subset \mathbb{R}^2$. This clearly implies (i).

As in Theorem 3.10, assertion (ii) is a consequence of the classical mountain pass theorem, which follows from Theorem 3.2 and the fact that φ_ℓ^a verifies $(PS)_c$.

For (iii) we note that if $c_\ell(a) = c_{\ell-1}(a)$ then the hypotheses in Corollary 4.24 are satisfied, and this means there exists $w \in X^a$ such that $d\varphi_\ell^a(w) = 0$, $\varphi_\ell^a(w) = c_\ell(a)$ and $\varphi_\ell^a(w) = \varphi_{\ell-1}^a(w)$. But the last relation implies that $w_\theta \equiv 0$ by (5), and hence $w \in \sum_a^1$. In other words $c_\ell(a) \in \varphi(\sum_a^1)$.

The rest of the proof consists of showing that, for a large enough a, we have that $c_1(a), c_2(a), ..., c_\ell(a) \notin \varphi(\sum_a^1)$ and therefore the critical levels are necessarily distinct: i.e.,

$$c_1(a) < c_2(a) < ... < c_\ell(a) < ...$$

To do that we need to analyse the asymptotic behaviour of $c(a)$ as $a \to +\infty$. We shall use the following notation: for positive quantities q and Q depending on a we write $q \preceq Q$ if $\liminf_{a \to +\infty} Q/q > 0$. We also write $q \sim Q$ if $q \preceq Q$ and $Q \preceq q$.

We first show the following:

Claim 2: If a family $\{u_a\}_{a \gg 1}$ with $u_a \in \sum_a^1$ satisfies $\varphi^a(u_a) = O(1)$ as $a \to \infty$, then necessarily $\varphi^a(u_a) = o(1)$.

Proof: First we note that the assumption $\varphi^a(u_a) = O(1)$ implies that

$$||\nabla u_a||_{L^2(\Omega_a)}^2 = O(1). \tag{6}$$

Indeed, condition (3) gives that

$$\varphi^a(u_a) = \int_{\Omega_a} \{\frac{1}{2}f(u_a)u_a - F(u_a)\}\, dx \geq \sigma \int_{\Omega_a} f(u_a)u_a\, dx$$

and hence

$$||\nabla u_a||_{L^2(\Omega_a)}^2 = \int_{\Omega_a} f(u_a)u_a dx = O(1) \text{ as } a \to +\infty$$

since $\varphi^a(u_a) = O(1)$. We need to prove that

$$||\nabla u_a||_{L^2(\Omega_a)}^2 = o(1). \tag{7}$$

We shall actually show that for $\varepsilon > 0$ and $\rho > 0$ given, there is a constant $C = C_{\varepsilon,\rho} > 0$ independent of $a \gg 1$ such that

$$||u||_{L^{p+1}(\Omega_a)}^{p+1} \leq \varepsilon||\nabla u||_{L^2(\Omega_a)}^{p+1} + Ca^{-\rho} \tag{8}$$

This inequality, combined with (6), implies (7), because

$$||\nabla u_a||_{L^2(\Omega_a)}^2 = \int_{\Omega_a} f(u_a)u_a\, dx$$

$$\leq K||u_a||_{L^{p+1}(\Omega_a)}^{p+1}$$

$$\leq K\{\varepsilon||\nabla u_a||_{L^2(A)}^{p+1} + Ca^{-\rho}\}.$$

To prove (8) we shall drop the dependence on a in our notation. Start by dividing the semi-annulus

$$\Omega^+ = \{(r,z); a^2 < r^2 + z^2 < (a+1)^2, r > 0\} \subset \mathbb{R}^2$$

into

$$\Omega_1^+ = \{(r,z) \in \Omega_+; \tan(|z|/r) > \varepsilon^\gamma\} \text{ and } \Omega_2^+ = \Omega \backslash \bar{\Omega}_1^+,$$

where $\gamma = 2/(5-p)$. Those portions correspond to the three dimensional domains $\Omega_1 = \{x \in \Omega; (r,z) \in \Omega_1^+\}$ and $\Omega_2 = \{x \in \Omega; (r,z) \in \Omega_2^+\}$ contained in Ω, respectively. In the polar coordinates $x = (\rho,\omega)$ for $\rho = |x|$ and $|\omega| = 1$, we have $\Omega_1 = (a, a+1) \times S_1$, where S_1 denotes a portion of the unit sphere S^2 in $\mathbb{R}^3$. From the definition we have that $|S_1| = C_1\varepsilon^{2\gamma}$, where $C_1 > 0$ is an absolute constant and $|S_1|$ denotes the area of S_1.

To control $u|_{\Omega_1}$, we use the Poincaré inequality on Ω_1,

$$||u||_{L^2(\Omega_1)}^2 \leq C_2|S_1|\, ||\nabla u||_{L^2(\Omega_1)}^2.$$

Therefore, for $p \in [1,5]$ we get

$$||u||_{L^{p+1}(\Omega_1)}^{p+1} \leq ||u||_{L^2(\Omega_1)}^{(5-p)/2}||u||_{L^6(\Omega_1)}^{3(p-1)/2}$$
$$\leq C_3\varepsilon||\nabla u||_{L^2(\Omega_1)}^{p+1}, \tag{9}$$

where $C_3 > 0$ is an absolute constant.

To estimate $\|\nabla u\|_{L^2(\Omega_2)}$, we fix $\varepsilon > 0$. We shall frequently use the following formulation of Poincaré's inequality: for $u \in X = H_0^1(\Omega)$,

$$\|u\|^2_{H^1(\Omega_2)} \sim \|\nabla u\|^2_{L^2(\Omega_2)}. \tag{10}$$

We start by showing that for $u = u(r, z) \in X^a$ and $a > 1$, we have

$$\|u\|^2_{L^q(\Omega_2)} \leq C_{\varepsilon,q} a^{2(\frac{6}{q}-1)} \|u\|^2_{H^1(\Omega_2)}, \tag{11}$$

where $1 < q < +\infty$. Here, $C_{\varepsilon,q} > 0$ is a constant depending on q and $\varepsilon > 0$.

To this end we introduce the transformation $z = a\hat{z}$ and $r = a\hat{r}$, which, for $a > 1$, sends Ω_2^+ into a set $\hat{\Omega}_2^+$ that is contained in

$$\{\varepsilon^\gamma < \hat{r} < 2, |\hat{z}| < 2\} \subset \mathbb{R}^2.$$

We will use the 2-dimensional Sobolev inequality for the transformed function $\hat{u}$ of u. Let

$$\hat{\Omega}_2 = \{(\hat{x}_1, \hat{x}_2, \hat{z}); (\hat{r}, \hat{z}) \in \hat{\Omega}_2^+ \text{ for } \hat{r} = (\hat{x}_1^2 + \hat{x}_2^2)^{1/2}\} \subset \mathbb{R}^3.$$

We have for $1 < q < +\infty$ that

$$\|\hat{u}\|^2_{L^q(\hat{\Omega}_2)} \sim \{\int_{\hat{\Omega}_2^+} |\hat{u}|^q \hat{r} \, d\hat{r} \, d\hat{r}\}^{2/q}$$

$$\sim \{\int_{\hat{\Omega}_2^+} |\hat{u}|^q \, d\hat{r} \, d\hat{z}\}^{2/q}$$

$$\preceq \int_{\hat{\Omega}_2^+} (\hat{u}_{\hat{r}}^2 + \hat{u}_{\hat{z}}^2 + \hat{u}^2) \, d\hat{r} \, d\hat{z}$$

$$\sim \int_{\hat{\Omega}_2^+} (\hat{u}_{\hat{r}}^2 + \hat{u}_{\hat{z}}^2 + \hat{u}^2)\hat{r} \, d\hat{r} d\hat{r}$$

$$\sim \|\hat{u}\|^2_{H^1(\hat{\Omega}_2)}.$$

Therefore, we get

$$\|u\|^2_{L^q(\Omega_2)} = a^{6/q}\|\hat{u}\|^2_{L^q(\hat{\Omega}_2)}$$

$$\preceq a^{6/q}\|\hat{u}\|^2_{H^1(\hat{\Omega}_2)}$$

$$\sim a^{6/q} \int_{\hat{\Omega}_2^+} (\hat{u}_{\hat{r}}^2 + \hat{u}_{\hat{z}}^2 + \hat{u}^2)\hat{r} \, d\hat{r} \, d\hat{z}$$

$$\preceq a^{6/q} \int_{\hat{\Omega}_2} (\hat{u}_{\hat{r}}^2 + \hat{u}_{\hat{z}}^2 + a^2\hat{u}^2)\hat{r} \, d\hat{r} \, d\hat{z}$$

$$= a^{(\frac{6}{q}-1)} \int_{\Omega_2^+} (u_r^2 + u_z^2 + u^2) r \, dr \, dz$$

$$\sim a^{(\frac{6}{q}-1)}\|u\|^2_{H^1(\Omega_2)}$$

which yields (11).

Now from the Poincaré inequality (10), we may write

$$||u||^2_{H^1(\Omega_2)} \leq C_4||\nabla u||^2_{L^2(\Omega_2)} \qquad (12)$$

with a constant $C_4 > 0$. On the other hand, there exists a constant C_5 such that

$$C_4||\nabla u||^2_{L^2(\Omega_2)} \leq C_5||\Delta u||^2_{L^2(\Omega_2)} + \frac{1}{4}||u||^2_{L^2(\Omega_2)} \qquad (13)$$

for $u \in H^2(\Omega_2) \equiv \{u \in H^1(\Omega_2); \nabla^2 u \in L^2(\Omega_2)\}$. Indeed, this inequality is true in $H^2(\mathbf{R}^3)$. On the other hand, by a standard reflection argument (see for instance [M]), we can take an extension $\hat{u} \in H^2(\mathbf{R}^3)$ of $u|_{\Omega_2}$ so that

$$||\hat{u}||_{H^2(\mathbf{R}^3)} \sim ||u||_{H^2(\Omega_2)}.$$

Now use assumption (1) to find a constant $C_6 > 0$ so that

$$C_5 f(u)^2 \leq u^2(C_6 u^q + \frac{1}{4}) \qquad (14)$$

for a given $q > 2p - 2$. Using that $u = u_a$ is a solution of the equation $(1)_a$, in addition to (12), (13) and (14), we obtain

$$||u||^2_{H^1(\Omega_2)} \leq C_6||u||^{q+2}_{L^{q+2}(\Omega_2)} + \frac{1}{2}||u||^2_{L^2(\Omega_2)}$$

and hence

$$||\nabla u||^2_{L^2(\Omega_2)} \leq 2C_6||u||^{q+2}_{L^{q+2}(\Omega_2)}.$$

Therefore, by (11), (10), and (6) we get that

$$||\nabla u||^2_{L^2(\Omega_2)} \leq C_\varepsilon a^{(6-(q+2))/2}||u||^{q+2}_{H^1(\Omega_2)} = O(a^{3-(q+2)/2}).$$

Let now $q \to +\infty$ to conclude that

$$||u||_{H^1(\Omega_2)} = O(a^{-\infty}).$$

From (11) with $q = p + 1$, this implies that

$$||u||^{p+1}_{L^{p+1}(\Omega_2)} = O(a^{-\infty}). \qquad (15)$$

The inequality (8) follows from (9) and (15), modulo replacing ε by constant $\times \varepsilon$ in the definition of Ω_1^+. The proof of Claim 2 is complete.

As in Theorem 3.10, the hypothesis allow us to set up the Mountain pass theorem for each φ^a. In other words, there exists $R_a > 0$ such that $\inf \varphi^a(B_{R_a}(0)) \geq \rho_a > 0$ and $e_a \in X^a \setminus B_{R_a}$ such that $\varphi^a(te_a) \leq 0$ for all $t \geq 1$. It then follows that $c(a) = c(\varphi^a; \Gamma^a(e_a))$ is a critical value for φ_a. We shall now study the behaviour of $c(a)$ as function of a. But first, we need to clear the ambiguity that might occur in the dependence on the endpoint e_a. For that we need to show that $c(a)$ does not depend on the endpoint e as long as the latter is taken in the set

$$E_{R_a} = \{e \in X^a \setminus B_{R_a}(0); e > 0, \varphi^a(te) \leq 0 \text{ for all } t \geq 1\}. \qquad (16)$$

Indeed, (3) gives that $\liminf_{t \to +\infty} F(t)/t^2 = +\infty$, hence $Re \in E_{R_a}$

holds for any $e > 0$ when $R > 0$ is large enough. Now take $e_1 \in E_{R_a}$ and let γ_1 be an arbitrary path connecting it to 0. Given another $e_2 \in E_{R_a}$, we connect it with e_1 in the following way: let C_1 and C_2 be the rays departing from 0 and leading to e_1 and e_2 respectively. Let C_3 be a circle with centre at the origin and with radius $R > 0$ sufficiently large. Then the points e_1 and e_2 are connected by those curves C_1, C_2, and C_3, crossing once from C_1 to C_3 and then from C_3 to C_2. When R is sufficiently large, φ^a is clearly negative on those curves. Combining them with γ_1, we then get a path γ_2 connecting 0 and e_2. Obviously, $\max_{0 \le t \le 1} \varphi^a(\gamma_1(t)) = \max_{0 \le t \le 1} \varphi^a(\gamma_2(t))$ and hence the critical levels corresponding to e_1 and e_2 are such that $c(a, e_1) \le c(a, e_2)$. By starting the reasoning with e_2, we get the reverse inequality and hence $c(a)$ does not depend on e in E_{R_a}.

Claim 3: There is a family of radii $R_a > 0$ with $0 < \liminf_{a \to \infty} R_a < \infty$, such that for $e_a \in E_{R_a}$, we have $0 < \liminf_{a \to \infty} c(a) < \infty$, where $c(a)$ is the mountain pass level between 0 and e_a.

Proof: In the notation introduced in Claim 2, we need to prove that there exists $R(a) \sim 1$ such that $c(a) \sim 1$.

By condition (2) we have
$$\int_\Omega F(v^+)dx \le C||v||_{L^{p+1}(\Omega)}^{p+1}.$$

On the other hand, the Poincaré inequality for $u \in X = H_0^1(\Omega)$ implies that
$$||u||_{L^2(\Omega)}^2 \preceq ||\nabla u||_{L^2(\Omega)}^2.$$
The Sobolev constant depends only on the dimension ($n = 3$) so that
$$||u||_{L^6(\Omega)}^2 \preceq ||\nabla u||_{L^2(\Omega)}^2.$$
This implies that
$$\int_\Omega F(u_+)\,dx \le C||\nabla u||_{L^2(\Omega)}^{p+1}$$
for $1 \le p \le 5$ with a constant $C > 0$ independent of $a \gg 1$. We now take the radius $R \sim 1$ of $B_R \subset X$ sufficiently small so that
$$\frac{1}{4}R^2 \ge C\,R^{p+1}. \tag{17}$$
We then get
$$\varphi(u) = \int_\Omega (\frac{1}{2}|\nabla u|^2 - F(u^+))\,dx \ge \frac{1}{2}||\nabla u||_{L^2(\Omega)}^2 - C||\nabla u||_{L^2(\Omega)}^{p+1} \ge \frac{1}{4}R^2$$
for $u \in \partial B_R$. Hence we can take as $\rho = \frac{1}{4}R^2$ so that
$$c^{-1} \le \rho^{-1} \sim 1.$$

To estimate c from above, we set

$$m(s) = \inf_{t \geq s} \frac{F(t)}{t^2}. \tag{18}$$

By assumption (2), we have that $\lim_{s \to +\infty} m(s) = +\infty$. Taking $t \geq 1$ and $s > 0$, we have for any positive function e that

$$\varphi(te) = t^2 \{ \frac{1}{2} \int_\Omega |\nabla e|^2 \, dx - \int_\Omega \frac{F(te)}{t^2} \, dx \}$$

$$\leq t^2 \{ \frac{1}{2} \int_\Omega |\nabla e|^2 \, dx - \int_{\{e \geq s\}} \frac{F(te)}{(te)^2} e^2 \, dx \}$$

$$\leq t^2 \{ \frac{1}{2} \int_\Omega |\nabla e|^2 \, dx - m(s) \int_{\{e \geq s\}} e^2 \, dx \}$$

and hence $\varphi(te) \leq 0$ for all $t \geq 1$ whenever $e \in E_s$, where

$$E_s = \{ e \in X; \, e > 0, \, \frac{\|\nabla e\|^2_{L^2(\Omega)}}{2 \int_{\{e \geq s\}} e^2 \, dx} \leq m(s) \}. \tag{19}$$

In other words, $E_s \setminus B \subset E$.

We shall now show that for an appropriate s, there exists $e \in E_s \setminus B$ satisfying

$$\|\nabla e\|_{L^2(\Omega)} \sim 1. \tag{20}$$

To this end we take a positive constant $s \sim 1$ sufficiently large so that

$$4s^2 \{ (\frac{1}{2})^n - (\frac{1}{4})^n \} \alpha_n \geq R^2, \tag{21}$$

as well as

$$8 \{ (\frac{1}{2})^n - (\frac{1}{4})^n \} / (\frac{1}{4})^n \leq m(s), \tag{22}$$

where α_n denotes the volume of the n-dimensional unit ball. We take a ball ω in Ω with radius $1/2$. Let ω_1 be the concentric ball of ω with radius $1/4$. Then a function $e \in H_0^1(\Omega)$ is defined in such a way that

$$e = \begin{cases} s & \text{in } \omega_1 \\ 0, & \text{in } \Omega \setminus \omega \end{cases} \tag{23}$$

and

$$|\nabla e| = 4s \text{ in } \omega \setminus \omega_1. \tag{24}$$

This function e satisfies

$$\|\nabla e\|^2_{L^2(\Omega)} = 16s^2 \{ (\frac{1}{2})^n - (\frac{1}{4})^n \} \alpha_n \geq 4R^2$$

$$\int_{\{e \geq s\}} e^2 \, dx = s^2 |\omega_1| = s^2 (\frac{1}{4})^n \alpha_n,$$

and hence

$$\frac{\|\nabla e\|^2_{L^2(\Omega)}}{\int_{\{e \geq s\}} e^2 \, dx} = 16\{(\frac{1}{2})^n - (\frac{1}{4})^n\}/(\frac{1}{4})^n \leq 2m(s).$$

Therefore, e belongs to $E_s \backslash B$ and also satisfies (20).

Let γ be the segment connecting 0 and e given above. Then we have

$$c \leq \max_{0 \leq t \leq 1} \frac{1}{2}\|\nabla\gamma(t)\|^2_{L^2} \leq \frac{1}{2}\|\nabla e\|^2_{L^2(\Omega)} \sim 1.$$

which finishes the proof of Claim 3.

To complete the proof of the theorem, we still need to prove Claim 2 for the critical values $c_\ell(a)$ corresponding to the functionals φ_ℓ^a. To this end, we first give an alternate formulation for these critical values.

Suppose that G is any one of the groups G_ℓ, and that e is a G-invariant function in $X_G^a = X_\ell^a$. We then define

$$\Gamma_G^a(e) = \{\gamma \in C([0,1]; X_G^a); \gamma(0) = 0, \gamma(1) = e\}.$$

It is then easy to see that for any ℓ, we have

$$c(\varphi_\ell, \Gamma^a(e)) = c(\varphi, \Gamma_\ell^a(e)) = c_\ell(a).$$

Hence, it is enough to show that Claim 2 holds with e_a being in X_ℓ^a. But looking back at the construction of e in (23) and (24) of that proof, we see that to make it G_ℓ-invariant, all that is needed is the existence (for a large enough) of a ball $\omega \subset \Omega^a$ with radius independent of $a \gg 1$ such that its iterates by the group G_ℓ are mutually disjoint. In that case, we set $e_\ell = \sum_{T \in G_\ell} T^* e$. It is easy to see that $e_\ell \in X_\ell^a$, $e_\ell \in E(s) \backslash B_R$ if s is chosen appropriately, $c_\ell(a) = c(\varphi, \Gamma_\ell^a(e_\ell))$ is independent of the choice of $e_\ell \in E_R \cap X_\ell$ while satisfying the asymptotic behaviour $c_\ell(a) \sim 1$.

If now N is fixed, we can also find the same $e \in X^a$ that is suitable for all groups G_ℓ with $1 \leq \ell \leq N$. Theorem 4.15 then follows from combining Claims 1, 2 and 3.

Notes and Comments: The concept of dual families was introduced by Ambrosetti–Rabinowitz in their seminal paper [A-R]. The strong form of the min-max principle and most of its (theoretical) implications that appear in this chapter are due to the author [G 1]. The case of the mountain pass theorem was established earlier by Ghoussoub-Preiss [G-P] to answer a question of Pucci-Serrin [P-S1] who had proved it in the case where *the separating mountain range F* had a *thickness*.

Multiplicity à la Ljusternik-Schnirelmann belongs to the folklore of the subject. The proof included here is very simple and it shows that the strong version of the min-max theorem is actually an existence and multiplicity statement. Corollaries 4.21 and 4.23 are new kinds of multiplicity results that are useful in limiting cases. See the examples in Chapter 8.

The notion of *uniform continuity of the gradient on pseudo-critical sequences* seems like a natural weakening of the Palais-Smale condition and can replace it in several instances, like in the result on the existence of good paths. The latter was proved earlier by various authors under the (PS) condition (See [B-N 3]). This notion will also reappear in Chapter 11. Theorem 4.25 is due to T. Suzuki [Su].

5

RELAXED BOUNDARY CONDITIONS

IN THE PRESENCE OF A DUAL SET

We start this chapter by establishing yet another refined version of the min-max principle. We then use it to obtain strengthened versions of various well known and widely used min-max theorems. We shall also present a few selected examples where the classical methods do not apply, but the refined versions do.

5.1 Another relaxation of the boundary condition

The refinement mentioned above will consist of relaxing the boundary condition $(F0)$ in two ways: first we will allow $\sup_B \varphi$ to take the critical value c. Secondly, we shall consider the case where B is not necessarily compact, nor contained in all the members of the class $\mathcal{F}$ and for that, we will introduce below the notion of *a homotopy-stable family with a generalized boundary B*.

We now discuss the first weakening of the boundary condition. Recall, that in the classical theorem, the critical point is obtained by taking a point that maximizes φ on an "optimal path", that is a set $A \in \mathcal{F}$ on which $\sup \varphi(A)$ is essentially equal to c. Condition (F0) makes sure that such a point is not on the boundary B. In general, not every point that maximizes φ on A is necessarily a critical point, however we will be able to prove an existence result by merely assuming that not all highest points on a given link A occur on the boundary; that is

$$M(A; \varphi) \setminus B \neq \emptyset \quad \text{for every } A \in \mathcal{F}.$$

where $M(A; \varphi) = \{x \in A; \varphi(x) = \sup \varphi(A)\}$.

Since necessarily $\sup \varphi(B) \leq c$ (when B is contained in all the members of $\mathcal{F}$), this new condition is equivalent to

(F'0) $A \cap \{\varphi \geq c\} \setminus B \neq \emptyset$ for every $A \in \mathcal{F}$.

The second improvement consists of relaxing the hypothesis on the homotopy-stable family in the following way.

Definition 5.1: Let B be a closed subset of X. We shall say that a class $\mathcal{F}$ of compact subsets of X is a *homotopy-stable family with extended boundary* B if for any set A in $\mathcal{F}$ and any $\eta \in C([0,1] \times X; X)$ satisfying $\eta(t,x) = x$ for all (t,x) in $(\{0\} \times X) \cup ([0,1] \times B)$ we have that $\eta(\{1\} \times A) \in \mathcal{F}$.

Here is the main result of this chapter. We prove it in the presence of a dual set F that is a priori smaller than $\{\varphi \geq c\}$ so that it contains Theorem 4.1.

Theorem 5.2: Let φ be a C^1-functional on a complete connected C^1-Finsler manifold X and consider a homotopy-stable family $\mathcal{F}$ with an extended closed boundary B. Set $c = c(\varphi, \mathcal{F})$ and let F be a closed subset of X satisfying

(F'1) $A \cap F \setminus B \neq \emptyset$ for every $A \in \mathcal{F}$

and

(F'2) $\sup \varphi(B) \leq c \leq \inf \varphi(F)$.

Then, for any sequence of sets $(A_n)_n$ in $\mathcal{F}$ such that $\limsup_{n} \varphi = c$, there exists a sequence $(x_n)_n$ in $X \setminus B$ such that

(i) $\lim_{n} \varphi(x_n) = c$
(ii) $\lim_{n} \|d\varphi(x_n)\| = 0$
(iii) $\lim_{n} \mathrm{dist}(x_n, F) = 0$.
(iv) $\lim_{n} \mathrm{dist}(x_n, A_n) = 0$.

In particular, if φ verifies $(PS)_{F,c}$ along a min-maxing sequence $(A_n)_n$, then the set $K_c \cap F \cap A_\infty$ is non empty.

Remark 5.3: Note that under condition $(F'2)$, the hypothesis $(F'1)$ is equivalent to

$$M(A \cap F; \varphi) \setminus B \neq \emptyset   \text{ for every } A \in \mathcal{F}.$$

Note also that $\sup \varphi(B) \leq c$ is automatically satisfied when B is assumed to be contained in all members of $\mathcal{F}$. It is then clear that conditions $(F'1)$ and $(F'2)$ are implied by $(F1)$ and $(F2)$ in the case of a homotopy stable family with a *real boundary* B.

Before proceeding with the proofs we give few immediate applications.

Corollary 5.4: *Let X, φ, B, $\mathcal{F}$ and $c = c(\varphi, \mathcal{F})$ be as in Theorem 5.2 and assume that φ verifies $(PS)_c$ along a min-maxing sequence $(A_n)_n$. Suppose that*

(F'0) $A \cap \{\varphi \geq c\} \setminus B \neq \emptyset$ *for every $A \in \mathcal{F}$.*

   and

(F''0). $\sup \varphi(B) \leq c$

Then the set $K_c \cap A_\infty$ is non empty.

Proof: Apply Theorem 5.2 with $F = \{\varphi \geq c\}$ and with boundary B.

Remark 5.5: The above corollary applies, in particular, if

(F'''0) $A \cap \{\varphi > c\} \neq \emptyset$ *for every $A \in \mathcal{F}$*

   and

(F''0). $\sup \varphi(B) \leq c$

In that case, the boundary is taken to be $B' = \{\varphi \leq c\}$.

Corollary 5.6: *Let X, φ, B, $\mathcal{F}$ and $c = c(\varphi, \mathcal{F})$ be as in Theorem 5.2, and suppose there exists a closed set F satisfying $(F'1)$ and*

(F3) $\sup \varphi(B) \leq \inf \varphi(F).$

   If φ verifies $(PS)_c$, then the set K_c is non empty.

Proof: Note that hypothesis $(F'1)$ always implies that $\inf \varphi(F) \leq c$. So either $\sup \varphi(B) < c$ or $\sup \varphi(B) = \inf \varphi(F) = c$. In both cases, condition $(F'0)$ and $(F''0)$ are satisfied. Corollary 5.4 then applies to get the claim. In the second case, a critical point can be found on F.

   For the sequel, we recall the following notation:

$$G_c = \{x \in X; \varphi(x) < c\}, \quad \varphi_c = \{x \in X; \varphi(x) \leq c\} \quad \text{and} \quad L_c = X \setminus G_c.$$

Corollary 5.7: *Under the hypothesis of Theorem 5.2, we have: for any $\delta > 0$ and any $\varepsilon > 0$,*
(1) there exists $A \in \mathcal{F}$ such that

$$\sup \varphi(A) < c + \varepsilon \quad \text{and} \quad A \subset (X \setminus F) \cup B \cup (F \cap K_c \cap A_\infty)^\delta.$$

(2) there exists $A' \in \mathcal{F}$ such that $\sup \varphi(A') < c + \varepsilon$,

$$A' \subseteq G_c \cup B \cup (K_c \cap A_\infty)^\delta \quad \text{and} \quad A' \cap L_c \cap K_c^\delta \setminus B \neq \emptyset.$$

Proof: (1) If not, then the set $F' = F \setminus (F \cap K_c \cap A_\infty)^\delta$ would satisfy $(F'1)$ and $(F'2)$ with respect to $\mathcal{F}$. By Theorem 5.2, this would imply that $F' \cap K_c \cap A_\infty \neq \emptyset$ which is absurd.
For (2), it is enough to apply part (1) to the sets $F' = \{\varphi \geq c\}$ and B.

Corollary 5.8: *Let X, φ, $\mathcal{F}$, c and $(A_n)_n$ be as in Theorem 5.2. Suppose F is a closed subset of X satisfying (F2) and*

$$A \cap F \cap \{\varphi > c\} \neq \emptyset \quad \text{for every } A \in \mathcal{F}.$$

Then, for any $\delta > 0$ and any $\varepsilon > 0$,
(1) there exists a set $A \in \mathcal{F}$ such that $\sup \varphi(A) < c + \varepsilon$ and

$$A \subseteq (X \backslash F) \cup \varphi_c \cup (F \cap K_c \cap A_\infty)^\delta.$$

(2) there exists a set $A' \in \mathcal{F}$ such that $\sup \varphi(A') < c + \varepsilon$ and

$$A' \subseteq \varphi_c \cup (K_c \cap A_\infty)^\delta.$$

Proof: It is enough to apply Corollary 5.4 with $B = \varphi_c$.

Remark 5.9: Under the classical condition $\sup \varphi(B) < c$, Corollary 5.7 gives the well known result about the existence of $A \in \mathcal{F}$ with $A \subseteq G_c \cup K_c^\delta$.

On the other hand, under conditions (F'''0) and (F''0), we may conclude that for any $\delta > 0$ and any $\varepsilon > 0$, there exists $A \in \mathcal{F}$ such that $\sup \varphi(A) < c + \varepsilon$ and $A \subseteq \varphi_c \cup K_c^\delta$. This follows from Corollary 5.8.

As in Theorem 4.1, we shall reduce the proof of Theorem 5.2 to the classical case, by using one more appropriately chosen perturbation of the function φ.

Proof of Theorem 5.2: Let $0 < \varepsilon$ and consider a set A in $\mathcal{F}$ such that

$$c \leq \sup \varphi(A) < c + \varepsilon^2/8. \tag{1}$$

We shall prove the existence of $x_\varepsilon \in X \backslash B$ such that
(i) $c - \varepsilon^2/8 \leq \varphi(x_\varepsilon) \leq c + 5\varepsilon^2/4$.
(ii) $\text{dist}(x_\varepsilon, F) \leq 3\varepsilon/2$.
(iii) $\|d\varphi(x_\varepsilon)\| \leq 10\varepsilon$.
(iv) $\text{dist}(x_\varepsilon, A) \leq \varepsilon/2$.

Let $F_\varepsilon = \{x \in X; \text{dist}(x, F) < \varepsilon\}$ and consider the subspace $\mathcal{L}$ of $C([0,1] \times X; X)$ consisting of all maps η such that

$$\eta(t, x) = x \text{ for all } (t, x) \in K_0 = (\{0\} \times X) \cup ([0,1] \times (A \setminus F_\varepsilon) \cup B)$$

and $\sup\{\rho(\eta(t, x), x); t \in [0,1], x \in X\} < +\infty$.

Since $(\{0\} \times X) \cup ([0,1] \times B) \subset K_0$, we get that $\eta(\{1\} \times A) \in \mathcal{F}$ for all η in $\mathcal{L}$. As in Chapter 4, the space $\mathcal{L}$ equipped with the uniform metric δ is a complete metric space. Consider now the following perturbations

$$\varphi_1(x) = \max\{0, \varepsilon^2 - \varepsilon \, \text{dist}(x, F)\}$$

and

$$\varphi_2(x) = \min\{\varepsilon^2/8, \varepsilon \, \text{dist}(x, (A \setminus F_\varepsilon) \cup B)\}.$$

Denote by ψ the function $\varphi + \varphi_1 + \varphi_2$, by K_1 the set $\{1\} \times A$ and define a lower semi-continuous function $I : \mathcal{L} \to \mathbb{R}$ by

$$I(\eta) = \sup\{\psi(\eta(1, x)); x \in A\} = \sup \psi(\eta(K_1))$$

Let $d = \inf\{I(\eta); \eta \in \mathcal{L}\}$ and note that since $\eta(K_1) \in \mathcal{F}$ for all $\eta \in \mathcal{L}$ and since $\varphi_1 = \varepsilon^2$ on F we get from conditions $(F'1)$ and $(F'2)$ that

$$I(\eta) \geq \sup\{(\varphi + \varphi_1)(x); x \in \eta(K_1) \cap F\} \geq c + \varepsilon^2.$$

Hence

$$d \geq c + \varepsilon^2. \tag{2}$$

Consider again the identity function $\bar{\eta}$ in $\mathcal{L}$ defined by $\bar{\eta}(t, x) = x$ for all (t, x) in $[0, 1] \times X$ and note that

$$
\begin{aligned}
d \leq I(\bar{\eta}) &= \sup\{(\varphi + \varphi_1 + \varphi_2)(x); x \in A\} \\
&< c + \varepsilon^2/8 + \varepsilon^2 + \varepsilon^2/8 = c + 5\varepsilon^2/4.
\end{aligned} \tag{3}
$$

Combine (2) and (3) to get that $\bar{\eta}$ is an element in $\mathcal{L}$ satisfying

$$I(\bar{\eta}) < c + 5\varepsilon^2/4 \leq d + \varepsilon^2/4 = \inf\{I(\eta); \eta \in \mathcal{L}\} + \varepsilon^2/4. \tag{4}$$

Apply Ekeland's theorem to get η_0 in $\mathcal{L}$ such that

$$I(\eta_0) \leq I(\bar{\eta}) \tag{5}$$

$$\delta(\eta_0, \bar{\eta}) \leq \varepsilon/2 \tag{6}$$

$$I(\eta) \geq I(\eta_0) - (\varepsilon/2)\delta(\eta, \eta_0) \quad \text{for all } \eta \text{ in } \mathcal{L}. \tag{7}$$

Let

$$C = \{x \in \eta_0(K_1); \psi(x) = \max \psi(\eta_0(K_1))\} \quad \text{and} \quad B' = (A \setminus F_\varepsilon) \cup B.$$

We need to show that

$$C \cap B' = \emptyset. \tag{8}$$

For that we shall prove that for some $\tau > 0$,

$$\sup \psi((A \setminus F_\varepsilon) \cup B) \leq \max \psi(\eta_0(K_1)) - \tau. \tag{9}$$

Indeed, by $(F'1)$ there exists $x_0 \in M(\eta_0(K_1) \cap F; \varphi) \setminus B$. Hence

$$\max \psi(\eta_0(K_1)) \geq \max \psi(\eta_0(K_1) \cap F) \geq c + \varepsilon^2 + \varphi_2(x_0).$$

Note that $\tau = \varphi_2(x_0) > 0$ since $x_0 \notin B \cup (A \setminus F_\varepsilon)$. On the other hand since $\varphi_1 = 0$ outside F_ε and $\varphi_2 = 0$ on $A \setminus F_\varepsilon$ we get from (1) that

$$\sup \psi(A \setminus F_\varepsilon) \leq \sup \varphi(A) < c + \varepsilon^2/8.$$

Moreover, since $\sup \varphi(B) \leq c$, we have

$$\sup \psi(B \cap F_\varepsilon) = \sup(\varphi + \varphi_1)(B \cap F_\varepsilon) \leq c + \varepsilon^2.$$

The last three displayed inequalities clearly imply (9). We shall now prove the following:

Claim: There exists $x_\varepsilon \in C = \eta_0(M)$ such that $\|d\varphi(x_\varepsilon)\| \leq 10\varepsilon$.

Before proving it, let us show how it implies Theorem 5.2. First note that since $x_\varepsilon \in C$, we have by (5) and (6) that $d \leq (\varphi + \varphi_1 + \varphi_2)(x_\varepsilon) \leq c + 5\varepsilon^2/4$. Since $0 \leq \varphi_1 \leq \varepsilon^2$ and $0 \leq \varphi_2 \leq \varepsilon^2/8$, we get from (2) that $c - \varepsilon^2/8 \leq \varphi(x_\varepsilon) \leq c + 5\varepsilon^2/4$ which is assertion (i).

For (ii) write $x_\varepsilon = \eta_0(1, x)$ where, in view of (8), x is necessarily in F_ε. Hence $\text{dist}(x, F) \leq \varepsilon$. On the other hand, by (6) we have $\rho(x_\varepsilon, x) = \rho(\eta_0(1, x), x) \leq \delta(\eta_0, \bar\eta) \leq \varepsilon/2$. Hence $\text{dist}(x_\varepsilon, F) \leq 3\varepsilon/2$. Note finally that (iv) is satisfied since $x \in A$.

Back to the claim. Suppose it is false. Fix $1 < k < 2$ and apply Lemma 3.7 to the sets C and B' to get $\alpha(t, x)$ satisfying the conclusion of that Lemma with a suitable function g and a time $t_0 > 0$.

For $0 < \lambda < t_0$, consider the function $\eta_\lambda(t, x) = \alpha(t\lambda, \eta_0(t, x))$. It belongs to $\mathcal{L}$ since it is clearly continuous on $[0, 1] \times X$ and since for all $(t, x) \in (\{0\} \times X) \cup ([0, 1] \times B')$ we have $\eta_\lambda(t, x) = \alpha(t\lambda, \eta_0(t, x)) = \alpha(t\lambda, x) = x$. Since $\delta(\eta_\lambda, \eta_0) < kt\lambda \leq k\lambda$, we get from (7) that

$$I(\eta_\lambda) \geq I(\eta_0) - \varepsilon k\lambda/2.$$

Since A is compact, let $x_\lambda \in A$ be such that $\psi(\eta_\lambda(1, x_\lambda)) = I(\eta_\lambda)$. We have

$$\psi(\eta_\lambda(1, x_\lambda)) - \psi(\eta_0(1, x)) \geq -\varepsilon k\lambda/2 \quad \text{for every } x \in A. \tag{10}$$

Since the Lipschitz constants of both φ_1 and φ_2 are less than ε we get

$$\varphi(\eta_\lambda(1, x_\lambda)) - \varphi(\eta_0(1, x_\lambda)) \geq -5\varepsilon\lambda k/2. \tag{11}$$

On the other hand, by (iii) of lemma 3.7, we have for all x

$$\begin{aligned} \varphi(\eta_\lambda(1, x_\lambda)) - \varphi(\eta_0(1, x_\lambda)) &= \varphi(\alpha(\lambda, \eta_0(1, x_\lambda))) - \varphi(\eta_0(1, x_\lambda)) \\ &\leq -5\varepsilon\lambda g(\eta_0(1, x_\lambda)). \end{aligned} \tag{12}$$

Combining (11) and (12) we get

$$-5\varepsilon k/2 \leq -5\varepsilon g(\eta_0(1, x_\lambda)). \tag{13}$$

If now x_0 is any cluster point of (x_λ) when $\lambda \to 0$, we have from (10) that $\eta_0(1, x_0) \in C$ and hence $g(\eta_0(1, x_0)) = 1$. Since $k < 2$, this clearly contradicts (13) and therefore the initial claim was true. The proof of the theorem is complete.

5.2 The relaxed mountain pass and saddle point theorems

We start with the following:

Corollary 5.10: *Let φ be a C^1-functional on a complete connected C^1-Finsler manifold X and let K be a compact space. Suppose γ_0 is a given continuous function from a closed subset K_0 of K into X and consider the family*

$$\Gamma = \{\gamma \in C(K; X) : \gamma = \gamma_0 \text{ on } K_0\}.$$

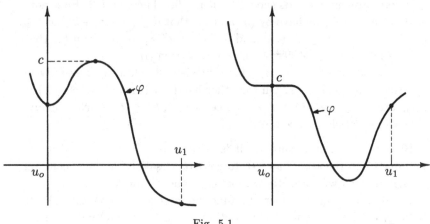

Fig. 5.1.

Set $c = \inf\limits_{\gamma \in \Gamma} \sup\limits_{x \in K} \varphi(\gamma(x))$ and let F be a closed subset of X satisfying

(F'1) $F \cap \gamma(K \setminus K_0) \neq \emptyset$ for all γ in Γ

and

(F3) $\sup \varphi(\gamma_0(K_0)) \leq \inf \varphi(F)$.

If φ verifies $(PS)_c$, then $K_c \neq \emptyset$. Moreover, if $c = \inf \varphi(F)$, then $K_c \cap F \neq \emptyset$.

The above result follows immediately from Theorem 5.2. Indeed, it is enough to notice that the class $\mathcal{F} = \{A; A = \gamma(K) \text{ for some } \gamma \in \Gamma\}$ is a homotopy stable class with boundary $B = \gamma_0(K_0)$.

Assume now that X is a Banach space and consider $\mathcal{F}_{u_0}^{u_1}$ to be the class of all continuous paths joining a point u_0 to another point u_1 in X. It is clear that $\mathcal{F}_{u_0}^{u_1}$ is homotopy-stable with boundary $B = \{u_0, u_1\}$ and that any sphere $S_\rho(u_0)$ centered at u_0 is dual to $\mathcal{F}_{u_0}^{u_1}$ provided $\rho < \|u_0 - u_1\|$. The following is well known when the inequality in hypothesis (2) below is supposed to be strict [A-R]. The novelty here is that no such assumption is needed. It follows immediately from Corollary 5.10.

Corollary 5.11 (Fig 5.1): *Let φ be a C^1-functional on X. Assume that for two points $\{u_0, u_1\}$ in X and a scalar a, the following hold:*
(1) *There is $0 < \rho < \|u_0 - u_1\|$ so that $\inf \varphi(S_\rho(u_0)) \geq a$*
(2) *$\max\{\varphi(u_0), \varphi(u_1)\} \leq a$.*
   Then $c = c(\varphi, \mathcal{F}_{u_0}^{u_1}) \geq a$ and if φ verifies $(PS)_c$, c is a critical value for φ. Moreover, if $c = a$ then $K_c \cap S_\rho(u_0) \neq \emptyset$.

Corollary 5.12: Let φ be a C^1-functional on X satisfying (PS). Assume that for two points u_0, u_1 in X, the following hold:
(1) u_0 is a local minimum for φ.
(2) $\varphi(u_1) \leq \varphi(u_0)$.

Then, either φ has a critical point which is not a local minimum at a level $c > \varphi(u_0)$, or for each $\rho > 0$ sufficiently small, φ has a local minimum at a point u_ρ with $\|u_0 - u_\rho\| = \rho$ and $\varphi(u_\rho) = \varphi(u_0)$.

The above holds in particular if both u_0 and u_1 are both local minima for φ.

Proof: It follows immediately from Corollary 1.6, Corollary 4.14 and Corollary 5.11. Note that we are in the first alternative if u_0 is supposed to be a strict local minimum.

Assume now that $X = Y \oplus Z$ where $\dim(Y) < +\infty$ and let U be an open bounded subset of Y. Let Γ be the class of all continuous functions $\gamma : \bar{U} \to X$ that are equal to the identity on the boundary ∂U. Consider the class $\mathcal{F}_U$ of all sets of the form $\gamma(\bar{U})$ with $\gamma \in \Gamma$. It is clear that $\mathcal{F}$ is homotopy-stable with boundary ∂U. Let us show that Z is dual to $\mathcal{F}_U$: Take any $\gamma \in \Gamma$ and let P be the projection of X onto Y. The map $P \circ \gamma : \bar{U} \to Y$ is continuous and is equal to the identity on the boundary of U. Brouwer's fixed point theorem then gives a point $x \in \bar{U}$ such that $P \circ \gamma(x) = 0$. But this means that $Z \cap \gamma(\bar{U}) \neq \emptyset$ and our claim is proved.

Again, the following result is well known when $\sup \varphi(\partial U) < \inf \varphi(Z)$ and Corollary 5.10 above shows that the result remains true if we have equality.

Corollary 5.13: Let φ be a C^1-functional satisfying (PS) on $X = Y \oplus Z$ with $\dim(Y) < \infty$. Assume there is a such that the following hold:
(1) $\inf \varphi(Z) \geq a$ and
(2) $\sup \varphi(\partial U) \leq a$, where U is an open bounded subset of Y.

Then, $c = c(\varphi, \mathcal{F}(U)) \geq a$ is a critical value for φ. Moreover, if $c = a$ then $K_c \cap Z \neq \emptyset$.

As an application of the above results, we give the following recent result of G. Tarantello [T 3]. It is an extension of Theorem 4.13.

5.3 The forced double pendulum

The equation of motion for a mechanical system with Lagrangian $\mathcal{L}(q, \xi, t)$, subject to the forcing $f = f(t) \in \mathbb{R}^N$ is given by

$$\frac{d}{dt} \frac{\partial \mathcal{L}}{\partial \xi}(q, \dot{q}, t) - \frac{\partial \mathcal{L}}{\partial q}(q, \dot{q}, t) = f(t). \tag{1}$$

where $q, \xi \in \mathbb{R}^N$ and $t \in \mathbb{R}$. We shall consider Lagrangians of the form

$$\mathcal{L}(q, \xi, t) = \frac{1}{2} A(t, q) \xi \cdot \xi - V(t, q) \qquad (2)$$

where for each (t, q), $A = A(t, q)$ is a symmetric positive definite $N \times N$ matrix, and the functions $V = V(t, q)$ and $f = f(t)$ are assumed to be time-periodic with the same period T. We shall also suppose that $(t, q) \to A(t, q)$ and $(t, q) \to V(t, q)$ are C^1-maps.

With the double pendulum equation in mind, we shall consider periodic potentials: that is for fixed numbers $T_k > 0, k = 1, ..., N$, we assume that for all $(t, q) \in \mathbb{R} \times \mathbb{R}^N$ and $k, k_s \in \mathbb{Z}, s = 1, ..., N$, we have

$$V(t + kT, q + (k_1 T_1, ..., k_N T_N)) = V(t, q) \qquad (3)$$

$$A(t + kT, q + (k_1 T_1, ..., k_N T_N)) = A(t, q) \qquad (4)$$

Theorem 5.14: *Let $V = V(t, q)$ and $A = A(t, q)$ satisfy (3) and (4) respectively and $f_k = f_k(t)$ be T-periodic continuous functions with $\int_0^T f_k = 0$ for $k = 1, ..., N$. There exist then constants $\lambda_* \leq 0 \leq \lambda^*$ (depending on V, A and f_k) so that the following hold:*
(a) *If $\lambda_* = \lambda^*$, then $\forall \xi \in \mathbb{R}$, problem (1) with $f(t) = (f_1(t), ..., f_N(t))$ admits a T-periodic solution $q_\xi(t) = (q_{1,\xi}(t), ..., q_{N,\xi}(t))$ such that $\frac{1}{T} \int_0^T q_{1,\xi} = \xi$;*
(b) *If $\lambda_* < \lambda^*$, then problem (1) with $f(t) = f_1(t) + \lambda, ..., f_N(t))$ admits at least two distinct solutions if $\lambda \in (\lambda_*, \lambda^*)$, and at least one solution if $\lambda = \lambda^*$ or $\lambda = \lambda_*$.*

Set $f(t) = (f_1(t) + \lambda, ..., f_N(t))$ with $f_k \in L^2[0, T], \int_0^T f_k = 0$ for $k = 1, ..., N$ and $\lambda \in \mathbb{R}$. We seek solutions for the following problem.

$$\begin{cases} \frac{d}{dt} \frac{\partial \mathcal{L}}{\partial \xi_1}(q, \dot{q}, t) - \frac{\partial \mathcal{L}}{\partial q_1}(q, \dot{q}, t) = f_1(t) + c \\ \frac{d}{dt} \frac{\partial \mathcal{L}}{\partial \xi_k}(q, \dot{q}, t) - \frac{\partial \mathcal{L}}{\partial q_k}(q, \dot{q}, t) = f_k(t), \quad k = 2, ..., N \qquad (1)_\lambda \\ q(0) = q(T), \quad \dot{q}(0) = \dot{q}(T) \end{cases}$$

Weak solutions of $(1)_\lambda$ are the critical points of the functional

$$\varphi_\lambda(q) = \frac{1}{2} \int_0^T A(t, q) \dot{q} \cdot \dot{q} - \int_0^T V(t, q) + \sum_{k=1}^N \int_0^T f_k q_k + \lambda \int_0^T q_1$$

defined on the Hilbert space

$$H = \{q = (q_1, ..., q_N) : q_k \in H^1([0, T]), q_k(0) = q_k(T), k = 1, ..., N\}$$

equipped with the scalar product

$$\langle q, Q \rangle = \int_0^T \dot{q} \cdot \dot{Q} + \int_0^T q \cdot Q, \quad q, Q \in H$$

and norm $\|q\| = (q,q)^{1/2}$. For $q = (q_1, ..., q_N) \in H$, set

$$\|q\|_2 = \left(\sum_{k=1}^{N} \|q_k\|_{L^2}^2 \right)^{1/2} \quad \text{and} \quad \|q\|_\infty = \sum_{k=1}^{N} \|q_k\|_{L^\infty}.$$

Since $A = A(t,q)$ is positive definite, let $A_0 > 0$ be such that $\forall t \in \mathbb{R}$, and $q, \xi \in \mathbb{R}^N$

$$A(t,q)\xi \cdot \xi \geq A_0 |\xi|^2$$

Note that if $\lambda = 0$, then φ_λ is bounded below on H and is T-periodic so that the method of Theorem 4.13 could handle the problem. However, for $\lambda \neq 0$, φ_λ is neither invariant nor bounded on H. To remedy the situation, we try to minimize φ_λ on sets of the form

$$\Lambda_\xi = \{q = (q_1, ..., q_N) \in H : \frac{1}{T} \int_0^T q_1(t)\,dt = \xi\}$$

over which φ_λ is bounded. Since it is a (mildly!) constrained minimization problem, we shall need the following Palais-Smale type condition.

Claim 1: For each λ, the functional φ_λ as well as its restriction $\varphi_{\lambda|_{\Lambda_\xi}}$ satisfy $(PS)_{F_N,c}$ for any $N \in \mathbb{N}$ and any $c \in \mathbb{R}$ where

$$F_N = \{u \in H; |\frac{1}{T}\int_0^T u(t)\,dt| \in [0,N]\}.$$

Proof of Claim 1: It suffices to show that a sequence $q_n \in H$ is relatively compact provided it satisfies the following conditions: for two positive constants C_1 and C_2, and for all $n \in \mathbb{N}$,

(i) $|\frac{1}{T}\int_0^T q_n| \leq C_1$

(ii) $\varphi_\lambda(q_n) \leq C_2$

(iii) $(\varphi'_\lambda(q_n) - \varphi'_\lambda(q_m))(q_n - q_m) \to 0$ as $n,m \to +\infty$.

Since the potential V is bounded, let

$$V_0 = \sup\{|V(t,q)|; (t,q) \in \mathbb{R} \times \mathbb{R}^N\}.$$

Set $f_0(t) = (f_1(t), ..., f_N(t))$ and $q_n = q_n^0 + \xi_n$ with $\xi_n = (\xi_{1,n}, ..., \xi_{N,n})$ in $\mathbb{R}^N$ and $\int_0^T q_n^0 = 0$. We have

$$C_2 \geq \varphi_\lambda(q_n) = \int_0^T (\frac{1}{2}A(t,q_n)\dot{q}_n \cdot \dot{q}_n - V(t,q_n) + f_0 \cdot q_n^0 + \lambda q_1)$$

$$\geq \frac{A_0}{2}\|\dot{q}_n\|_2^2 - \|f_0\|_2\|q_n^0\|_2 - (V_0 + |\lambda|C_1)T$$

$$\geq \frac{A_0}{2}\|\dot{q}_n\|^2 - \frac{T\|f_0\|_2}{2\pi}\|\dot{q}_n\|_2 - (V_0 + |\lambda|C_1)T.$$

This yields that the sequence $\|\dot{q}_n\|_2$ is uniformly bounded. This fact together with (i) imply that for a subsequence of q_n (which we still call

q_n) and $q \in H$ we have

$$\|q_n - q\|_\infty \to 0 \quad \text{as} \quad n \to +\infty.$$

Set

$$A_n(t) = \left(\frac{\partial A}{\partial q_1}(t, q_n)\dot{q}_n \cdot \dot{q}_n, ..., \frac{\partial A}{\partial q_N}(t, q_n)\dot{q}_n \cdot \dot{q}_n \right) \in \mathbb{R}^N.$$

For a suitable constant $K_1 > 0$, we have for all $n \in \mathbb{N}$,

$$\sum_{k=1}^{N} \left\| \frac{\partial A}{\partial q_k}(t, q_n)\dot{q}_n \cdot \dot{q}_n \right\|_{L^1} \le K_1.$$

Furthermore, by the Lipschitz continuity of $A(t,q)$ in q (uniformly in t) we have

$$(\varphi_\lambda'(q_n) - \varphi_\lambda'(q_m))(q_n - q_m)$$

$$= \int_0^T A(t, q_n)(\dot{q}_n - \dot{q}_m) \cdot (\dot{q}_n - \dot{q}_m)$$

$$+ \int_0^T (A(t, q_n) - A(t, q_m))\dot{q}_m \cdot (\dot{q}_n - \dot{q}_m)$$

$$+ \frac{1}{2} \int_0^T (A_n(t) - A_m(t)) \cdot (q_n - q_m)$$

$$+ \int_0^T \left(\frac{\partial V}{\partial q}(t, q_n) - \frac{\partial V}{\partial q}(t, q_m) \right) \cdot (q_n - q_m)$$

$$\ge A_0 \|\dot{q}_n - \dot{q}_m\|_2^2 - K_2 \|q_n - q_m\|_\infty,$$

where K_2 is a positive constant. Consequently $\|\dot{q}_n - \dot{q}_m\|_2 \to 0$ as $n, m \to +\infty$, and therefore $q_n \to q$ in H.

We now can prove the following:

Claim 2: For every $\lambda \in \mathbb{R}$, φ_λ is bounded from below in $\wedge_\xi$. Furthermore, for every $\xi \in \mathbb{R}$, there exists $q_\xi \in \wedge_\xi$ such that for any $\lambda \in \mathbb{R}$,

$$\varphi_\lambda(q_\xi) = \inf_{\wedge_\xi} \varphi_\lambda = \varphi_0(q_\xi) + \lambda T\xi \qquad (5)$$

Proof of Claim 2: Set $m_\xi = \inf_{\wedge_\xi} \varphi_0$ and let $q_n = (q_{1,n}, ..., q_{N,n}) \in \wedge_\xi$ be a minimizing sequence for φ_0. Since $\varphi_0(q) = \varphi_0(q + (k_1T_1, ..., k_0T_n))$ for all $k_s \in \mathbb{Z}, s = 1, ..., N$, we can assume that $\frac{1}{T}\int_0^T q_{k,n} \in [0, T_k]$ for $k = 2, ..., N$. As in the proof of Claim (1), this implies that for some constant $C > 0, \|q_n - q_m\| \le C$ for all n, m.

If we set $\frac{1}{T}\int_0^T \varphi_0'(q_n) = (\sigma_{1,n}, ..., \sigma_{N,n})$ we get that

$$\lim_n \|\varphi_0'(q_n) - (\sigma_{1,n}, 0, ..., 0)\| \to 0.$$

Since $\int_0^T (q_{1,n} - q_{1,m}) = 0$ for all $n, m \in \mathbb{N}$, we conclude that

$$|(\varphi_0'(q_n) - \varphi_0'(q_m))(q_n - q_m)|$$

$$= \left| \left((\varphi_0'(q_n) - (\sigma_{1,n}, 0, ..., 0)) - (\varphi_0'(q_m) - (\sigma_{1,m}, 0, ..., 0)) \right) (q_n - q_m) \right|$$

$$\leq \left(\|\varphi_0'(q_n) - (\sigma_{1,n}, 0, ..., 0)\| + \|\varphi_0'(q_{1,m}) - (\sigma_m, 0, ..., 0)\| \right) \|q_n - q_m\|.$$

Consequently $|(\varphi_0'(q_n) - \varphi_0'(q_m))(q_n - q_m)| \to 0$ as $n, m \to +\infty$. So Claim 1 applies to the sequence q_n and yields the conclusion.

For each $\xi \in \mathbb{R}$, set

$$\Gamma_\xi = \{q \in \Lambda_\xi : \varphi_0(q) = \inf_{\Lambda_\xi} \varphi_0\}.$$

By Claim (2) we know that $\Gamma_\xi \neq \emptyset$ for all $\xi \in \mathbb{R}$. Moreover, for $q_\xi \in \Gamma_\xi$ we have $\varphi_0'(q_\xi) = -T(\psi(q_\xi), 0, ..., 0) \in \mathbb{R}^N$, where

$$\psi(q) = -\frac{1}{T} \left(\frac{1}{2} \int_0^T \frac{\partial A}{\partial q_1}(t, q)\dot{q} \cdot \dot{q} - \int_0^T \frac{\partial V}{\partial q_1}(t, q) \right).$$

We shall find a first solution by minimizing φ_λ over the set

$$\Lambda_{\xi_1, \xi_2} = \left\{ q = (q_1, ..., q_N) \in H : \xi_1 \leq \frac{1}{T} \int_0^T q_1 \leq \xi_2 \right\}$$

where ξ_1, ξ_2 are appropriately chosen real numbers. To do this, we need the following two estimates. In the sequel L will be the Lipschitz constant of V, i.e., for all $t \in \mathbb{R}, q, Q \in \mathbb{R}^N$,

$$|V(t, q) - V(t, Q)| \leq L|q - Q|. \tag{6}$$

For the sequel, we also set $f_0(t) = (f_1(t), ..., f_N(t))$.

Claim 3: The following estimates hold:

(i) If $q_\xi \in \Gamma_\xi$, then $\|\dot{q}_\xi\|_2 \leq \frac{T}{\pi A_0}(\|f_0\|_2 + \sqrt{T}L)$.

(ii) For every $\lambda \in \mathbb{R}$, there exists $L_\lambda > 0$ such that for all $\sigma_i \in \mathbb{R}^N, \xi_i \in \mathbb{R}$ and $q_{\xi_i} \in \Gamma_{\xi_i}$, $i = 1, 2$, we have

$$|\varphi_\lambda(q_{\xi_1} + \sigma_1) - \varphi_\lambda(q_{\xi_2} + \sigma_2)| \leq L_\lambda(\|q_{\xi_1} - q_{\xi_2}\| + |\sigma_1 - \sigma_2|)$$

Proof of Claim 3: (i) Set $q_\xi = q_\xi^0 + (\xi, \sigma_2, ..., \sigma_N)$ with $\int_0^T q_\xi^0 = 0$. Since

$$TV(\xi, \sigma_2, ..., \sigma_N) = \varphi_0(\xi, \sigma_2, ..., \sigma_N) \geq \varphi_0(q_\xi)$$

we have

$$0 \geq \frac{1}{2} \int_0^T A(t, q_\xi)\dot{q}_\xi \cdot \dot{q}_\xi - \int_0^T (V(q_\xi^0 + (\xi, \sigma_2, ..., \sigma_N) - V(\xi, \sigma_2, ..., \sigma_N))$$

$$+ \int_0^T f_0 \cdot q_\xi^0$$

$$\geq \frac{A_0}{2} \|\dot{q}_\xi\|_2^2 - L\sqrt{T}\|q_\xi^0\|_2 - \|f_0\|_2 \|q_\xi^0\|_2.$$

Therefore $\|\dot{q}_\xi\|_2 \left(\frac{A_0}{2} \|\dot{q}_\xi\|_2 - \frac{T}{2\pi} \|f_0\|_2 + L\sqrt{T} \right) \leq 0$ and the first estimate is verified.

ii) Let L' be the (uniform) Lipschitz constant corresponding to $A(t,q)$ and let M_0 be such that for all $t \in \mathbb{R}$ and $q, \xi, \eta \in \mathbb{R}^N$,

$$|A(t,q)\xi \cdot \eta| \leq M_0 |\xi||\eta|.$$

Set $\hat{q}_{\xi_i} = q_{\xi_i} + \sigma_i, i = 1, 2$. We have

$$|\varphi_\lambda(\hat{q}_{\xi_1}) - \varphi_\lambda(\hat{q}_{\xi_2})|$$

$$\leq \frac{1}{2} \int_0^T |A(t, \hat{q}_{\xi_1})\dot{q}_{\xi_1} \cdot \dot{q}_{\xi_1} - A(t, \hat{q}_{\xi_2})\dot{q}_{\xi_2} \cdot \dot{q}_{\xi_2}|$$

$$+ \int_0^T |V(t, \hat{q}_{\xi_1}) - V(t, \hat{q}_{\xi_2})|$$

$$+ \int_0^T |f_0 \cdot (q_{\xi_1}^0 - q_{\xi_2}^0)| + T|\lambda|(|\xi_1 - \xi_2| + |\sigma_1 - \sigma_2|)$$

$$\leq \frac{1}{2} \int_0^T |(A(t, \hat{q}_{\xi_1}) - A(t, \hat{q}_{\xi_2}))\dot{q}_{\xi_1} \cdot \dot{q}_{\xi_1}|$$

$$+ \frac{1}{2} \int_0^T |A(t, \hat{q}_{\xi_2})\dot{q}_{\xi_1} \cdot (\dot{q}_{\xi_1} - \dot{q}_{\xi_2})|$$

$$+ \frac{1}{2} \int_0^T |A(t, \hat{q}_{\xi_2})\dot{q}_{\xi_2} \cdot (\dot{q}_{\xi_1} - \dot{q}_{\xi_2})| + L\sqrt{T}\|\hat{q}_{\xi_1} - \hat{q}_{\xi_2}\|_2$$

$$+ \frac{T}{2\pi}\|f_0\|_2\|\dot{q}_{\xi_1} - \dot{q}_{\xi_2}\|_2 + |\lambda|T(|\xi_1 - \xi_2| + |\sigma_1 - \sigma_2|)$$

$$\leq \frac{L'}{2}\|\hat{q}_{\xi_1} - \hat{q}_{\xi_2}\|_\infty \|\dot{q}_{\xi_1}\|_2^2 + \frac{M_0}{2}\|\dot{q}_{\xi_1}\|_2\|\dot{q}_{\xi_1} - \dot{q}_{\xi_2}\|_2$$

$$+ \frac{M_0}{2}\|\dot{q}_{\xi_2}\|_2\|\dot{q}_{\xi_1} - \dot{q}_{\xi_2}\|_2 + \sqrt{T}(L + |\lambda|)\|q_{\xi_1} - q_{\xi_2}\|_2$$

$$+ \frac{T}{2\pi}\|f_0\|_2\|\dot{q}_{\xi_1} - \dot{q}_{\xi_2}\|_2 + T(|\lambda| + L)|\sigma_1 - \sigma_2|.$$

Hence by the estimate in (i), we conclude

$$|\varphi_\lambda(q_{\xi_1}) - \varphi_\lambda(q_{\xi_2})|$$

$$\leq \frac{L'}{2} \left(\frac{T}{\pi A_0}(\|f_0\|_2 + \sqrt{T}L) \right)^2 \|q_{\xi_1} - q_{\xi_2}\|_\infty$$

$$+ |\sigma_1 - \sigma_2|) + \left(M_0 \frac{T}{\pi A_0}(\|f_0\|_2 + \sqrt{T}L) + \frac{T}{2\pi}\|f_0\|_2 \right) \|\dot{q}_{\xi_1} - \dot{q}_{\xi_2}\|_2$$

$$+ \sqrt{T}(L + |\lambda|)(\|q_{\xi_1} - q_{\xi_2}\|_2 + \sqrt{T}|\sigma_1 - \sigma_2|)$$

$$\leq L_\lambda(\|q_{\xi_1} - q_{\xi_2}\| + |\sigma_1 - \sigma_2|)$$

for some suitable constant $L_\lambda > 0$ independent of ξ_i and $\sigma_i, i = 1, 2$.

Proof of Theorem 5.14: For $q \in H$, recall that

$$\psi(q) = -\frac{1}{T}\left(\frac{1}{2}\int_0^T \frac{\partial A}{\partial q_1}(t,q)\dot{q}\cdot\dot{q} - \int_0^T \frac{\partial V}{\partial q_1}(t,q)\right).$$

Set $\lambda_* = \inf_{\xi \in \mathbb{R}} \inf_{q \in \Gamma_\xi} \psi(q)$ and $\lambda^* = \sup_{\xi \in \mathbb{R}} \sup_{q \in \Gamma_\xi} \psi(q)$.

By Claim 3.(i), we have that $-\infty < \lambda_* \leq \lambda^* < +\infty$. Furthermore, φ_0 is bounded from below in H, and by Claim 2, its minimum is achieved at some point $q_0 \in H$. In particular $\psi(q_0) = 0$. So $\lambda_* \leq \psi(q_0) = 0 \leq \lambda^*$. For all $\xi \in \mathbb{R}$ and $q_\xi \in \Gamma_\xi$ we have $\varphi_0'(q_\xi) = -T(\psi(q_\xi), 0, ..., 0) \in \mathbb{R}^N$. We distinguish three cases:

Case (1): Assume $\lambda_* = \lambda^* = 0$. Then for every $\xi \in \mathbb{R}, q_\xi \in \Gamma_\xi$ (which is not empty by Claim 2) would be a solution for $(1)_{\lambda=0}$ and $\frac{1}{T}\int_0^T q_{1,\xi} = \xi$.

Case (2): If $\lambda_* < \lambda^*$ and $\lambda \in (\lambda_*, \lambda^*)$, there exists then $\xi_1, \xi_2 \in \mathbb{R}$ such that $\psi(q_{\xi_2}) < \lambda < \psi(q_{\xi_1})$ for all $q_{\xi_1} \in \Gamma_{\xi_1}$ and $q_{\xi_2} \in \Gamma_{\xi_2}$. Since $\psi(q_{\xi+kT_1}) = \psi(q_\xi)$ and $q_\xi \in \Gamma_\xi$ if and only if $q_{\xi+kT_1} \in \Gamma_{\xi+kT_1}$ for all $k \in \mathbb{Z}$, we can always assume $0 < \xi_2 - \xi_1 < T_1$.

Notice that φ_λ is bounded from below in $\wedge_{\xi_1,\xi_2}$. We shall obtain our first solution by showing that $\inf_{\wedge_{\xi_1,\xi_2}} \varphi_\lambda$ is achieved at an interior point of $\wedge_{\xi_1,\xi_2}$.

Set $d = \inf_{\wedge_{\xi_1,\xi_2}} \varphi_\lambda$ and let $q_n = (q_{1,n}, ..., q_{N,n}) \in \wedge_{\xi_1,\xi_2}$ be a minimizing sequence, that is $\lim_{n\to+\infty} \varphi_\lambda(q_n) = d$. If $\xi_n = \frac{1}{T}\int_0^T q_{1,n} \in [\xi_1, \xi_2]$, without loss of generality we can assume $q_n = q_{\xi_n} \in \Gamma_{\xi_n}$ and $\lim_{n\to+\infty} \xi_n = \xi_0$ with $\xi_1 \leq \xi_0 \leq \xi_2$. In addition, given $q_{\xi_0} \in \Gamma_{\xi_0}$, we get from Claim 3.(ii) the following estimate

$$d = \inf_{\wedge_{\xi_1,\xi_2}} \varphi_\lambda \leq \varphi_\lambda(q_{\xi_0}) \leq \varphi_\lambda(q_{\xi_n} + (\xi_0 - \xi_n, 0, ..., 0))$$

$$= \varphi_\lambda(q_{\xi_n} + (\xi_0 - \xi_n, 0, ..., 0)) - \varphi_\lambda(q_{\xi_n}) + \varphi_\lambda(q_{\xi_n})$$

$$\leq L_\lambda|\xi_0 - \xi_n| + \varphi_\lambda(q_{\xi_n}) \to d \quad \text{as} \quad n \to \infty.$$

Therefore $\varphi_\lambda(q_{\xi_0}) = d$.

We will be done if we show that $\xi_1 < \xi_0 < \xi_2$. To this purpose set $\ell_1(s) = \varphi_\lambda(q_{\xi_1} + (s, 0, ..., 0))$ and note that

$$\ell_1'(0) = \frac{1}{2}\int_0^T \frac{\partial A}{\partial q_1}(t, q_{\xi_1})\dot{q}_{\xi_1}\cdot\dot{q}_{\xi_1} - \int_0^T \frac{\partial V}{\partial q_1}(t, q_{\xi_1}) + \lambda T$$

$$= -T(\psi(q_{\xi_1}) - \lambda) < 0.$$

Similarly, if $\ell_2(s) = \varphi_\lambda(q_{\xi_2} + (s, 0, ..., 0))$, then $\ell_2'(0) = -T(\psi(q_{\xi_2}) - \lambda) > 0$. So for small $\varepsilon > 0$,

$$\varphi_\lambda(q_{\xi_0}) \leq \varphi_\lambda(q_{\xi_1} + (\varepsilon, 0, ..., 0)) < \varphi_\lambda(q_{\xi_1})$$

and

$$\varphi_\lambda(q_{\xi_0}) \le \varphi_\lambda(q_{\xi_2} - (\varepsilon, 0, ..., 0)) < \varphi_\lambda(q_{\xi_2}).$$

Consequently $\xi_0 \ne \xi_1$ and $\xi_0 \ne \xi_2$, which yields $\varphi'_\lambda(q_{\xi_0}) = 0$. So q_{ξ_0} is a solution for $(1)_\lambda$ which is a local minimum for φ_λ.

To get a second solution, we shall use Corollary 5.12. Indeed, note that

$$\varphi_\lambda(q) = \varphi_\lambda(q + (0, k_2 T_2, ..., k_N T_N)) \quad \text{for all } k_s \in \mathbf{Z}, s = 2, ..., N,$$

hence q_{ξ_0} and $p = q_{\xi_0} + (0, k_2 T_2, ..., k_N T_N)$ are two local minima for φ. Since φ_λ verifies $(PS)_{F_N,c}$ for any $N \in \mathbf{N}$ and any $c \in \mathbf{R}$, then φ_λ has a critical point different from $q_{\xi_0} + (0, k_2 T_2, ..., k_N T_N)$ for all $k_s \in \mathbf{Z}, s = 2, ..., N$.

Case (3): If $\lambda = \lambda_*$ or $\lambda = \lambda^*$, take a sequence $\{\lambda_n\}$ in (λ_*, λ^*) such that $\lambda_n \to \lambda_*$ (or $\lambda_n \to \lambda^*$) as $n \to +\infty$. By the previous arguments, for every $n \in \mathbf{N}$ there exists $q_n = (q_{1,n}, ..., q_{N,n}) \in H$ such that

(a) $\dfrac{1}{T}\displaystyle\int_0^T q_{k,n} \in [0, T_k]$ for all $n \in \mathbf{N}$ and for all $k = 1, ..., N$;

(b) $\|\dot{q}_n\|_2 \le \dfrac{T}{A_0 \pi}(\|f_0\|_2 + \sqrt{T}L)$

(c) $0 = \varphi'_{\lambda_n}(q_n) = \varphi'_{\lambda_*}(q_n) + \lambda_n - \lambda_*$.

Therefore $|(\varphi'_{\lambda_*}(q_n) - \varphi'_{\lambda_*}(q_m))(q_n - q_m)| = |\lambda_n - \lambda_m|\|q_n - q_m\| \to 0$ as $n, m \to +\infty$, since (a) and (b) necessarily imply that for some constant $C > 0$, $\|q_n - q_m\| \le C$ for all $n, m \in \mathbf{N}$. Hence as in Claim 1, we obtain a subsequence $\{q_{n_k}\}$ of $\{q_n\}$ and $q \in H$ such that $q_{n_k} \to q$ in H. Obviously $\varphi'_{\lambda_*}(q) = 0$, and Theorem 5.14 is proved.

5.4 A relaxed boundary condition in the presence of linking

Corollary 5.15 (Fig 5.2): *Let φ be a C^1-functional satisfying (PS) on a Banach space $X = Y \oplus Z$ with dim $(Y) < \infty$. Assume that for a fixed unit vector z in Z, there is a continuous map γ_0 from the boundary of the half ball $K = \{u = y + sz; y \in Y, s \ge 0, \|u\| \le R\}$ of $Y \oplus \mathbf{R}z$ into X with the following properties:*

(i) $\gamma_0(y) = y$ for $y \in Y$, $\|y\| \le R$ and $\|\gamma_0(u)\| \ge r > 0$ for $\|u\| = R$.
(ii) $\varphi(\gamma_0(u)) \le a$ for all $u \in \partial K$.
(iii) For some positive $\rho < r$, inf $\varphi(Z \cap S_\rho) \ge a$.

Then φ has a critical point at a level $c \ge a$. Moreover, if $c > a$ then K_c contains a point which is not a local minimum, and if $c = a$ then $K_c \cap Z \cap S_\rho \ne \emptyset$.

In particular, if $\varphi(0) = a$, then φ has a non-zero critical point.

Proof: Let Γ be the class of all continuous extensions of γ_0 to all of K.

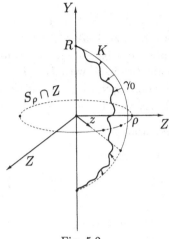

Fig. 5.2.

The result will follow immediately from Corollary 5.10 if we show the following

Claim: The set $Z \cap S_\rho$ is dual to the class Γ.

Indeed, assume without loss of generality that $R = 1$ and consider any continuous map $\gamma : K \to X$ satisfying

$$\gamma(y) = y \quad \text{if} \quad y \in Y \text{ and } \|y\| \le 1$$
$$\|\gamma(u)\| \ge r > 0 \quad \text{if} \quad u \in K \text{ and } \|u\| = 1. \tag{1}$$

Let P be the projection onto Y along Z. We shall prove that there exists $\bar{u} \in K$ such that

$$P \circ \gamma(\bar{u}) = 0 \quad \text{and} \quad \|\gamma(\bar{u})\| = \rho. \tag{2}$$

Let $Y_1 = Y \oplus \mathbb{R}z$ and consider the map $T : K \to Y_1$,

$$T(u) = P \circ \gamma(u) + \|(\text{Id} - P) \circ \gamma(u)\| z.$$

To prove (2), it suffices to show that for some point $\bar{u} \in K$, $T(\bar{u}) = \rho z$. To do that, we will use degree theory (see Appendix D). Since $\rho < r$, it follows from (i) that $T(u) \ne \rho z$ for all $u \in \partial K$. Consequently $\deg(T, K, \rho z)$ is defined. To get the desired result, we need to show that it equals 1. Now since the degree depends only on the boundary values of T, we may consider only $T|_{\partial K}$. Set

$$A = \{(y, 0); y \in Y, \|y\| \le 1\} \text{ and } A' = \{u \in K; \|u\| = 1\}.$$

We have $\partial K = A \cup A'$. Clearly $Tu = u$ for $u \in A$ and $\|Tu\| \ge r > 0$ for $u \in A'$. On ∂K define

$$\tilde{T}u = \begin{cases} u & \text{if } u \in A \\ Tu/\|Tu\| & \text{if } u \in A'. \end{cases}$$

Using (1) we see that T and $\tilde{T}$ are homotopic in $Y_1\backslash\{\rho z\}$ via $T_t u = tTu + (1-t)\tilde{T}u, t \in [0,1]$. Note that $\tilde{T}(A') \subset A'$ and $\tilde{T} = I$ on $\partial A'$. Since A' is homeomorphic to a ball, there is a continuous deformation $\tilde{T}_t$ connecting $\tilde{T}$ to the identity in A' with $\tilde{T}_t = I$ on $\partial A'$ for all $t \in [0,1]$. It follows that $T|_{\partial K}$ is homotopic to the identity in $Y_1\backslash\{\rho z\}$ and thus $\deg(T, K, \rho z) = \deg(\mathrm{Id}, K, \rho z) = 1$.

5.5 Second order systems

We now prove the following:

Theorem 5.16: *Consider a system of $\mathbb{R}^N$-valued functions $x(t)$ satisfying*

$$\ddot{x} + V_x(t,x) = 0 \tag{1}$$

with V a real smooth function defined in $\mathbb{R} \times \mathbb{R}^n$ and periodic in t of period 2π, such that for some θ $(0 < \theta < \frac{1}{2})$, it satisfies

$$0 < V(t,x) \le \theta x \cdot V_x(t,x) \quad \text{for } |x| \text{ large and all } t. \tag{2}$$

Also assume that

$$V(t,x) \ge 0 \tag{3}$$

and

$$L := \max_{\substack{t \\ |x| \le 1}} V(t,x) \le \frac{3}{2\pi^2}. \tag{4}$$

Then (1) has a nonconstant periodic solution with period 2π.

Proof: As usual, the variational expression is given by

$$\varphi(x) = \int_0^{2\pi} [\frac{1}{2}|\dot{x}|^2 - V(t,x(t))]\, dt \tag{5}$$

on the Banach space $X = H^1$ of vector valued 2π-periodic functions. A critical point x of (5) is a 2π-periodic solution of (1).

For $u \in H^1$, write $v = u - \frac{1}{2\pi}\int_0^{2\pi} u(t)\, dt$ and $w = \frac{1}{2\pi}\int_0^{2\pi} u(t)\, dt$. We shall equip H^1 with the following equivalent norm:

$$\|u\| = \|\dot{v}\|_2 + \|w\|_\infty.$$

Essentially the same proof as of Theorem 3.10, gives that φ is of class C^1 on $X = H^1$ and satisfies the (PS) condition.

To apply Corollary 5.15, let Y be the N-dimensional subspace of X consisting of the constant vectors and let Z be the space of functions with mean zero. Consider the following $(N+1)$-dimensional subspace of X:

$$Y_1 = \{x(t) = x_0 + \beta e_1 \cos t; x_0 \in \mathbb{R}^N, \beta \in \mathbb{R}, e_1 = (1, ..., 0)\}$$

and let K_R be the open cylinder

$$K_R = \{x(t) = x_0 + \beta e_1 \cos t; |x_0| < R, 0 < \beta < R\}$$

for some R to be chosen later. Note that for x in Y_1 we have

$$\varphi(x) = \frac{1}{2}\beta^2 \int_0^{2\pi} \sin^2 t\, dt - \int_0^{2\pi} V(t, x(t))\, dt = \frac{\pi}{2}\beta^2 - \int_0^{2\pi} V(t, x(t))\, dt.$$

Claim 1: For R large enough, we have $\varphi \leq 0$ on ∂K_R.

Proof: Note that condition (2) implies that for some $C, C' > 0$,

$$V(t, x) \geq C|x|^{1/\theta} - C'. \tag{2'}$$

On the base of the cylinder, $\beta = 0$, hence $\varphi(x) \leq 0$ there, since $V \geq 0$. For the function $x(t) = x_0 + \beta e_1 \cos t$ in Y_1, we have

$$\int_0^{2\pi} |x(t)|^2\, dt = 2\pi|x_0|^2 + \pi\beta^2.$$

In addition, by (2)',

$$\int_0^{2\pi} V(t, x(t))\, dt \geq C \int_0^{2\pi} |x(t)|^{1/\theta}\, dt - C' \geq C_1 \left(\int_0^{2\pi} |x(t)|^2\, dt \right)^{1/2\theta} - C'$$

by Hölder's inequality - with a suitable constant C_1. Hence

$$\varphi(x) \leq \frac{\pi}{2}\beta^2 - C_1(2\pi|x_0|^2 + \pi\beta^2)^{1/2\theta} + C'. \tag{6}$$

Thus on the lateral surface of the cylinder, i.e., where $|x_0| = R$ and $0 \leq \beta \leq R$, we have for large R,

$$\varphi(x) \leq \frac{\pi}{2}R^2 - C_1(2\pi R^2)^{1/2\theta} + C' \leq 0.$$

The claim is proved.

Consider now the set $F = \{y \in Z; \|y\| = \sqrt{\frac{6}{\pi}}\}$. We have the following

Claim 2: $\inf \varphi(F) \geq \frac{3}{\pi} - 2\pi L \geq 0$.

Indeed, if $y \in F$, we have by Sobolev's embedding theorem in one dimension (Appendix A) that $\|y\|_\infty \leq 1$. This combined with assumption (4) gives

$$\varphi(y) = \int_0^{2\pi} [\frac{1}{2}|\dot{y}|^2 - V(t, y(t))]dt \geq \frac{3}{\pi} - 2\pi L \geq 0$$

Hence we can use Corollary 5.15 to conclude that φ has a non-zero critical point at a level $c \geq 0$. Note that if $c > 0$, the corresponding critical point must be nonconstant since for any constant critical point x we have $\varphi(x) = -\int V(t, x)\, dt \leq 0$. On the other hand, if $c = 0$, the fact that x is non-zero again guarantees that it is necessarily non constant. Actually, in this case x belongs to F and the proof is complete.

Remark 5.17: Note that in the case where $c = \frac{3}{\pi} - 2\pi L$, what is needed is only that the functional φ verifies $(PS)_{F,c}$. This holds without assumption (2).

5.6 Critical points in the presence of splitting

We now give a proof of the following result:

Theorem 5.18: *Let φ be a C^1-functional satisfying (PS) on a Banach space $X = Y \oplus Z$ with $k = \dim Y < \infty$. Assume $\varphi(0) = 0$ and that for some $R > 0$, the following hold:*

(a) $\varphi(u) \geq 0$ for $u \in Z$ with $\|u\| \leq R$,
(b) $\varphi(y) \leq 0$ for $y \in Y$ with $\|y\| \leq R$,
(c) φ is bounded below and $\inf_X \varphi < 0$.

Then φ has at least two non-zero critical points. Furthermore, either φ has a non-zero critical point which is not a local minimum, or φ has a continum of local minima at level zero.

We shall need the following Lemma. As in Chapter 3, we denote by v a continuous pseudo-gradient vector field associated to φ on the set $\{u \in X; \varphi'(x) \neq 0\}$.

Lemma 5.19: *Let φ be a C^1 function satisfying (PS) on a Banach space X and suppose that $u_0 \in X$ is a unique global minimum for φ on X. Let y be such that $\varphi'(y) \neq 0$, and φ has no critical value in the interval $(\varphi(u_0), \varphi(y))$. Then the "negative gradient flow" starting at y, defined by*

$$\frac{dx}{dt} = -\frac{v(x)}{\|v(x)\|^2}, \quad x(0) = y, \tag{1}$$

exists for a maximal finite time $0 \leq t \leq T(y)$ and $x(T(y)) = u_0$.

Proof: On the integral curve of (1) we have that $\frac{d\varphi}{dt} \leq -\frac{1}{4}$. Thus the solution $x(t)$ of (1) exists on a maximal open interval $(0, T)$ with $T \leq 4\varphi(y)$, and $\varphi(u_0) < \varphi(x(t)) < \varphi(y)$ on $(0, T)$. We need to prove that $x(t) \to u_0$ as $t \to T$. It will be enough to show that there exists a sequence $t_j \to T$ such that $\|\varphi'(x(t_j))\| \to 0$, since the (PS) condition then yields a subsequence that converges to a critical point of φ; it can only be u_0, therefore $\lim_{t \to T} \varphi(x(t)) = \varphi(u_0)$ and by Corollary 1.4, $x(t) \to u_0$ as $t \to T$.

Assume now that $\|\varphi'(x(t))\| \geq \delta$ on $(0, T)$ for some $\delta > 0$. Then $\|v(x(t))\| \geq \|\varphi'(x(t))\| \geq \delta$, and $\int_0^T \frac{dx}{dt} dt$ exists, and so $\lim_{t \to T} x(t) = x(T)$ exists; $x(T)$ is necessarily u_0 for otherwise the solution $x(t)$ could be continued beyond $t = T$. This is a contradiction since $\varphi'(u_0) = 0$.

Proof of Theorem 5.18: Clearly φ achieves its minimum at some point u_0. Supposing that 0 and u_0 are the only critical points for φ, we shall look for a contradiction. We consider 3 cases:

Case (1): $\dim(Y) > 0$ and $\dim(Z) > 0$.

We first claim that there exists $\delta > 0$ such that

$$\{u : F(u) < F(u_0) + \delta\} \subset \{u : \|u - u_0\| < \|u_0\|/2\}.$$

Indeed, if not, there exists then a minimizing sequence $(u_n)_n$ such that $\|u_n - u_0\| \geq \|u_0\|/2$. By Ekeland's Theorem and as in Corollary 1.6, we can also assume that $\varphi'(x_n) \to 0$. Since φ verifies (PS), we then obtain a critical point that is different from 0 and u_0, contrary to our assumption.

Back to the proof, we suppose without loss of generality, that $R = 1 < \|u_0\|$. Since at every point $y \in Y$ with $\|y\| = 1$, $\varphi'(y) \neq 0$, we may apply the preceding Lemma to conclude that the flow starting at y, described by (1), exists on a maximal open interval $0 < t < T(y) < -4\varphi(u_0)$ and that $x(t) \to u_0$ as $t \to T(y)$. Moreover, by choosing δ sufficiently small, we obtain that for a unique value $t = t(y) < T(y)$ we have $\varphi(x(t(y))) = \varphi(u_0) + \delta$. This uniqueness property necessarily implies that $t(y)$ is continuous in y.

For z_0, a unit vector in Z, let K denote the set

$$K = \{u = sz_0 + y; y \in Y, s \geq 0 \text{ and } \|u\| \leq 1\}. \tag{2}$$

We now define a continuous map γ_0 of ∂K into X in the following way:

$$\gamma_0(y) = y \text{ if } y \in Y, \|y\| \leq 1, \text{ and } \gamma_0(z_0) = u_0. \tag{3}$$

Moreover, any $u \neq z_0$ on ∂K, with $\|u\| = 1$, has the unique representation

$$u = sz_0 + \sigma y \tag{4}$$

with $0 \leq s \leq 1, y \in Y, \|y\| = 1$ and $0 < \sigma \leq 1$ (i.e., s, σ, y are unique). For such a u, we define

$$\gamma_0(sz_0 + \sigma y) = x(2st(y)) \text{ for } 0 \leq s \leq \frac{1}{2} \tag{5}$$

where $x(t)$ is our solution of (1). So $\gamma_0(\frac{1}{2}z_0 + \sigma y) = x(t(y))$ and it lies in the set $\{x : \|x - u_0\| < \frac{1}{2}\|u_0\|\}$. Finally define

$$\gamma_0(sz_0 + \sigma y) = (2s - 1)u_0 + (2 - 2s)x(t(y)) \text{ for } \frac{1}{2} \leq s < 1. \tag{6}$$

As s goes from $\frac{1}{2}$ to 1, the right hand side traverses the straight segment from $x(t(y))$ to u_0, and so for $s \geq 1/2$, $\|\gamma_0(sz_0 + \sigma y) - u_0\| \leq \|u_0\|/2$.

The mapping γ_0 is clearly continuous and $\varphi \leq 0$ on its image $\gamma_0(\partial K)$. In addition we see that for some $r \leq 1$, we have $\|\gamma_0(u)\| \geq r >$

0 for $\|u\| = 1$. Fix $0 < \rho < r$ and use Corollary 5.15 to find a non-zero critical point u_1 for φ at a nonnegative level c. If $c > 0$, then Corollary 4.14 applies to conclude that u_1 is not a local minimum. If $c = 0$ and u_1 is a local minimum, Corollary 5.12 applies between 0 and u_1 and we get the existence of a continuum of local minima.

Case (2): $\dim(Y) = 0$ and $\dim(Z) > 0$.

In that case 0 is a local minimum for φ and u_0 is a global minimum. Corollary 5.12 then applies and we get the claim.

Case (3): $\dim(Y) > 0$ and $\dim Z = 0$.

We may even assume in this case that $\dim(Y) = \infty$. Note that now 0 is a local minimum for $-\varphi$ and since φ verifies (PS), we get from Corollary 1.5 that φ is coercive, that is $-\varphi(u) \to -\infty$ as $\|u\| \to \infty$. The claim follows again from Corollary 5.12.

5.7 Multiple solutions for a boundary value problem

A typical situation where the splitting required by Theorem 5.18 occurs around a critical point u_0 is when the latter is a non-degenerate critical point for φ (i.e., $d^2\varphi(u_0)$ invertible) with a finite *Morse index*. In this case the splitting is assured by the Morse Lemma. (See Chapter 9 for details). Theorem 5.18 then yields the following:

Corollary 5.20: *Assume φ is a bounded below C^2-functional that verifies (PS) and suppose u_0 is a non-degenerate critical point with a finite Morse index such that $\inf_X \varphi < \varphi(u_0)$. Then φ has at least two critical points that are distinct from u_0.*

As an application of Corollary 5.20, we give the following example:

Let Ω be a bounded domain in $\mathbb{R}^N$, and consider the problem

$$\begin{cases} -\Delta u = g(u) & \text{in } \Omega, \\ u = 0 & \text{on } \partial\Omega. \end{cases} \tag{1}$$

Let $\{\lambda_i\}_{i=1}^\infty$ be the spectrum $\sigma(-\Delta)$ of $-\Delta$ in $H_0^1(\Omega)$ with homogeneous boundary value conditions.

Theorem 5.21: *Let g be a function in $C^1(\mathbb{R})$ that verifies $g(0) = 0$ and*

$$\limsup_{|s|\to\infty} \frac{g(s)}{s} < \lambda_1. \tag{2}$$

Assume also that

$$g'(0) > \lambda_1 \quad \text{and} \quad g'(0) \notin \sigma(-\Delta). \tag{3}$$

Then, equation (1) has at least two non trivial solutions.

Proof: Suppose first that $\tilde{g} : \mathbb{R} \to \mathbb{R}$ is a C^1-function such that for some constants $C > 0$ and $0 < \mu < \lambda_1$,

$$\tilde{g}(t) \leq \mu t + 2C \text{ on } \mathbb{R}^+ \text{ and } \tilde{g}(t) \geq \mu t - 2C \text{ on } \mathbb{R}^-.$$

If u is a C^2-solution of $-\Delta u = \tilde{g}(u)$ in Ω and $u = 0$ on $\partial\Omega$, then the maximum principle yields that $u_- \leq u \leq u_+$ where u_+ and u_- are the solutions of

$$-\Delta u - \mu u = \pm 2C \text{ in } \Omega \text{ and } u = 0 \text{ on } \partial\Omega.$$

Note now that assumption (2) yields constants $C > 0$ and $0 < \mu < \lambda_1$, such that

$$g(t) \leq \mu t + C \text{ on } \mathbb{R}^+ \text{ and } g(t) \geq \mu t - C \text{ on } \mathbb{R}^-.$$

It follows from the above discussion that we can replace g by a C^1-modification $\tilde{g}$ with $|\tilde{g}'(t)|$ uniformly bounded on $\mathbb{R}$ and such that the solutions of the corresponding boundary value problem are exactly the solutions of the initial one.

In other words, assumption (2) combined with the maximum principle allow us to assume that $g(t) = bt + c$ with $b < \lambda_1$, for $|t|$ large. But in this case, the functional

$$\varphi(u) = \frac{1}{2} \int_\Omega |\nabla u|^2 - \int_\Omega G(u)$$

is a C^2-functional on $X = H_0^1$ that verifies the (PS) condition. Moreover, its critical points correspond to the solutions of (1). We claim that φ satisfies the conditions of Corollary 5.20. Indeed, (2) implies that it is bounded below. On the other hand, since $\varphi''(0) = -\Delta - g'(0)$, it follows from condition (3) that 0 is a nondegenerate critical point with finite index larger than 1. Hence 0 is not the global minimum and Corollary 5.15 applies to yield the result.

In the next chapter, we shall be able to remove the condition $g'(0) \notin \sigma(-\Delta)$.

We could have avoided the use of the Morse lemma in the above example as we could have written the splitting explicitly. We shall do just that in the next example borrowed from Brezis-Nirenberg [B-N3].

5.8 Multiple solutions for a nonlinear eigenvalue value problem

Let Ω be a smoothly bounded domain in $\mathbb{R}^N$, $N > 2$, $2^* = \frac{2N}{N-2}$ and consider the problem

$$\begin{cases} -\Delta u + a(x)u = \lambda g(u) & \text{in } \Omega, \\ u = 0 & \text{on } \partial\Omega. \end{cases} \tag{1}$$

Theorem 5.22: *Assume that $a \in L^\infty$, g is smooth with*

$$g(0) = g'(0) = 0, \tag{2}$$

and

$$\limsup_{|t|\to\infty} \frac{g(t)}{t} < 0. \tag{3}$$

In addition, assume that for some t_0,

$$G(t_0) = \int_0^{t_0} g(s)\, ds > 0.$$

In case 0 is an eigenvalue of $-\Delta + a$, we also suppose that for some small $\delta > 0$,

$$G(t) \geq 0 \quad \text{for } |t| \leq \delta. \tag{4}$$

Then, for every λ sufficiently large, there are at least two nontrivial solutions of (1).

Proof: As in the preceding example, assumption (3) allows us to modify g in such a way that $g(u) = -bu + c$ with $b > 0$, for $|u|$ large and the solutions of the modified problem are also solutions of the original problem. The functional

$$\varphi_\lambda(u) = \frac{1}{2}\int_\Omega |(\nabla u|^2 + au^2) - \lambda \int_\Omega G(u)$$

is then C^1 and verifies the (PS)-condition on $X = H_0^1$.

We now show that φ_λ verifies the conditions of Theorem 5.18. First, since $G(t_0) > 0$ for some t_0, there exists a function $u_0 \in H_0^1$ such that $\int_\Omega G(u_0)\, dx > 0$, hence $\inf_X \varphi_\lambda < 0$ for λ large enough. On the other hand, we may use (3) to deduce easily that $\inf_X \varphi_\lambda > -\infty$ for λ large.

To construct the appropriate splitting for X around the critical point 0, choose for Y the finite dimensional space spanned by the eigenfunctions corresponding to the nonpositive eigenvalues of $-\Delta + a$, and let Z be its orthogonal complement in X. Using (2) and the fact that $G(t) < 0$ for $|t|$ large we get that for any $\varepsilon > 0$, there exists a constant $C_\varepsilon > 0$ such that for all t,

$$G(t) \leq \varepsilon t^2 + C_\varepsilon |t|^{2^*}. \tag{5}$$

We deduce that there is an $\alpha > 0$ such that on Z,

$$\varphi_\lambda(u) \geq \alpha \|u\|^2 - \frac{\alpha}{2}\|u\|_2^2 - C\|u\|^{2^*} \geq 0$$

for $\|u\| \leq R$ small. On Y, which is finite dimensional, we have in view of (5) — and (4) in case 0 is an eigenvalue of $-\Delta + a$ — that

$$\varphi_\lambda(u) \leq 0 \quad \text{for } \|u\| \text{ small.}$$

Theorem 5.18 now yields the desired conclusion.

An example of a function g satisfying the conditions of Theorem 5.22 is $g(u) = u^3 + u^4 - u^5$.

Notes and Comments: In the case where $F = \{\varphi \geq c\}$, the min-max principle with the relaxed boundary condition (Theorem 5.2) is due to Brezis-Nirenberg [B-N 3]. The general case and the corollaries that follow, appeared in [G 2]. Corollary 5.11 was known to P. Rabinowitz [R 1] who used a dual class for locating the critical level. Theorem 5.14 is a recent result of G. Tarantello [T 3]. Corollary 5.15 is an extension of a theorem of Rabinowitz [R 1] due to Brezis-Nirenberg [B-N 3]. Theorem 5.16 is a personal communication of L. Nirenberg. The proof given above of Theorem 5.18 is due to Brezis-Nirenberg [B-N 3]. This theorem is related to results of K.C. Chang [Ch] whose arguments rely on Morse theory while the above use the strong form of the min-max principle established in [G 1]. Theorem 5.22 is also from [B-N 3].

6

THE STRUCTURE OF THE CRITICAL SET

IN THE MOUNTAIN PASS THEOREM

In this chapter, we shall use the information obtained in Theorem 4.1 about the location of min-max critical points, to describe the structure of the critical set in the mountain pass theorem. The interest of this chapter stems from the fact that no Morse theory is needed and that the functionals are only supposed to be C^1.

To state Thorem 4.1 in the case of the mountain pass theorem, we need the following:

Definition 6.1: A closed subset F of a Banach space X is said to *separate two points u and v in X*, if u and v belong to two disjoint connected components of $X \setminus F$.

In the terminology of Chapter 4, it is equivalent to say that the set F is dual to the class Γ_u^v of all continuous paths joining the points u and v in X, that is

$$\Gamma_u^v = \{g \in C([0,1]; X) : g(0) = u \text{ and } g(1) = v\}$$

where $C([0,1]; X)$ is the space of all X-valued continuous functions on $[0,1]$.

Theorem 4.1, applied to the class Γ_u^v, gives the following:

Theorem 6.2: Let $\varphi : X \to \mathbb{R}$ be a C^1-function on a Banach space X and assume that for two points u and v in X, the number $c = \inf_{g \in \Gamma_u^v} \max_{0 \le t \le 1} \varphi(g(t))$ is finite. Suppose also that F is a closed subset of $\{\varphi \ge c\}$ that separates u and v. If φ verifies $(PS)_{F,c}$, then the set $F \cap K_c$ is non empty.

6.1 Saddle points of mountain pass type

To classify the various types of critical points, we shall use the following notation:

$$G_c = \{x \in X; \varphi(x) < c\} \quad \text{and} \quad L_c = \{x \in X; \varphi(x) \geq c\}$$

$$M_c = \{x \in K_c; x \text{ is a local minimum of } \varphi\}$$

$$P_c = \{x \in K_c; x \text{ is a proper local maximum of } \varphi\}$$

$$S_c = \{x \in K_c; x \text{ is a saddle point of } \varphi\}$$

Recall that x is said to be *a saddle point* if in each neighborhood of x there exist two points y and z such that $\varphi(y) < \varphi(x) < \varphi(z)$.

Theorem 6.3: *Under the hypothesis of Theorem 6.2, assume $F \cap P_c$ contains no compact set that separates u and v, then either $F \cap M_c \neq \emptyset$ or $F \cap S_c \neq \emptyset$.*

Proof: Suppose that $F \cap M_c = \emptyset = F \cap S_c$. Since F separates u and v, we can use a result of Whyburn (see [Ku] ch. VIII, S57, III, theorem 1)) to find a closed connected subset $\hat{F} \subset F$ that also separates u and v. Note that $\hat{F} \cap K_c = \hat{F} \cap P_c$ and the latter is relatively open in $\hat{F}$ while $\hat{F} \cap K_c$ is closed. Since $\hat{F}$ is connected, then either $\hat{F} \cap P_c = \emptyset$ or $\hat{F} \cap P_c = \hat{F}$. But the first case is impossible since by Theorem 6.2 we have $\hat{F} \cap P_c = \hat{F} \cap K_c \neq \emptyset$. Hence $\hat{F} \subset P_c$ which is impossible by assumption and the claim is proved.

Corollary 6.4: *Under the hypothesis of Theorem 6.2, assume φ verifies $(PS)_c$ and that $u, v \notin \overline{M}_c$. If P_c contains no compact subset that separates u and v, then $S_c \neq \emptyset$.*

In particular, if $\max\{\varphi(u), \varphi(v)\} < c$ and X is infinite dimensional, then $S_c \neq \emptyset$.

Proof: First observe that K_c is the disjoint union of S_c, M_c and P_c. By the (PS) condition, we know that K_c is compact. Suppose $S_c = \emptyset$. For each $x \in M_c$, there exists a $B(x, \varepsilon_x)$ such that $B(x, \varepsilon_x) \subseteq L_c$. Let $N = \bigcup_{x \in M_c} B(x, \varepsilon_x)$. Then $M_c \subseteq N \subseteq L_c$. Since $u, v \notin \overline{M}_c$ and $\overline{M}_c$ is compact, we may assume that $u, v \notin \overline{N}$. Now put $F_0 = (F \backslash N) \cup \partial N$. It is clear that $\inf_{x \in F_0} \varphi(x) \geq c$ and that F_0 separates u, v. Moreover, $F_0 \cap (M_c \cup S_c) = \emptyset$. By Theorem 6.3, $P_c \cap F_0$ and hence P_c must contain a compact subset that separates u, v. A contradiction that completes the proof.

Definition 6.5: Say that a point x in K_c is *of mountain-pass type* if

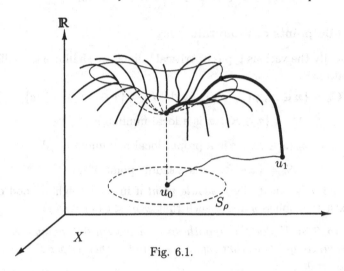

Fig. 6.1.

for any neighborhood N of x, the set $\{x \in N; \varphi(x) < c\}$ is nonempty and not path connected. We denote by H_c the set of critical points of mountain pass type at the level c.

Theorem 6.6 (Fig 6.1): *With the hypothesis of Theorem 6.2, the following hold:*
(1) Either $F \cap \overline{M_c} \neq \emptyset$ or $F \cap H_c \neq \emptyset$.
(2) If $F \cap P_c$ contains no compact set that separates u and v, then either $F \cap \overline{M_c} \neq \emptyset$ or $F \cap K_c$ contains a saddle point of mountain pass type.

Before we prove it, we give the following immediate corollary.

Corollary 6.7: *With the hypothesis of Theorem 6.2, assume that $\max\{\varphi(u), \varphi(v)\} < c$, then the following hold:*
(1) Either $\overline{M_c} \setminus M_c \neq \emptyset$ or $H_c \neq \emptyset$.
(2) If X is infinite dimensional, then either $\overline{M_c} \setminus M_c \neq \emptyset$ or K_c contains a saddle point of mountain pass type.

Proof: It is enough to apply Theorem 6.6 to the set $F = \partial\{\varphi \geq c\}$ and to notice that no local minimum can be on such a set.

To prove Theorem 6.6, we shall need the following topological lemma.

Lemma 6.8: *Let F_0 be a closed subset of X that separates two distinct points u and v. Let Z_i $(i = 1, 2, \cdots, n)$ be n mutually disjoint open subsets of X such that $u, v \notin \cup_{i=1}^{n}\overline{Z_i}$. Let G be an open subset of $X \setminus F_0$ and denote by $Y_i = Z_i \setminus G$. Then the following hold:*
(i) The set $F_1 = [F_0 \setminus (\cup_{i=1}^{n} Z_i)] \cup (\cup_{i=1}^{n} \partial Y_i)$ separates u and v.

(ii) If A_i $(i = 1, 2, \cdots, n)$ are n nonempty connected components of G and T_i $(i = 1, 2, \cdots, n)$ are subsets of $Z_i \cap \partial A_i$ that are relatively open in ∂Y_i and such that $T_i \cap \partial L = \emptyset$ for any connected component L of G with $L \neq A_i$, then the set

$$F_2 = [F_0 \backslash (\cup_{i=1}^{n} Z_i)] \cup (\cup_{i=1}^{n} \partial Y_i \backslash T_i)$$

also separates u and v.

Proof: (i) Since $G \subseteq X \backslash F_0$, we have

$$F_1 = [F_0 \backslash (\cup_{i=1}^{n} Y_i)] \cup (\cup_{i=1}^{n} \partial Y_i). \tag{1}$$

Clearly F_1 is closed and $u, v \notin F_1$. We need only to show that for any $g \in \Gamma_u^v$, $g([0, 1])$ intersects F_1. If $g([0, 1]) \cap (F_0 \backslash \cup_{i=1}^{n} Y_i) \neq \emptyset$, we are done. Otherwise $g([0, 1]) \cap (\cup_{i=1}^{n} Y_i) \cap F_0 \neq \emptyset$ so that if $g([0, 1]) \cap (\cup_{i=1}^{n} \partial Y_i) = \emptyset$, then $g([0, 1]) \subseteq \cup_{i=1}^{n} Y_i \subseteq \cup_{i=1}^{n} Z_i$ which contradicts that $u, v \notin \cup_{i=1}^{n} \overline{Z_i}$.

To establish (ii), we first prove the following:

Claim 1: For $i, j = 1, 2, \cdots, n$, we have
(a) $T_i \subseteq Y_i \cap \partial Y_i$, $T_i \cap G = \emptyset$ and $A_i \cap F_2 = \emptyset$.
(b) $T_j \cap \overline{Y_i} = \emptyset$ and $T_i \cap T_j = \emptyset$ if $i \neq j$.
(c) $Z_i \cap (\partial G \backslash T_i) \subseteq \partial Y_i \backslash T_i$.

Proof of Claim 1: (a) Since G is open, it is clear from the definition of T_i that $T_i \subseteq Z_i \cap \partial G$ so that $T_i \subseteq Y_i \cap \partial Y_i$ and $T_i \cap G = \emptyset$ for $i = 1, 2, \cdots, n$. On the other hand, $A_i \cap \overline{Y_j} \subseteq A_i \cap \overline{(Z_j \backslash G)} \subseteq A_i \cap (\overline{Z_j} \backslash G)$, hence $A_i \cap F_2 = \emptyset$.

(b) If $i, j = 1, 2, \cdots, n$ and $i \neq j$, then $T_j \cap \overline{Y_i} \subseteq T_j \cap \overline{Z_i} \subseteq Z_j \cap \overline{Z_i} = \emptyset$ and $T_i \cap T_j \subseteq Z_i \cap Z_j = \emptyset$.

(c) Since G is open, we have that for any $x \in Z_i \cap \partial G \backslash T_i$, $x \notin G$, hence $x \in Z_i \backslash G$ and $x \in Y_i$. Moreover, for any $x \in \partial G \backslash T_i$ and any $\varepsilon > 0$ there is $y \in B(x, \varepsilon) \cap G$. Clearly $y \notin Y_i$ so that $x \in \partial Y_i$. Since $T_i \cap Z_i \cap (\partial G \backslash T_i) = \emptyset$, we have that $x \in \partial Y_i \backslash T_i$.

Back to the proof of the Lemma, we note first that, in view of (a), the set F_2 is closed and is equal to

$$F_2 = [F_0 \backslash (\cup_{i=1}^{n} Y_i)] \cup (\cup_{i=1}^{n} \partial Y_i \backslash T_i). \tag{2}$$

Clearly $u, v \notin F_2$ and we need only to show that for any $g \in \Gamma_u^v$,

$$g([0, 1]) \cap F_2 \neq \emptyset.$$

Suppose not, and take $g_0 \in \Gamma_u^v$ such that $g_0([0, 1]) \cap F_2 = \emptyset$. We shall work towards a contradiction.

First by (1), we have $g_0([0, 1]) \cap (\cup_{i=1}^{n} T_i) \neq \emptyset$. Let i_1 be the first i in

$\{1, ..., n\}$ such that $g_0([0,1]) \cap T_i \neq \emptyset$. We shall find a $g_{i_1} \in \Gamma_u^v$ such that

$$g_{i_1}([0,1]) \cap F_2 = \emptyset, \quad g_{i_1}([0,1]) \cap T_i = \emptyset \quad \text{for } 1 \le i \le i_1. \tag{3}$$

To do this, we define the following times:

$$s_1 = \inf\{t \in [0,1]; g_0(t) \in Z_{i_1}\}, \quad s_2 = \inf\{t \in [0,1]; g_0(t) \in Y_{i_1}\}, \tag{4}$$

$$t_1 = \sup\{t \in [0,1]; g_0(t) \in Y_{i_1}\}, \quad t_2 = \sup\{t \in [0,1]; g_0(t) \in Z_{i_1}\}. \tag{5}$$

Claim 2: The following properties hold:
(d) $0 < s_1 < s_2 < t_1 < t_2 < 1$.
(e) $g_0(t_1)$ and $g_0(s_2)$ belong to T_{i_1}.
(f) $g_0(t) \in A_{i_1}$ for $t \in (s_1, s_2) \cup (t_1, t_2)$.

Indeed, it is clear that $0 \le s_1 \le s_2 \le t_1 \le t_2$. Since $u, v \notin \cup_{i=1}^n \bar{Z}_i$, we have $0 < s_1$ and $t_2 < 1$. On the other hand, $g_0(t_2) \notin Z_{i_1}$ since the latter is open, while $g_0(t_1) \in \partial Y_{i_1} \cap T_{i_1}$ since $g_0([0,1]) \cap F_2 = \emptyset$, hence (a) yields that $g_0(t_1) \in \partial Y_{i_1} \cap T_{i_1} = T_{i_1} \subset Z_{i_1}$. Modulo a similar reasoning for s_1, s_2, (d) and (e) are therefore verified.

To prove (f), we note first that $g_0(t) \in G$ for $t \in (s_1, s_2) \cup (t_1, t_2)$, since otherwise $g_0(t) \in Y_{i_1}$ which contradicts (4) and (5). So, for any $t \in (t_1, t_2)$, $g_0(t) \in U$ for some connected component U of G. If $U \neq A_{i_1}$, we have that $T_{i_1} \cap \partial U = \emptyset$ and since $g_0(t_1) \in T_{i_1}$, we see that $g_0(t_1) \notin \partial U$. Hence there must exist $t_3 \in (t_1, t)$ such that $g_0(t_3) \in \partial U \subseteq \partial G \backslash T_{i_1}$. By (c) we see that $g_0(t_3) \in F_2$ which is a contradiction. So $U = A_{i_1}$ and consequently, $g_0(t) \in A_{i_1}$ for all $t \in (t_1, t_2)$, and (f) is proved.

To finish the proof of the Lemma, note that since A_{i_1} is path connected, then for $s_1 < s^{i_1} < s_2$, $t_1 < t^{i_1} < t_2$, we can use a path in A_{i_1} to join $g_0(s^{i_1})$ and $g_0(t^{i_1})$. In this way, we get a path $g_{i_1} \in \Gamma_u^v$ such that $g_{i_1}([0,1]) \cap T_{i_1} = \emptyset$ and $g_{i_1}([0,1]) \cap T_i = \emptyset$ for $1 \le i \le i_1$, since by (a), $A_{i_1} \cap T_i = \emptyset$ for all $i = 1, 2, \cdots, n$. On the other hand, since $A_{i_1} \cap F_2 = \emptyset$, we get that $g_{i_1}([0,1]) \cap F_2 = \emptyset$ and (3) is established.

Next, let i_2 be the first $i \in \{1, ..., n\}$ such that $g_{i_1}([0,1]) \cap T_i \neq \emptyset$. Clearly $i_1 < i_2 \le n$. In the same way, we can construct $g_{i_2} \in \Gamma_u^v$ such that for $1 \le i \le i_2$,

$$g_{i_2}([0,1]) \cap F_2 = \emptyset \quad \text{and} \quad g_{i_2}([0,1]) \cap T_i = \emptyset.$$

By iterating a finite number of times, we will get a $g_n \in \Gamma_u^v$ such that for $1 \le i \le n$,

$$g_n([0,1]) \cap F_2 = \emptyset \quad \text{and} \quad g_n([0,1]) \cap T_i = \emptyset.$$

But this contradicts assertion (i) and the lemma is proved.

Proof of Theorem 6.6: (1) Suppose $F \cap K_c$ contains no critical points of mountain-pass type and $F \cap \overline{M}_c = \emptyset$. We claim that there exist finitely many components of G_c, say $C_1, \ldots, C_p$ and $\varepsilon_1 > 0$ such that

$$G_c \cap \{x; \operatorname{dist}(x, F \cap K_c) < \varepsilon_1\} \subset C_1 \cup C_2 \ldots \cup C_p. \tag{1}$$

Indeed, otherwise we could find a sequence x_i in $F \cap K_c$ and a sequence $(C_i)_i$ of different components of G_c such that $\operatorname{dist}(x_i, C_i) \to 0$. But then any limit point of the sequence x_i would be a critical point for φ of Mountain-pass type belonging to F, thus contradicting our initial assumption. Hence (1) is verified.

Let now $M_i = F \cap K_c \cap \bar{C}_i$. Since any point of $M_i \cap \left(\bigcup_{j \neq i} \bar{C}_j \right)$ would be a critical point of Mountain-pass type, we may find for each $i = 1, \ldots, p$ an open set N_i such that

$$M_i \subset N_i, \text{ and } \overline{N}_i \cap \overline{N}_j = \emptyset \text{ for } i \neq j. \tag{2}$$

Since $F \cap \overline{M}_c = \emptyset$, we may also assume that

$$(\cup_{i=1}^p \overline{N}_i) \cap \overline{M}_c = \emptyset \text{ and } u, v \notin \cup_{i=1}^p \overline{N}_i. \tag{3}$$

Now for each i ($1 \leq i \leq n$) and any $x \in M_i$, there must be $\varepsilon_x > 0$ such that $B(x, \varepsilon_x) \cap U = \emptyset$ for any component U of G_c with $U \neq C_i$. Put

$$T_i = \cup_{x \in M_i} B(x, \varepsilon_x/2) \cap \partial C_i \cap N^i \text{ and } Y_i = N_i \setminus G_c.$$

Note that $T_i \subset N^i \cap \partial C_i$ and is relatively open in ∂Y_i. By Lemma 6.8, the set

$$\tilde{F} = [F \setminus (\cup_{i=1}^p \overline{N}_i)] \cup [(\cup_{i=1}^p (\partial Y_i \setminus T_i)$$

separates u and v, hence $\tilde{F} \cap K_c \neq \emptyset$ by Theorem 6.2, which is clearly a contradiction since $M_i \subset T_i$ and $F \cap \overline{M}_c = \emptyset$.

2) Since F separates u and v, we can again use the result of Whyburn mentioned above to get a closed connected subset $\hat{F}$ also separating u and v and such that $\hat{F} = \partial U = \partial V$ where U and V are two components of $X \setminus \hat{F}$ containing u and v respectively. Assume $F \cap \overline{M}_c = \emptyset$. The set $K = \tilde{F} \cap P_c$ is an open subset relative to $\tilde{F}$. If K is not closed, then any $x \in \overline{K} \setminus K$ is a saddle point since $\hat{F} \cap M_c = \emptyset$. Moreover, if H is any open neighborhood of x not intersecting M_c and such that $\varphi \leq c$ on H, then both sets $U \cap H$ and $V \cap H$ meet the set $\{\varphi < c\}$. This shows that x is a saddle point of mountain pass type.

Assume now K is closed. Then it is a closed and open set in the connected space $\tilde{F}$. Hence either $K = \tilde{F}$ or $K = \emptyset$. In the first case $\tilde{F}$ is then contained in P_c and since it separates u and v, we get a contradiction. In the second case, note that by part (1), $\tilde{F} \cap K_c$ contains a point of mountain pass type. Such a point is necessarily a saddle point since $\tilde{F} \cap M_c$ and $\tilde{F} \cap P_c$ are both empty. This clearly finishes the proof of Theorem 6.6.

6.2 A theorem of G. Fang

For the sequel, we shall need the following concept.

Definition 6.9: For A, B two disjoint subsets of X and any nonempty subset C of X, we say that A, B are connected through C if there is no set $F \subseteq C \cup A \cup B$ that is both relatively closed and open such that $A \subseteq F$ and $F \cap B = \emptyset$.

When A and B are connected through C, we also say that C connects A and B or that the space $C \cup A \cup B$ is connected between A and B. We refer to Kuratowski ([K], p.142–148) for details.

Theorem 6.10 (G. Fang): *With the hypothesis of Theorem 6.2, we further assume $u, v \notin K_c$ and that φ verifies $(PS)_c$. Then one of the following three assertions concerning the set K_c must hold:*

  (i) *P_c contains a compact subset that separates u and v;*

  (ii) *K_c contains a saddle point of mountain-pass type;*

  (iii) *There are finitely many components of G_c, say C_i ($i = 1, 2, \cdots, n$) such that $S_c = \cup_{i=1}^{n} S_c^i$, $S_c^i \cap S_c^j = \emptyset$ for $i \neq j, 1 \leq i, j \leq n$) where $S_c^i = S_c \cap \overline{C}_i$. Moreover, there are at least two of them $S_c^{i_1}, S_c^{i_2}$ ($i_1 \neq i_2, 1 \leq i_1, i_2 \leq n$) such that the sets $\overline{M}_c \cap S_c^{i_1}, \overline{M}_c \cap S_c^{i_2}$ are nonempty and connected through M_c.*

Corollary 6.11: *With φ, u and v as in Theorem 6.2, assume that φ verifies $(PS)_c$ and that $\max\{\varphi(u), \varphi(v)\} < c$. If K_c does not separate u and v, then at least one of the following two cases occurs:*

  (α) *K_c contains a saddle point of mountain-pass type;*

  (β) *$\overline{M}_c$ intersects at least two components of S_c.*

  Moreover, if $\overline{M}_c$ has only a finite number of components, then either

  (α') *K_c contains a saddle point of mountain-pass type, or*

  (β') *at least one component of $\overline{M}_c$ intersects two or more components of S_c.*

To prove Theorem 6.10, we shall need the following two topological lemmas.

Lemma 6.12: *Let M be a subset of a metric space (X, d). Suppose $M = M_1 \cup M_2$ and $M_1 \cap M_2 = \emptyset$. If M_1 is both open and closed relative to the subspace M, then there exist open sets D_1, D_2 of X such that $M_1 \subseteq D_1$, $M_2 \subseteq D_2$ and $D_1 \cap D_2 = \emptyset$.*

Proof: Since M_1 is both relatively open and closed, so is M_2. Hence there exist open sets E_1, E_2 such that $M_1 \subseteq E_1, M_2 \subseteq E_2$ and $(E_1 \cap E_2) \cap M = \emptyset$. Set $H_1 = E_1 \backslash (E_1 \cap E_2)$ and $H_2 = E_2 \backslash (E_1 \cap E_2)$. Then

$M_1 \subseteq H_1$, $M_2 \subseteq H_2$ and $H_1 \cap H_2 = \emptyset$. Since E_1 is open, for each $x \in \partial E_2 \cap E_1, \exists\, \varepsilon_x > 0$ such that the ball $B(x, \varepsilon_x)$ centered at x with radius ε_x is contained in E_1. Let $\varepsilon'_x = \text{dist}(x, \partial E_1 \cap E_2)$. Then $\varepsilon'_x \geq \varepsilon_x > 0$. Set

$$D_1 = H_1 \cup \Big(\bigcup_{x \in \partial E_2 \cap E_1} B(x, \varepsilon'_x/4) \Big) \quad \text{and}$$

$$D_2 = H_2 \cup \Big(\bigcup_{y \in \partial E_1 \cap E_2} B(y, \varepsilon'_y/4) \Big).$$

Clearly, $M_1 \subseteq D_1$, $M_2 \subseteq D_2$ and D_1, D_2 are open. We now claim that $D_1 \cap D_2 = \emptyset$. Indeed, if not, say $z \in D_1 \cap D_2$, then there exist $B(x, \varepsilon'_x/4), B(y, \varepsilon'_y/4)$ such that $z \in B(x, \varepsilon'_x/4) \cap B(y, \varepsilon'_y/4)$. Then

$$d(x, y) \leq d(x, z) + d(z, y) \leq \varepsilon'_x/4 + \varepsilon'_y/4 < \max(\varepsilon'_x, \varepsilon'_y).$$

On the other hand, $\varepsilon'_x \leq d(x, y)$ and $\varepsilon'_y \leq d(x, y)$ which imply that $d(x, y) \geq \max(\varepsilon'_x, \varepsilon'_y)$. A contradiction which completes the proof of the lemma.

Lemma 6.13: *Let S^i ($i = 1, 2, \cdots, n$) be n mutually disjoint compact subsets of a Banach space X and let M be any nonempty subset of X. If for all i, j ($i \neq j$ $i, j = 1, 2, \cdots, n$), the sets $S^i \cap \overline{M}$ and $S^j \cap \overline{M}$ are not connected through M, then there are n mutually disjoint open sets N^i ($i = 1, 2, \cdots, n$) such that*

$$M \subseteq \cup_{i=1}^n N^i \quad \text{and} \quad S^i \subseteq N^i \quad \text{for all} \quad i = 1, 2, \cdots, n. \tag{1}$$

Proof: For each i ($i = 1, 2, \cdots, n$), we denote by M^i the compact set $S^i \cap \overline{M}$. Since by assumption none of the pairs M^i, M^j ($i \neq j$ $i, j = 1, 2, \cdots, n$) are connected through M, there exist by Lemma 6.9 open sets O_{ij} and P_{ij} such that for $i \neq j$ $i, j = 1, 2, \cdots, n$) we have: $O_{ij} = P_{ji}, M^i \subseteq O_{ij}, M^j \subseteq P_{ij}, O_{ij} \cap P_{ij} = \emptyset$ and $M^i \cup M \cup M^j \subseteq O_{ij} \cup P_{ij}$. For each i, ($i = 1, 2, \cdots, n$), let

$$O_i = \bigcap_{\substack{j=1 \\ j \neq i}}^n O_{ij}, \qquad P_i = \bigcup_{\substack{j=1 \\ j \neq i}}^n P_{ij} \tag{2}$$

and

$$M_s = \bigcup_{i=1}^n M^i, \qquad \tilde{M}^i = \bigcup_{\substack{j=1 \\ j \neq i}}^n M^j. \tag{3}$$

Then

$$M^i \subseteq O_i, \qquad \tilde{M}^i \subseteq P_i \tag{4}$$

and

$$O_i \cap P_i = \emptyset, \qquad M_s \cup M \subseteq O_i \cup P_i. \tag{5}$$

Put for each i $(i = 1, 2, \cdots, n)$

$$O^i = O_i \cap \left(\bigcap_{\substack{j=1 \\ j \neq i}}^{n} P_j \right). \tag{6}$$

Then by (2)—(5), we have

$$M^i \subseteq O^i, \qquad O^i \cap O^j = \emptyset \qquad (i \neq j \quad i, j = 1, 2, \cdots, n). \tag{7}$$

It is not generally true that $M_s \cup M \subseteq \cup_{i=1}^n O^i$. In order to prove the lemma, we let

$$M' = (M_s \cup M) \backslash (\cup_{i=1}^n O^i), \qquad M'' = (M_s \cup M) \cap (\cup_{i=1}^n O^i).$$

Then

$$M_s \cup M = M' \cup M'', \qquad M' \cap M'' = \emptyset. \tag{8}$$

By (5) and (6), we see that M'' is both open and closed relative to $M_s \cup M$. Again by Lemma 6.9, there exist two open sets D' and D'' such that

$$M' \subseteq D', \qquad M'' \subseteq D'', \qquad D' \cap D'' = \emptyset. \tag{9}$$

Now for each i $(i = 1, 2, \cdots, n)$ put $O_D^i = O^i \cap D''$. By (7) and (9), we have

$$D' \cap (\cup_{i=1}^n O_D^i) = \emptyset, \qquad M^i \subseteq O_D^i, \qquad O_D^i \cap O_D^j = \emptyset \qquad (i \neq j). \tag{10}$$

By the compactness of S^i and M^i, we may introduce

$$a_i = \text{dist}(M^i, X \backslash O_D^i) > 0, \quad \delta_1 = \frac{1}{2} \min\{\text{dist}(S^i, S^j); i \neq j\} > 0. \tag{11}$$

Let $\delta_2 = \frac{1}{4} \min\{a_i, \delta_1; i = 1, 2, \cdots, n\}$ and

$$Q_i = \{x \in X; \text{dist}(x, M^i) < \delta_2\} \qquad S_q^i = S^i \backslash Q_i. \tag{12}$$

Then $Q_i \subseteq O_D^i$ and $S_q^i \cap \overline{M} = \emptyset$. By (11), we see that

$$\text{dist}(S_q^i, Q_j) \geq \text{dist}(S_q^i, S_q^j) - \delta_2 \geq 3\delta_2.$$

By the compactness of S_q^i, we may also introduce

$$b_i = \frac{1}{4} \text{dist}(S_q^i, M) > 0 \quad \text{and} \quad \delta_3 = \min\{b_i, \delta_2; i = 1, 2, \cdots, n\} > 0.$$

Put

$$P = \{x \in X; \text{dist}(x, M) < \delta_3\} \tag{13}$$

and

$$N_i = Q_i \cup (O_D^i \cap P), \qquad R' = D' \cap P. \tag{14}$$

Then

$$M' \subseteq R', \qquad M'' \subseteq \cup_{i=1}^n N_i. \tag{15}$$

By (10), we have that $R' \cap (\cup_{i=1}^n N_i) = \emptyset$ and $N_i \cap N_j = \emptyset$ if $i \neq j$, $(i, j = 1, 2, \cdots, n)$. Furthermore

$$\text{dist}(S_q^i, P) \geq \text{dist}(S_q^i, M) - \delta_3 \geq 3\delta_3. \tag{16}$$

Hence
$$\text{dist}(S_q^i, R') \geq \text{dist}(S_q^i, P) \geq \text{dist}(S_q^i, M) - \delta_3 \geq 3\delta_3. \tag{17}$$
By (14) and (16), we also have that

$$\text{dist}(S_q^i, N_j) \geq \min\{\text{dist}(S_q^i, Q_j), \text{dist}(S_q^i, P)\}$$
$$\geq \min(3\delta_2, 3\delta_3) \geq 3\delta_3. \tag{18}$$

Now let $N^1 = N_1 \cup \{x \in X; \text{dist}(x, S_q^1) < \delta_3\} \cup R'$ and
$$N^i = N_i \cup \{x \in X; \text{dist}(x, S_q^i) < \delta_3\} \quad (i \neq 1 \quad i = 1, 2, \cdots, n).$$
By (8), (12) and (15) it follows that
$$S^i \subseteq N^i \quad \text{and} \quad M \subseteq \cup_{i=1}^n N^i. \tag{19}$$
By (10), (17) and (18), we see that
$$N^i \cap N^j = \emptyset \quad (i \neq j \quad i, j = 1, 2, \cdots, n). \tag{20}$$
So (19) and (20) imply that N^i satisfy (1). This completes the proof of the lemma.

Proof of Theorem 6.10: Suppose assertions (ii) and (iii) are not true and let us prove (i). The critical set K_c is the disjoint union of S_c, M_c and P_c. Also by the $(PS)_c$ condition, K_c is compact. It is also clear that S_c is closed and compact. We will assume that $S_c \neq \emptyset$ since otherwise we may conclude by Corollary 6.4. We start with the following:

Claim 1: There exist finitely many components of G_c, say C_i ($i = 1, 2, \cdots, n$) and $\eta_1 > 0$ such that
$$G_c \cap \{x; \text{dist}(x, S_c) < \eta_1\} \subseteq \cup_{i=1}^n C_i. \tag{1}$$

Indeed, if not, we could find a sequence x_i in S_c and a sequence $(C_i)_i$ of different components of G_c such that $\text{dist}(x_i, C_i) \to 0$. But then any limit point of the sequence x_i would be a saddle point of mountain-pass type for φ, thus contradicting our assumption that assertion (ii) is false. Claim 1 is hence proved. We clearly may assume that $C_i \neq \emptyset$ for all $i = 1, 2, \cdots, n$.

Next, for each $i = 1, 2, \cdots, n$, let $S_c^i = S_c \cap \overline{C_i}$. They all are compact and mutually disjoint. Also we have that
$$S_c = \cup_{i=1}^n S_c^i. \tag{2}$$

Claim 2: There are n mutually disjoint open sets N^i ($i = 1, 2, \cdots, n$) such that $u, v \notin \cup_{i=1}^n \overline{N^i}$ and
$$S_c \cup M_c \subseteq \cup_{i=1}^n N^i \quad \text{and} \quad S_c^i \subseteq N^i \quad \text{for all} \quad i = 1, 2, \cdots, n. \tag{3}$$

Indeed, we have two cases to consider.

Case 1: $M_c = \emptyset$. This is a trivial case. Since $u, v \notin K_c$, for each i $(i = 1, 2, \cdots, n)$ there exists an open neighborhood N^i of S_c^i such that $u, v \notin N^i$. Since the S_c^i's are mutually disjoint compact sets, we may take the N^i's in such a way that they are also mutually disjoint.

Case 2: $M_c \neq \emptyset$. In this case we are in a situation where we have n mutually disjoint compact sets S_c^i $(i = 1, 2, \cdots, n)$ and a nonempty set M_c. Moreover all the pairs $S_c^i \cap \overline{M}_c, S_c^j \cap \overline{M}_c$ $(i \neq j$ $i, j = 1, 2, \cdots, n)$ are not connected through M_c since assertion (3) is assumed false. Applying Lemma 6.13, we can then find n mutually disjoint open sets N^i such that (3) is verified. Since $u, v \notin K_c$, we may clearly assume that $u, v \notin \cup_{i=1}^{n} \overline{N}_i$. Claim 2 is therefore proved in both cases.

In order to finish the proof of Theorem 6.10, we still need the following:

Claim 3: There exists a closed set $\hat{F}$ such that $\hat{F}$ separates u, v while
$$\inf_{x \in \hat{F}} \varphi(x) \geq c \quad \text{and} \quad \hat{F} \cap (S_c \cup M_c) = \emptyset. \tag{4}$$
Indeed, for each i $(i = 1, 2, \cdots, n)$, let
$$Y_i^c = N^i \backslash G_c. \tag{5}$$
Observe that for each i $(1 \leq i \leq n)$ and any $x \in S_c^i$ there must be a ball $B(x, \varepsilon_x)$ such that for any connected component U of G_c with $C_i \neq U$, $B(x, \varepsilon_x) \cap U$ is necessarily empty, since otherwise x is a saddle point of mountain pass type. Set
$$T_i^c = \cup_{x \in S_c^i} B(x, \varepsilon_x/2) \cap \partial C_i \cap N^i \cap \{x \in X; \operatorname{dist}(x, S_c^i) < \eta_1\}. \tag{6}$$
Clearly
$$S_c^i \subseteq T_i^c, \qquad T_i^c \subseteq N^i \cap \partial C_i \tag{7}$$
and T_i^c is open relative to ∂Y_i. Also $T_i^c \cap \partial U = \emptyset$ for any component U of G_c with $U \neq C_i$. Now let
$$\hat{F} = [F \backslash (\cup_{i=1}^{n} N^i)] \cup (\cup_{i=1}^{n} \partial Y_i^c \backslash T_i^c).$$
Clearly, $\inf_{x \in \hat{F}} \varphi(x) \geq c$. Since F separates u, v and in view of Claim 1, Claim 2, (5) and (7), we see that we can apply Lemma 6.8 with $A_i = C_i, G = G_c, Z_i = N^i, Y_i = Y_i^c, T_i = T_i^c$ for all $i = 1, 2, \cdots, n$ to conclude that $\hat{F}$ separates u, v. On the other hand, since $M_c \cap (\overline{G_c} \backslash G_c) = \emptyset$, we have by (3) and (5), that $\partial Y_i^c \cap M_c = \emptyset$. Therefore by (2) and (6), we have $\cup_{i=1}^{n} (\partial Y_i^c \backslash T_i^c) \cap (S_c \cup M_c) = \emptyset$. Hence $\hat{F} \cap (M_c \cup S_c) = \emptyset$ and Claim 3 is thus proved.

Finally by Theorem 6.3, we see that $\hat{F} \cap P_c$ and hence P_c must contain a compact subset that separates u, v which implies assertion (1). This finishes the proof of the theorem.

Remark 6.14: Theorem 6.10 remains true if instead of $(PS)_c$, φ verifies only $(PS)_{(F \cup K_c)_\varepsilon, c}$. The proof also shows that the condition $u, v \notin K_c$ can be replaced by $u, v \notin S_c \cup \overline{M_c}$.

We have the following surprising corollary concerning the cardinality of the critical set K_c generated by the mountain pass theorem.

Corollary 6.15: *Suppose* $\dim(X) \geq 2$. *Then under the hypothesis of Theorem 6.10, one of the following three assertions must be true:*
(1) K_c *has a saddle point of mountain-pass type.*
(2) K_c *contains a continuum of local maxima.*
(3) K_c *contains a continuum of local minima.*

Proof: If K_c does not contain a saddle point of mountain-pass type, then either assertion (i) or assertion (iii) in Theorem 6.10 is true. Let us first assume that assertion (iii) is true. Then there exist two disjoint nonempty closed subsets of K_c, say, M_c^1 and M_c^2 which are connected through M_c. Clearly $\text{dist}(M_c^1, M_c^2) = d > 0$. For any $0 < \sigma < d$, let $M_\sigma = \{x \in X; \text{dist}(x, M_c^1) < \sigma\}$. Then $\overline{M_\sigma} \cap M_c^2 = \emptyset$, $M_c^1 \subseteq M_\sigma$. We claim that $\partial M_\sigma \cap M_c \neq \emptyset$. Otherwise, there will be two disjoint open sets M_σ and $X \backslash \overline{M_\sigma}$ such that

$$M_c^1 \subseteq M_\sigma, \quad M_\sigma \cap M_c^2 = \emptyset \text{ and } M_c \cup M_c^1 \cup M_c^2 \subseteq M_\sigma \cup (X \backslash \overline{M_\sigma}).$$

This contradicts that M_c^1, M_c^2 are connected through M_c. Now let $m_\sigma \in \partial M_\sigma \cap M_c$. Then we have a map $f : \sigma \in (0, d) \longrightarrow m_\sigma \in M_c$. Clearly f is injective which implies (3).

In the case where assertion (i) of Theorem 6.10 is true, then we can argue similarly that assertion (2) in Corollary 6.15 is true. Indeed, let $d = \text{dist}(u, v)$. Then for all σ $(0 < \sigma < d)$, we have $\partial B(u, \sigma) \cap P_c \neq \emptyset$. Otherwise, as above, we will see that P_c will not separate u, v. Define a map f from $(0, d)$ to P_c as follows: $f : \sigma \in (0, d) \longrightarrow m_\sigma \in P_c \cap \partial B(u, \sigma)$. Again f is injective which implies assertion (2) of the corollary.

The following result is to be compared to Corollaries 5.12 and 5.20 of the last chapter.

Corollary 6.16: *Suppose* φ *has a local maximum and a local minimum on a Banach space* X. *If* φ *satisfies* (PS) *and if* $\dim(X) \geq 2$, *then necessarily* φ *has a third critical point.*

Proof: Suppose u_1 is a local maximum and u_2 is a local minimum. If φ is not bounded below, then we have a mountain pass situation with u_2 as an initial point and Corollary 6.15 applies to give either an infinite number of critical points or a saddle point of mountain pass type which is necessarily distinct from u_1 and u_2.

If, on the other hand, φ is bounded below then, since it satisfies $(PS)_c$,

Corollary 1.5 yields that φ cannot be bounded above. Hence we have a mountain pass situation for $-\varphi$ with u_1 as an initial point. Again Corollary 6.15 applies to yield our claim.

6.3 A bifurcation problem

Let Ω be a bounded domain in $\mathbb{R}^N$, $(N > 2)$ and consider the boundary value problem

$$\begin{cases} -\Delta u = g(u) & \text{in } \Omega, \\ u = 0 & \text{on } \partial\Omega. \end{cases} \tag{1}$$

Recall that $\{\lambda_i\}_{i=1}^{\infty}$ is the spectrum $\sigma(-\Delta)$ of $-\Delta$ in $H_0^1(\Omega)$.

Theorem 6.17: Let g be a function in $C^1(\mathbb{R})$ that verifies $g(0) = 0$ and

$$\limsup_{|s| \to \infty} \frac{g(s)}{s} < \lambda_1. \tag{2}$$

(i) If $g'(0) > \lambda_1$, then equation (1) has at least two nontrivial solutions.

(ii) If $g'(0) > \lambda_2$ and $g'(0) \notin \sigma(-\Delta)$, then equation (1) has at least three nontrivial solutions.

Proof: As in Theorem 5.21, assumption (2) combined with the maximum principle allow us to assume that $g(t) = bt + c$ with $b < \lambda_1$, for $|t|$ large. It follows that the functional

$$\varphi(u) = \frac{1}{2} \int_\Omega |\nabla u|^2 - \int_\Omega G(u)$$

is a C^2-functional on $X = H_0^1$ that verifies the (PS) condition. Moreover, its critical points correspond to the solutions of (1).

Use (2) to find $\alpha \in (0, \lambda_1)$ and a constant $C > 0$ such that if $t > 0$, then

$$g(t) \le \alpha t + C \quad \text{and} \quad g(-t) \ge -\alpha t - C. \tag{3}$$

Let e_0 be the solution of the following boundary value problem

$$\begin{cases} -\Delta u = \alpha u + C & \text{in } \Omega, \\ u = 0 & \text{on } \partial\Omega. \end{cases} \tag{4}$$

By the maximum principle, $e_0 > 0$. It follows that $-e_0$ (resp. e_0) is a subsolution (resp. supersolution) of (1).

Let now $e_1 > 0$ be the first eigenfunction for $-\Delta$, normalized in such a way that $\max_{\bar\Omega} e_1 = 1$. Using the fact that $g'(0) > \lambda_1$, we may choose $\varepsilon > 0$ small enough in such a way that $\varepsilon e_1 \le e_0$ and εe_1 is a subsolution of (1). It follows that $-e_0 < -\varepsilon e_1$ and $\varepsilon e_1 < e_0$ are two pairs of sub-super solutions for (1). Hence, we may apply Theorem 1.16 and Remark 1.17 to find two distinct local minima $\{u_1, u_2\}$ for φ such that $-e_0 \le u_1 \le -\varepsilon e_1$ and $\varepsilon e_1 \le u_2 \le e_0$.

ii) Consider now the mountain pass theorem between u_1 and u_2 and note that

$$\inf_{\gamma \in \Gamma_{u_1}^{u_2}} \max_{0 \le t \le 1} \varphi(\gamma(t)) \ge \max\{\varphi(u_1), \varphi(u_2)\}.$$

If the number of critical points of φ is finite, Corollary 6.15 yields then $u_3 \in K_c$ which is of mountain pass type. But if $g'(0) > \lambda_2$ and $g'(0) \notin \sigma(-\Delta)$, then 0 is a nondegenerate critical point and the set where φ is less than $\varphi(0) = 0$ on a suitable deleted neighborhood of 0 is connected. It follows that 0 is not a critical point of mountain pass type and hence u_3 is not the trivial solution.

Notes and Comments: The idea of studying the structure of min-max critical points without Morse theory and for C^1-functions was started by Hofer [H 1,2] and by Pucci-Serrin [P-S 1,2,3]. The approach and methods of proofs adopted in this chapter were initiated by Ghoussoub-Preiss [G-P] and later further developed by Ghoussoub [G 1] and Fang [F 1].

Theorems 6.3 and Corollary 6.4 improve earlier results by Pucci and Serrin [P-S 1] and they are due to Ghoussoub-Preiss [G-P] and Fang [F 1] respectively. Theorem 6.6 appears in [G-P] and improve the results of Hofer [H 1] and Pucci-Serrin [P-S 2] that appear in Corollary 6.7. Theorem 6.10 is due to Fang [F 1] while the following Corollary 6.11 was established by Pucci and Serrin [P-S 3].

Theorem 6.16 was considered by many authors. By using different methods, Hofer [H 1] actually proves the existence of at least 4 nontrivial solutions for that equation.

7

GROUP ACTIONS AND MULTIPLICITY

OF CRITICAL POINTS

In this chapter, we consider the case where the function φ is invariant under the action of a group G acting on the manifold. In this setting, one can often define a topological index ind_G, associated with the group, which permits to measure the size of the critical set. We shall obtain two types of multiplicity results: one is standard and is in the spirit of the Ljusternik-Schnirelmann theory, i.e., estimates can be obtained when two different classes give the same critical value. The second type of multiplicity result is more surprising and gives a quantitative version of Theorem 4.1. Roughly speaking, it says that the size of the set of critical points on a dual set F is at least as large as the intersections of the sets in the class $\mathcal{F}$ with F.

7.1 The equivariant min-max principles

Let $\{T_g; g \in G\}$ be a representation of a topological group G over a complete metric space X. Say that a function $\varphi : X \to \mathbb{R}$ is G-invariant if $\varphi(T_g x) = \varphi(x)$ for all $x \in X$ and $g \in G$. A subset A of X is G-invariant if its characteristic function is. A function $F : X \to X$ is said to be G-equivariant if $f(T_g x) = T_g f(x)$ for $x \in X$ and $g \in G$. A homotopy $\eta : [0, 1] \times X \to X$ will be called G-equivariant if for each $t \in [0, 1]$, the function $\eta(t, \cdot)$ is equivariant from X into itself.

Definition 7.1: Let B be a closed subset of X. We shall say that a class $\mathcal{F}$ of compact subsets of X is a G-homotopy stable family with boundary B provided
(a) every set in $\mathcal{F}$ is G-invariant,

(b) every set in $\mathcal{F}$ contains B,

(c) for any set A in $\mathcal{F}$ and any G-equivariant $\eta \in C([0,1] \times X; X)$ satisfying $\eta(t,x) = x$ for all (t,x) in $(\{0\} \times X) \cup ([0,1] \times B)$ we have that $\eta(\{1\} \times A) \in \mathcal{F}$.

The following is the equivariant counterpart of Theorem 4.5.

Theorem 7.2: *Let G be a compact Lie group acting differentiably (i.e. locally linear) on a complete connected C^{2^-}-Finsler manifold X without boundary. Let φ be a G-invariant C^1-functional on X and consider a G-homotopy stable family $\mathcal{F}$ with a closed boundary B. Set $c = c(\varphi, \mathcal{F})$ and let F be a closed G-invariant subset of X satisfying*

(F1) $\qquad\qquad F \cap B = \emptyset$ *and* $F \cap A \neq \emptyset$ *for all A in $\mathcal{F}$*

and

$(F2)_\delta$ $\qquad\qquad\qquad \inf \varphi(F) \geq c - \delta.$

Suppose $0 < \delta < \max\left\{ \frac{1}{32}\mathrm{dist}^2(B,F); \frac{1}{8}[\inf\varphi(F) - \sup\varphi(B)] \right\}$, then for any A in $\mathcal{F}$ satisfying $\max\varphi(A) \leq c + \delta$, there exists $x_\delta \in X$ such that

(i) $c - \delta \leq \varphi(x_\delta) \leq c + 9\delta$
(ii) $\|d\varphi(x_\delta)\| \leq 18\sqrt{\delta}$
(iii) $\mathrm{dist}(x_\delta, F) \leq 5\sqrt{\delta}$
(iv) $\mathrm{dist}(x_\delta, A) \leq 3\sqrt{\delta}.$

The following corollaries of Theorem 7.2 are the equivariant versions of Theorem 4.1 and Corollary 4.7.

Corollary 7.3: *Under the hypothesis of Theorem 7.2, let F be a closed G-invariant subset of X satisfying (F1) and (F2). Assume φ verifies $(PS)_{F,c}$ along a min-maxing sequence $(A_n)_n$, then the set $F \cap K_c \cap A_\infty$ is non empty.*

Corollary 7.4: *Let G, X, φ and $\mathcal{F}$ be as in Corollary 7.3, and consider a family of G-invariant sets $\mathcal{F}^*$ that is dual to $\mathcal{F}$. Assume that*

$$\sup_{F \in \mathcal{F}^*} \inf_{x \in F} \varphi(x) = \inf_{A \in \mathcal{F}} \max_{x \in A} \varphi(x) = c,$$

and that φ satisfies $(PS)_c$ along a min-maxing sequence $(A_n)_n$ in $\mathcal{F}$ and a suitable max-mining sequence $(F_n)_n$ in $\mathcal{F}^*$. Then the set $A_\infty \cap F_\infty \cap K_c$ is non empty.

We also have the equivariant version of Theorem 5.2.

Theorem 7.5: *Let G, X, φ be as in Theorem 7.2 and consider a G-homotopy stable family $\mathcal{F}$ of compact subsets of X with an extended* *closed boundary* B. *Set* $c = c(\varphi, \mathcal{F})$ *and let* F *be a closed* G-*invariant subset of* X *satisfying* $(F'1)$ *and* $(F'2)$. *Assume that* φ *verifies* $(PS)_{F,c}$ *along a min-maxing sequence* $(A_n)_n$, *then the set* $F \cap K_c \cap A_\infty$ *is non empty* .

The proofs of Theorems 7.2 and 7.5 above are identical to those of Theorems 4.5 and 5.2 provided one proves an equivariant version of Lemma 3.7 which is the following

Lemma 7.6: *Let* φ *be a* G-*invariant* C^1-*functional on a complete connected* C^{2^-}-*Finsler manifold* X *equipped with a metric* ρ *and let* B *and* C *be two closed and disjoint subsets of* X. *Suppose that* C *is compact,* G-*invariant and that* $\|d\varphi(x)\| > 2\varepsilon > 0$ *for every* $x \in C$. *Then for each* $k > 1$, *there exist a positive continuous function* g *on* X *and an equivariant deformation* $\alpha \in C([0,1] \times X; X)$ *such that for some* $t_0 > 0$, *the following hold for every* $t \in [0, t_0)$:

 (i) $\alpha(t, x) = x$ *for every* $x \in B$.
 (ii) $\rho(\alpha(t, x), x)) \leq kt$ *for every* $x \in X$.
 (iii) $\varphi(\alpha(t, x)) - \varphi(x) \leq -\varepsilon g(x)t$ *for every* $x \in X$.
 (iv) $g(x) = 1$ *for all* $x \in C$.

For the proof we recall some facts about group actions. Suppose that G is a compact Lie group acting differentially on X. By using, if necessary, an averaging procedure, we can suppose –modulo changing the metric– that each operation of G is an isometry. Similarly, if v is a continuous pseudo-gradient vector field associated to a G-invariant functional φ on X, one can construct a G-equivariant one $\tilde{v}(x)$ with the same properties by just taking

$$\tilde{v}(x) = \int dg(x)^{-1} v(g(x))\, dg$$

where dg is the Haar measure on G.

On the other hand, the action of the group can be linearized locally in the following way (See [Pa 2] and [Mo] for details).

For any $x \in X$, there is an invariant neighborhood (tube) U_x of the orbit Gx that is generated by a *slice through* x. That is, there exists an equivariant diffeomorphism f_x from U_x onto $S_x \times G/G_x$ where:

 — S_x is a disc in a Banach space E_x on which the isotropy group $G_x = \{h \in G; hx = x\}$ acts linearly,
 — G_x acts on $S_x \times G$ via $h(s, g) = (hs, gh^{-1})$ for every $h \in G_x$,
 — G acts on $S_x \times G/G_x$ by $\bar{g}[(s, g)] = [(s, \bar{g}g)]$ for every $\bar{g} \in G$ and where $[(s, g)]$ denotes the equivalent class of (s, g) in $S_x \times G/G_x$.

Suppose now that φ is a G-invariant functional, we shall indicate the changes to be made in the construction of Lemma 3.7 in order to obtain the G-equivariant deformation $\alpha(t, x)$.

First, as mentioned above, the set U_i can be taken to be a neighborhood of the orbit Gx_i that is generated by a slice. Let $f_i : U_i \to S_i \times G/G_{x_i}$ be the associated equivariant diffeomorphism and let $\tilde{w}_i$ be a G-equivariant pseudo-gradient for $\varphi \circ f_i^{-1}$. Define $\tilde{v}_i([s, g]) = \tilde{w}_i([0, g])$. It is easy to see that it is well defined (i.e., it only depends on the equivalent classes) and that it is G-equivariant.

Since we can suppose the group action to be isometric, we can construct the family $(V_i)_{i=1}^m$, its subordinate partition of unity $(\chi_i)_{i=1}^m$ and the function ℓ to be all G-invariant. Define now on $S_i \times G/G_{x_i}$ the functions $\tilde{\chi}_i([s, g]) = (\chi_i.\ell)(f_i^{-1}([s, g]))$. Starting with $\alpha_0(t, x) = x$, we define by induction on j $(1 \leq j \leq m)$ the functions

$$\alpha_j(t, x) = \begin{cases} f_j^{-1}(\tilde{\alpha}_j(t, f_j(x))) & \text{if } x \in W_j \\ \alpha_{j-1}(t, x) & \text{otherwise} \end{cases}$$

where $W_j = \{x \in U_j; \alpha_{j-1}(t, x) \in U_j\}$ and where the function $\tilde{\alpha}_j$ is defined on $f_j(W_j)$ by

$$\tilde{\alpha}(t, [s, g]) = \left(s' - t\tilde{\chi}_i[s, g]\frac{\tilde{v}_i[s, g]}{\|\tilde{v}_i[s, g]\|}, g'\right)$$

(s', g') being a representative of the class $f_j(\alpha_{j-1}(t, f_j^{-1}(s, t)))$. It is tedious but straightforward to show that all these functions are well defined, continuous and equivariant. The rest of the proof is now identical to the group action-free case.

Remark 7.7: Since X is now supposed to be a C^{2^-}–Finsler manifold, one can then find a locally Lipschitz pseudo-gradient v for φ and hence the deformation α can be obtained by solving a Cauchy problem on the manifold. Indeed, let D be a neighborhood of C such that $\overline{D} \cap B = \emptyset$ and $\overline{D} \subset R$ (the set of regular points), the required deformation α can be obtained as the (unique) solution of the Cauchy problem

$$\begin{cases} \dot{\sigma} = F(\sigma) \\ \sigma(0) = x \end{cases}$$

where

$$F(x) = \begin{cases} -g(x)\frac{v(x)}{\|v(x)\|} & \text{if } x \in D \\ 0 & \text{otherwise} \end{cases}$$

and

$$g(x) = \frac{\text{dist}(x, D^c)}{\text{dist}(x, D^c) + \text{dist}(x, C)}.$$

7.2 Indices associated to a group and multiplicity results

In the following, we deal with the size of K_c i.e., the critical set of a function φ at a level c. For that, we need the notion of a *topological index* Ind_G associated to a compact group G acting on the manifold X. It is defined as a mapping from the closed G-invariant subsets of X into $\mathbb{N} \cup \{+\infty\}$ satisfying the following properties:

(I1) $\mathrm{Ind}_G(A) = 0$ if and only if $A = \emptyset$.

(I2) $\mathrm{Ind}_G(A_2) \geq \mathrm{Ind}_G(A_1)$ if there is a G-equivariant continuous map from A_1 to A_2.

(I3) If K is compact invariant, there exists a closed invariant neighborhood $K^\delta = \{x; \mathrm{dist}(x, K) \leq \delta\}$ of K so that $\mathrm{Ind}_G(K^\delta) = \mathrm{Ind}_G(K)$.

(I4) $\mathrm{Ind}_G(A_1 \cup A_2) \leq \mathrm{Ind}_G(A_1) + \mathrm{Ind}_G(A_2)$ for all closed G-invariant sets A_1, A_2.

If $I(G)$ denotes the set $\{x; \exists g \neq e, gx = x\}$ that is the set of all points x with a non-trivial isotropy group G_x, we shall sometimes need an index satisfying also the following properties:

(I5) If K is compact invariant with $K \cap I(G) = \emptyset$, then K contains at least n orbits provided $\mathrm{Ind}_G(K) \geq n$.

(I6) If K is compact invariant with $K \cap I(G) = \emptyset$, then $\mathrm{Ind}_G(K) < +\infty$.

By now, there are many examples of topological indices associated with the classical groups $\mathbb{Z}_p$ and S^1. Other cohomological indices associated with more general Lie groups were introduced in [F-R 1-2], [F-H-R] and [F-H]. They will be discussed briefly below. However, in the sequel, we shall insist on the simplest examples since they tell most of the story. But first, we state and prove various results that are common to all indices as long as they satisfy the above axioms. We start with the following:

Corollary 7.8: *Under the hypothesis of Corollary 7.3, and if* Ind_G *is an index associated to the group* G, *then*

$$\mathrm{Ind}_G(K_c \cap F \cap A_\infty) \geq \inf\{\mathrm{Ind}_G(A \cap F); A \in \mathcal{F}\}.$$

Remark 7.9: If $\sup \varphi(B) < c$, we can then apply Corollary 7.8 to $F = \{\varphi \geq c\}$ to obtain

$$\mathrm{Ind}_G(K_c \cap A_\infty) \geq \inf\{\mathrm{Ind}_G(A \cap \{\varphi \geq c\}); A \in \mathcal{F}\}.$$

Proof of Corollary 7.8: Let $n = \inf\{\mathrm{Ind}_G(A \cap F); A \in \mathcal{F}\}$. By property (I3) of the index, there exists an invariant neighborhood U of $K_c \cap F \cap A_\infty$ such that $\mathrm{Ind}_G(\overline{U}) = \mathrm{Ind}_G(K_c \cap F \cap A_\infty)$. Let $F' = X \setminus U$. It is clearly closed and invariant. By property (I4) of the index, we have for A in $\mathcal{F}$

$$\operatorname{Ind}_G(A \cap F) \leq \operatorname{Ind}_G(A \cap F \setminus U) + \operatorname{Ind}_G(\overline{U})$$
$$= \operatorname{Ind}_G(A \cap F \cap F') + \operatorname{Ind}_G(K_c \cap F \cap A_\infty).$$

It follows that if $\operatorname{Ind}_G(K_c \cap F \cap A_\infty) \leq n - 1$, then $\operatorname{Ind}_G(A \cap F \cap F') \geq 1$ and in particular $A \cap F \cap F' \neq \emptyset$ for all A in $\mathcal{F}$ by property (I1). On the other hand $\inf \varphi(F \cap F') \geq \inf \varphi(F) \geq c$. Hence Corollary 7.3 applies to the dual set $F \cap F'$ and we get that $K_c \cap F \cap F' \cap A_\infty \neq \emptyset$ which is clearly a contradiction.

An intersection index

Given two closed disjoint G-invariant subsets B and F of X, we can define an *F-intersection index* Int_B^F on the class Σ_B of all closed G-invariant sets containing B, in the following fashion:

$$\operatorname{Int}_B^F(A) = \inf\{\operatorname{Ind}_G(\eta(A) \cap F); \eta \in \mathcal{L}_B\}$$

where

$$\mathcal{L}_B = \{\eta \in C(X, X); \ \eta \ G\text{-invariant and } \eta(x) = x \text{ for } x \in B\}.$$

It is easy to check that $\operatorname{Int}_B^F : \Sigma_B \to \mathbb{N} \cup \{+\infty\}$ satisfies the following properties:

(T1) $\operatorname{Int}_B^F(A_1) \leq \operatorname{Int}_B^F(A_2)$ if there exists $\eta \in \mathcal{L}_B$ with $\eta(A_1) \subset A_2$.

(T2) $\operatorname{Int}_B^F(A_1 \cup A_2) \leq \operatorname{Int}_B^F(A_1) + \operatorname{Ind}_G(A_2)$.

(T3) If $\operatorname{Int}_B^F(A) \geq 1$, then $A \cap F \neq \emptyset$.

Note that the intersection-index coincides with the index if $F = X$ and $B = \emptyset$.

The definition of the intersection index is custom-made to insure that for any G-homotopy stable class $\mathcal{F}$ with boundary B and any set F the (possibly empty) classes

$$\mathcal{F}_n(F) = \{A \in \mathcal{F}; \operatorname{Int}_B^F(A) \geq n\}$$

are also G-homotopy stable with boundary B for each $n \geq 1$. Let $c_n(F) = c(\varphi, \mathcal{F}_n(F))$ and note that if (F1) is satisfied then

$$\mathcal{F} = \mathcal{F}_1 \supset \mathcal{F}_2 \supset \cdots \supset \mathcal{F}_n \supset \mathcal{F}_{n+1} \supset \cdots$$

and $c = c_1 \leq c_2 \leq \cdots \leq c_n \leq \cdots (c_n = \infty$ if $\mathcal{F}_n$ is empty). The following is a reformulation of Corollary 7.8

Corollary 7.10:   *Under the hypothesis of Corollary 7.8, we have*

$$\operatorname{Ind}_G(K_c \cap F) \geq \sup\{n; c_n(F) = c\}$$

The intersection-index will be used in a more crucial way in Corollary 7.17 below. We now prove the following multiplicity result.

Corollary 7.11:   *Under the hypothesis of Corollary 7.4 and if* Ind_G *is an index associated to the group* G, *then*

$$\text{Ind}_G(A_\infty \cap F_\infty \cap K_c) \geq \inf\{\text{Ind}_G(A \cap F); A \in \mathcal{F}, F \in \mathcal{F}^*\}.$$

Proof: Indeed, if not, we proceed as above and we find an invariant neighborhood U of $K_c \cap F_\infty \cap A_\infty$ such that $\text{Ind}_G(\overline{U}) = \text{Ind}_G(K_c \cap F_\infty \cap A_\infty)$. Let $F' = X \setminus U$. It is clearly closed and invariant. The same computation as in Corollary 7.8 shows that for any A in $\mathcal{F}$ and any $F \in \mathcal{F}^*$ we have that $\text{Ind}_G(A \cap F \cap F') \geq 1$ and in particular $A \cap F \cap F' \neq \emptyset$ for all A in $\mathcal{F}$ and $F \in \mathcal{F}^*$. In other words, the family

$$\mathcal{F}_1^* = \{F \cap F'; F \in \mathcal{F}^*\}$$

is also dual to $\mathcal{F}$. Note now that $\lim_n \inf \varphi(F_n \cap F') = c$ which implies by Corollary 7.4 that $K_c \cap F_\infty \cap F' \cap A_\infty \neq \emptyset$. A contradiction.

A dual index

One can associate to any topological index Ind_G on a manifold X, the following *dual index* Ind_G^*. If F is any closed G-invariant subset of X, we let

$$\text{Ind}_G^*(F) = \sup\{n; \exists A \text{ compact } G\text{-invariant, } \text{Ind}_G(A) \geq n$$
$$\text{and } A \cap F = \emptyset\}$$

It is clear that for each $n \geq 1$, the class

$$\mathcal{F}_n^* = \{F; \ F \text{ closed } G\text{-invariant and } \text{Ind}_G^*(F) \leq n - 1\}$$

is dual to the G-homotopy class

$$\mathcal{F}_n = \{A; \ A \text{ compact } G\text{-invariant and } \text{Ind}_G(A) \geq n\}$$

The dual index does not necessarily satisfy the properties of an index. However, one can prove the following

Corollary 7.12:   *Let* X, G *and* φ *be as in Theorem 7.2 and suppose that* Ind_G *is an index associated to* G. *Then, for each* k *such that* $c_k = \inf_{A \in \mathcal{F}_k} \max_{x \in A} \varphi(x)$ *and* $c_k^* = \sup_{F \in \mathcal{F}_k^*} \inf_{x \in F} \varphi(x)$ *are finite, we have* $c_k = c_k^*$.

Moreover, if φ verifies $(PS)_{c_k}$, then for every min-maxing sequence $(A_n)_n$ in $\mathcal{F}_k$ and every suitable max-mining sequence $(F_k)_k$ in $\mathcal{F}_k^*$, we have that $K_{c_k} \cap A_\infty \cap F_\infty$ is non-empty.

Proof: Since $A \cap F \neq \emptyset$ for every $A \in \mathcal{F}_k$ and $F \in \mathcal{F}_k^*$, we clearly have $c_k^* \leq c_k$. On the other hand, if $A = \{x \in X; \varphi(x) \leq c_k - \varepsilon\}$ and $U = \{x \in X; \varphi(x) < c_k - \varepsilon/2\}$, then necessarily, $\text{Ind}_G(A) \leq \text{Ind}_G(\overline{U}) < k$. Hence the set $F = X \setminus U$ intersects every set K' with $\text{Ind}_G(K') \geq k$ since

$$\text{Ind}_G(F \cap K') \geq \text{Ind}_G(K') - \text{Ind}_G(A) \geq 1.$$

This means that $F \in \mathcal{F}_k^*$. Hence, $c_k^* \geq \inf_F \varphi \geq c_k - \varepsilon$ and therefore $c_k = c_k^*$. The rest of the claim follows now from Corollary 7.11.

Other types of multiplicity results similar to those in the Ljusternik-Schnirelmann theory can also be deduced from Theorem 7.2. They are concerned with the case where two different homotopy-stable classes yield the same critical value. We summarize them in the following:

Theorem 7.13: *Let G and X be as in Theorem 7.2 and let φ be a G-invariant C^1-functional on X satisfying (PS). Let $(\mathcal{F}_j)_{j=1}^N$ be a decreasing sequence of G-homotopy stable families with boundaries $(B_j)_{j=1}^N$ and satisfying the following excision property with respect to an index Ind_G:*

(E) For every $1 \leq j \leq j + p \leq N$, any A in $\mathcal{F}_{j+p}$ and any U open and invariant such that $\overline{U} \cap B_j = \emptyset$ and $\mathrm{Ind}_G(\overline{U}) \leq p$, we have $A \setminus U \in \mathcal{F}_j$.

Let F be a closed invariant set that is dual to $\mathcal{F}_j$ while $\sup \varphi(B_j) \leq d := \inf \varphi(F)$ for each $1 \leq j \leq N$. Set $c_j = c(\varphi, \mathcal{F}_j)$ and let $M = \sup\{k \geq 0; c_k = d\}$. Then

(a) $\mathrm{Ind}_G(K_{c_M} \cap F \cap A_\infty) \geq M$ for every min-maxing sequence $(A_n)_n$ in $\mathcal{F}_M$.

(b) For very $M < j \leq j+p \leq N$ such that $c_j = c_{j+p}$ and for every min-maxing sequence $(A_n)_n$ in $\mathcal{F}_{j+p}$, we have $\mathrm{Ind}_G(K_{c_j} \cap A_\infty) \geq p+1$.

In particular, if $I(G) \subset (X \setminus F) \cap \{\varphi \leq d\}$ then

(c) φ has at least N distinct critical orbits.

(d) φ has an unbounded sequence of critical values provided the sequence $(\mathcal{F}_j)_{j=1}^N$ is infinite $(N = +\infty)$.

Proof: (a) Assume $1 \leq M$ since otherwise there is nothing to prove. In view of Corollary 7.8, it is enough to show that $\mathrm{Ind}_G(A \cap F) \geq M$ for every A in $\mathcal{F}_M$. Suppose it was false for some A in $\mathcal{F}_M$. By property (I3) of the index and since $F \cap B_1 = \emptyset$ we can find U open and invariant such that $A \cap F \subset U$, $\overline{U} \cap B_1 = \emptyset$ and $\mathrm{Ind}_G(\overline{U}) = \mathrm{Ind}_G(A \cap F) \leq M - 1$. The excision property implies that $A \setminus U \in \mathcal{F}_1$ and hence that $(A \setminus U) \cap F \neq \emptyset$ which clearly contradicts that F is dual to $\mathcal{F}_1$.

(b) Suppose now $M < j < j + p \leq N$ such that $c_j = c_{j+p}$. If we suppose $\mathrm{Ind}_G(K_{c_j} \cap A_\infty) \leq p$ and since $\sup \varphi(B_j) \leq \inf \varphi(F) < c_j$, we can find, as above, an open invariant neighborhood U of $K_{c_j} \cap A_\infty$ such that $\overline{U} \cap B_j = \emptyset$ and $\mathrm{Ind}_G(\overline{U}) \leq p$. By property (E) we have $A \setminus U \in \mathcal{F}_j$ for every A in $\mathcal{F}_{j+p}$. Hence $(A \setminus U) \cap \{\varphi \geq c_j\} \neq \emptyset$ for such an A. It follows that $F' = \{\varphi \geq c_{j+p}\} \setminus U = \{\varphi \geq c_j\} \setminus U$ verifies (F1) and (F2) with respect to $\mathcal{F}_{j+p}$. By Theorem 7.2, $K_{c_j} \cap A_\infty \setminus U = K_{c_{j+p}} \cap A_\infty \cap F' \neq \emptyset$ which is clearly a contradiction.

(c) Follows immediately from the above estimates and property (I5) of the index since $K_{c_i} \cap F$ is disjoint from $I(G)$ (when $1 \le i \le M$) and $K_{c_i} \cap I(G) = \emptyset$ if $M + 1 \le i \le N$ since then $d < c_i$.

(d) Suppose now that $(\mathcal{F}_j)_j$ is an infinite sequence of G-homotopy stable classes. We note first that M must be finite. Indeed, we have by (a) that $\mathrm{Ind}_G(K_d \cap F) \ge M$. Since $K_d \cap F$ is compact and disjoint from $I(G)$ we cannot have $\mathrm{Ind}_G(K_d \cap F) = +\infty$ by property (I6) of the index. The same reasoning shows that the sequence $(c_j)_{j>M}$ cannot become stationary. It remains to show that $c = \lim_j c_j$ must be $+\infty$. Suppose not and consider

$$K = \{x \in X; d\varphi(x) = 0 \text{ and } c_{M+1} \le \varphi(x) \le c\}.$$

Since it is invariant, compact and disjoint from $I(G)$, we have $\mathrm{Ind}_G(K) = q < +\infty$. Let U be an open set containing K, disjoint from $\{\varphi \le d\}$ and such that $\mathrm{Ind}_G(\overline{U}) = q$. We shall show the existence of a sequence $(x_n)_n$ such that $\lim_n \varphi(x_n) = c$, $\lim_n d\varphi(x_n) = 0$ and $\lim_n(\mathrm{dist}(x_n, X \setminus U) = 0$. Since φ has (PS), we get then, that $K_c \cap X \setminus U \ne \emptyset$ which is clearly a contradiction.

To get the sequence of approximate critical points, fix $\delta > 0$ such that $c_{M+1} < c - \delta$ and find j large enough so that $c_{M+1} < c - \delta \le c_j \le c_{j+q} \le c$. By the excision property, we have for every A in $\mathcal{F}_{j+1}$ that $A \setminus U \in \mathcal{F}_j$. Hence if we set $F' = \{\varphi \ge c_j\} \setminus U$ we get that $F' \cap A \ne \emptyset$ for all A in $\mathcal{F}_{j+q}$ and $F' \cap B_{j+q} = \emptyset$ since $\sup \varphi(B_{j+q}) \le d < c_{M+1} < c_{j+q}$. Moreover, $\inf \varphi(F') \ge c_j \ge c - \delta \ge c_{j+q} - \delta$. It follows from Theorem 7.2 that if we choose δ initially to be less than $\frac{1}{8}(c_{M+1} - d) \le \frac{1}{8}[\inf \varphi(F') - \sup \varphi(B_{j+q})]$ then there is $x_\delta \in X$ so that

(i) $c_{j+q} - \delta \le \varphi(x_\delta) \le c_{j+q} + 9\delta$, hence $c - 2\delta \le \varphi(x_\delta) \le 10\delta$

(ii) $\|d\varphi(x_\delta)\| \le 18\sqrt{\delta}$

(iii) $\mathrm{dist}(x_\delta, F') \le 5\sqrt{\delta}$, hence $\mathrm{dist}(x_\delta, X \setminus U) \le 5\sqrt{\delta}$.

Letting $\delta \to 0$ we get the required contradiction.

Remark 7.14: The above result remains true if every class $\mathcal{F}_j$ is replaced by a union $\widetilde{\mathcal{F}}_j = \bigcup_\alpha \mathcal{F}_j^\alpha$ of G-homotopy stable classes with boundaries $(B_j^\alpha)_\alpha$ provided the condition on F becomes for every j, $F \cap B_j^\alpha = \emptyset$, $\sup \varphi(B_j^\alpha) \le \inf \varphi(F)$ for each α and $F \cap A \ne \emptyset$ for all A in $\mathcal{F}_j$.

The simplest nested sequence of G-homotopy-stable families satisfying the excision property mentioned above, can be obtained from the index itself. Indeed, let $(B_n)_n$ be an increasing sequence of G-invariant compact sets and define for each integer n, the following class

$$\textstyle\sum_n^{B_n} = \{A; A \text{ compact invariant containing } B_n \text{ and } \mathrm{Ind}_G(A) \ge n\}.$$

It is clearly G-homotopy stable with boundary B_n and for each $m \le n$, we have

$$\sum_n^{B_n} \subset \sum_m^{B_m} \quad \text{and} \quad c_m := c(\varphi, \sum_m^{B_m}) \leq c_n := c(\varphi, \sum_n^{B_n})$$

provided, of course, the classes are non-empty.

Suppose now that $\sup \varphi(B_n) < c_n$, Theorem 7.2 then gives that the sets K_{c_n} are non empty. Moreover, if $c_n = c_m$, then Theorem 7.13.b gives the classical estimate $\mathrm{Ind}_G(K_{c_n}) \geq n - m + 1$. In the following, we use Theorems 7.2 and 7.5 to give multiplicity results under the relaxed boundary condition $\sup \varphi(B_n) \leq c_n$.

If $\mathcal{F}$ is a G-homotopy-stable family of sets, we shall write

$$\mathrm{Ind}_G(\mathcal{F}) = \inf\{\mathrm{Ind}_G(A); A \in \mathcal{F}\}.$$

Proposition 7.15: *Let G, φ, B and $\mathcal{F}$ be as in Theorem 7.2. Let C be a closed invariant set with $C \subseteq B$ such that for some $m \geq 1$,*

(F4) $$c = c(\varphi, \mathcal{F}) = c(\varphi, \sum_m^C).$$

Let F be a closed invariant set satisfying (F1) and (F2) with respect to the class $\sum_m^C$ whose boundary is C. If φ satisfies $(PS)_{F,c}$ then, for any min-maxing sequence $(A_n)_n$ in $\mathcal{F}$, we have

$$\mathrm{Ind}_G(K_c \cap F \cap A_\infty) \geq \mathrm{Ind}_G(\mathcal{F}) - \mathrm{Ind}_G(F \cap B) - m + 1$$

In particular,

(i) *if $\mathcal{F} = \sum_{m+p}^B$, then $\mathrm{Ind}_G(K_c \cap F) \geq p + 1 - \mathrm{Ind}_G(B \cap F)$.*

(ii) *if $\mathcal{F} = \sum_{m+p}^B$ and $\sup \varphi(C) < c$, then*

$$\mathrm{Ind}_G(K_c) \geq p + 1 - \mathrm{Ind}_G(B \cap \{\varphi \geq c\}).$$

Proof: Assume $\mathrm{Ind}_G(K_c \cap F \cap A_\infty) \leq \mathrm{Ind}_G(\mathcal{F}) - \mathrm{Ind}_G(F \cap B) - m$. Since $F \cap C = \emptyset$, we can use property (I3) of the index to find an invariant neighborhood U of $K_c \cap F \cap A_\infty$ that is disjoint from C and such that $\mathrm{Ind}_G(\overline{U}) = \mathrm{Ind}_G(K_c \cap F \cap A_\infty)$. Similarly, find an invariant neighborhood V of $F \cap B$ that is disjoint from C and such that $\mathrm{Ind}_G(\overline{V}) = \mathrm{Ind}_G(F \cap B)$.

For any A in $\mathcal{F}$, we have by property (I4) of the index

$$\mathrm{Ind}_G(A) \leq \mathrm{Ind}_G(A \setminus U \setminus V) + \mathrm{Ind}_G(K_c \cap F \cap A_\infty) + \mathrm{Ind}_G(B \cap F).$$

It follows that $\mathrm{Ind}_G(A \setminus U \setminus V) \geq m$, hence $A \setminus U \setminus V \in \sum_m^C$ and $(A \setminus U \setminus V) \cap F \neq \emptyset$, for all A in $\mathcal{F}$. In other words, if we set $F' = F \setminus U \setminus V$, then F' verifies (F1) and (F2) with respect to the class $\mathcal{F}$. It follows that $F' \cap K_c \cap A_\infty \neq \emptyset$ which is a contradiction.

Corollary 7.16 : *Let $G, \varphi, \mathcal{F}$ and B be as in Theorem 7.2, with B invariant, $F = \{\varphi \geq c\}$ and φ satisfying $(PS)_c$. Suppose that $c(\varphi, \mathcal{F}) = c(\varphi, \sum_m^B)$ for some $m \geq 1$. Then, we have*

$$\mathrm{Ind}_G(\overline{K_c \setminus B}) \geq \mathrm{Ind}_G(\mathcal{F}) - \mathrm{Ind}_G(B \cap \{\varphi \geq c\}) - m + 1.$$

In particular, if $\mathcal{F} = \sum_{m+p}^{B}$, then
$$\mathrm{Ind}_G(\overline{K_c \setminus B}) \geq p + 1 - \mathrm{Ind}_G(B \cap \{\varphi \geq c\}).$$

Proof: Let U be an open invariant neighborhood of $\overline{K_c \setminus B}$ with the same index. Let also V be an open invariant neighborhood of $B \cap \{\varphi \geq c\}$ with the same index. If the assertion were false, then we get that for any $A \in \mathcal{F}$, the set $A \setminus U \setminus V \in \sum_m^B$. This means that the set $F = \{\varphi \geq c\} \setminus U \setminus V$ and B satisfy conditions $(F'1)$ and $(F'2)$ with respect to the class $\mathcal{F}$. By Theorem 7.5, $F' \cap K_c \neq \emptyset$ which is a contradiction.

Here is a useful reformulation of Theorem 7.13, in terms of the intersection index.

Corollary 7.17: *Let G, X and φ be as in Theorem 7.2 and suppose $(B_j)_{j=1}^N$ is an increasing family of G-invariant closed sets. Let F be a closed G-invariant set such that*
$$(F3) \qquad\qquad \sup \varphi(B_N) \leq \inf \varphi(F)$$
and let A be a compact G-invariant set containing B_N such that
$$(F5) \qquad\qquad \mathrm{Int}_{B_N}^F(A) \geq N$$
There exist then critical values c_i $(1 \leq i \leq N)$ for φ such that if we denote by $d = \inf \varphi(F)$ and $M = \sup\{k \geq 0; c_k = d\}$ then
(a) $\mathrm{Ind}_G(K_{c_M} \cap F) \geq M$.
(b) For very $M < j \leq j + p \leq N$ such that $c_j = c_{j+p}$ we have $\mathrm{Ind}_G(K_{c_j}) \geq p + 1$.
In particular, if $I(G) \subset (X \setminus F) \cap \{\varphi \leq d\}$ then
(c) φ has at least N distinct critical orbits.
(d) φ has an unbounded sequence of critical values if the hypothesis hold for arbitrarily large N.

Proof: It is enough to notice that for every $1 \leq j \leq N$, the classes
$$\mathcal{F}_j = \{A; \ A \text{ compact } G\text{-invariant containing } B_j \text{ and } \mathrm{Int}_{B_j}^F(A) \geq j\}$$
are non-empty and they satisfy the hypothesis of Theorem 7.13.

Before we give the examples, we present one more notion.

Relative indices

A *relative index* will be an $\mathbb{N} \cup \{+\infty\}$-valued function $\mathrm{Ind}_G(\ ,\ )$ defined on pairs (A, B) of invariant paracompact sets in X with $A \supset B$ and satisfying the following properties:
(R1) $\mathrm{Ind}_G(\ ,\emptyset)$ verifies properties (I1)–(I6) of the index and will be denoted $\mathrm{Ind}_G(\ )$.
(R2) If $f : (A_1, B) \to (A_2, B)$ is equivariant and $f_{|B}$ is a homeomorphism, then $\mathrm{Ind}_G(A_1, B) \leq \mathrm{Ind}_G(A_2, B)$.

(R3) $\mathrm{Ind}_G(A_1 \cup A_2, B) \leq \mathrm{Ind}_G(A_1, B) + \mathrm{Ind}_G(A_2)$.

The existence of such a relative index in various cohomological settings is established in [F-H].

Corollary 7.18: *Let G and X be as in Theorem 7.2 and let φ be a G-invariant C^1-functional on X satisfying (PS). Let B and F be two disjoint closed and invariant subsets of X such that*
(1) $k = \mathrm{Ind}_G(X \setminus F, B) < \mathrm{Ind}_G(X, B) = n$
(2) $\sup \varphi(B) \leq \inf \varphi(F)$
(3) $I(G) \subset B$.
Then φ has at least $n - k$ distinct critical orbits. Moreover, if $\mathrm{Ind}_G(X, B) = +\infty$ then φ has an unbounded sequence of critical values.

Proof: For each j, $(k < j \leq n)$ let

$$\Sigma_j(B) = \{A; A \text{ compact invariant containing } B \text{ and } \mathrm{Ind}_G(A, B) \geq j\}.$$

By (R3), the family $\{\Sigma_j(B); k < j \leq n\}$ verifies the excision property E. By (R2), each $\Sigma_j(B)$ is G-homotopy stable with boundary B. On the other hand F verifies (F1) and (F3) with respect to each $\Sigma_j(B)$ $(k < j \leq n)$. Indeed, if $F \cap A = \emptyset$ for some A in $\Sigma_j(B)$, then $(A, B) \subset (X \setminus F, B)$ and by (R2), $\mathrm{Ind}_G(X \setminus F, B) \geq \mathrm{Ind}_G(A, B) \geq j > k$ which contradicts (1). The corollary now follows from Theorem 7.13 since by (3), $I(G) \subset B \subset (X \setminus F) \cap \{\varphi \leq \inf \varphi(F)\}$.

We shall give one more example to show the flexibility of Theorem 7.2 in proving multiplicity results.

Let $\mathcal{F}$ be G-homotopy stable with boundary B and we would like to compare $c = c(\varphi, \mathcal{F})$ to the min-max value of φ on another family Σ in order to obtain an estimate for K_c. Several choices for Σ are possible. Many of them can be covered by the following setting.

Let D and C be two closed invariant sets such that $D \subseteq C \subseteq B$ and consider the following class:

$$\Sigma_n^C(D) = \{A; A \text{ compact invariant containing } C \text{ and } \mathrm{Ind}_G(A, D) \geq n\}.$$

It is clearly G-homotopy stable with boundary C. We shall give an estimate for K_c in the case where $c(\varphi, \mathcal{F}) = c(\varphi, \Sigma_n^C(D))$. Note that the case $B = \emptyset$ is the most classical, while the case where $D = \emptyset$ was covered earlier in this section. We shall write

$$\mathrm{Ind}_G(\mathcal{F}, D) = \inf\{\mathrm{Ind}_G(A, D); A \in \mathcal{F}\}.$$

Proposition 7.19: *Let G, φ, B and $\mathcal{F}$ be as in Theorem 7.2. Let D and C be two closed invariant sets with $D \subseteq C \subseteq B$ such that for some*

$m \geq 1$,

(F4)
$$c = c(\varphi, \mathcal{F}) = c(\varphi, \Sigma_m^C(D))$$

Let F be a closed invariant set satisfying (F1) and (F2) with respect to the class $\Sigma_m^C(D)$ whose boundary is C. Then, for any min-maxing sequence $(A_n)_n$ in $\mathcal{F}$, we have

$$\mathrm{Ind}_G(K_c \cap F \cap A_\infty) \geq \mathrm{Ind}_G(\mathcal{F}, D) - \mathrm{Ind}_G(F \cap B) - m + 1.$$

Proof: Assume $\mathrm{Ind}_G(K_c \cap F \cap A_\infty) \leq \mathrm{Ind}_G(\mathcal{F}, D) - \mathrm{Ind}_G(F \cap B) - m$. Since $F \cap C = \emptyset$, we can use property (I3) of the index to find an invariant neighborhood U of $K_c \cap F \cap A_\infty$ that is disjoint from C and such that $\mathrm{Ind}_G(\overline{U}) = \mathrm{Ind}_G(K_c \cap F \cap A_\infty)$. Similarly, find an invariant neighborhood V of $F \cap B$ that is disjoint from C and such that $\mathrm{Ind}_G(\overline{V}) = \mathrm{Ind}_G(F \cap B)$.

For any A in $\mathcal{F}$, we have by property (R3) of the relative index

$$\mathrm{Ind}_G(A, D) \leq \mathrm{Ind}_G(A \setminus U \setminus V), D) + \mathrm{Ind}_G(K_c \cap F \cap A_\infty) + \mathrm{Ind}_G(B \cap F).$$

It follows that $\mathrm{Ind}_G(A \setminus U \setminus V) \geq m$, hence $A \setminus U \setminus V \in \Sigma_m^C(D)$ and $(A \setminus U \setminus V) \cap F \neq \emptyset$, for all A in $\mathcal{F}$. In other words, if we set $F' = F \setminus U \setminus V$, then F' verifies (F1) and (F2) with respect to the class $\mathcal{F}$. It follows that $F' \cap K_c \cap A_\infty \neq \emptyset$ which is a contradiction.

7.3 The $\mathbb{Z}_2$-index of Krasnoselski

Suppose the symmetry group $\mathbb{Z}_2$ is acting on a smooth manifold X. Denote by Σ the class of all closed symmetric subsets X. The $\mathbb{Z}_2$-index of Krasnoselski is defined on Σ in the following way:

$$\gamma_{\mathbb{Z}_2}(A) = \inf\{k; \text{there exists } f : A \to \mathbb{R}^k \setminus \{0\} \text{ odd and continuous}\}.$$

If no such a finite k exists, we set $\gamma_{\mathbb{Z}_2}(A) = \infty$. We also let $\gamma_{\mathbb{Z}_2}(\emptyset) = 0$.

Lemma 7.20: *Suppose $\mathbb{Z}_2$ acts on a C^1-Finsler manifold X, then the following hold:*

(a) *$\gamma_{\mathbb{Z}_2}$ verifies the properties (I1)-(I6) of an index.*

(b) *If there exists an odd homeomorphism $h \in C(A, \partial\Omega)$ where $A \in \Sigma$ and Ω is a bounded neighborhood of 0 in $\mathbb{R}^k$, then $\gamma_{\mathbb{Z}_2}(A) = k$.*

(c) *If X is a Banach space and Z is a subspace of X with finite codimension, then for every $A \in \Sigma$,*

$$\gamma_{\mathbb{Z}_2}(A \cap Z) \geq \gamma_{\mathbb{Z}_2}(A) - \mathrm{codim}(Z).$$

Proof: (a) Note that if C is closed and $C \cap (-C) = \emptyset$, then necessarily $\mathrm{Ind}_{\mathbb{Z}_2}(C \cup -C) = 1$ since the function

$$f(x) = \begin{cases} 1 & \text{on } B \\ -1 & \text{on } -B \end{cases}$$

is an odd function in $C(B \cup (-B); \mathbb{R} \setminus \{0\})$. It follows that if $x \neq 0$, then

$\gamma_{\mathbb{Z}_2}(\{x, -x\}) = 1$. More generally, if $0 \notin A$ and $\gamma_{\mathbb{Z}_2}(A) > 1$, then A contains infinitely many distinct points, since otherwise $A = C \cup (-C)$ with $C \cap (-C) = \emptyset$. (I1) and (I5) are proved.

For the rest, we shall assume that $\gamma_{\mathbb{Z}_2}(A_1)$ and $\gamma_{\mathbb{Z}_2}(A_2)$ are finite since otherwise there is nothing to prove. For (I2), assume $\gamma_{\mathbb{Z}_2}(A_2) = n$ which means that there exists an odd f in $C(A_2; \mathbb{R}^n \backslash \{0\})$. If now $\varphi : A_1 \to A_2$ is odd and continuous, we have that $f \circ \varphi \in C(A_1; \mathbb{R}^n \backslash \{0\})$ and hence $\gamma_{\mathbb{Z}_2}(A_1) \leq n$.

For (I3), suppose $\gamma_{\mathbb{Z}_2}(A_1) = m$ and $\gamma_{\mathbb{Z}_2}(A_2) = n$ and consider odd mappings $f_1 \in C(A_1; \mathbb{R}^m \backslash \{0\})$ and $f_2 \in C(A_2; \mathbb{R}^n \backslash \{0\})$. By the Tietze extension theorem, there exist $\tilde{f}_1 \in C(X, \mathbb{R}^m)$ and $\tilde{f}_2 \in C(X, \mathbb{R}^n)$ such that $\tilde{f}_1|A_1 = f_1$ and $\tilde{f}_2|A_2 = f_2$. Modulo an obvious symmetrization, we can clearly assume that $\tilde{f}_1$ and $\tilde{f}_2$ are also odd functions. Note now that $f = (\tilde{f}_1, \tilde{f}_2)$ is an odd function in $C(A_1 \cup A_2; \mathbb{R}^{m+n} \backslash \{0\})$, hence $\gamma_{\mathbb{Z}_2}(A_1 \cup A_2) \leq m + n = \gamma_{\mathbb{Z}_2}(A_1) + \gamma_{\mathbb{Z}_2}(A_2)$.

To prove (I4), use the compactness of K to cover it by $\cup_{i=1}^k T_{x_i}$ where

$$T_{x_i} = \{y \in X; \|y - x_i\| < \|x_i\|/2 \text{ or } \|y + x_i\| < \|x_i\|/2\}$$

and $x_i \in K$. By the proof of (I1) we have $\gamma_{\mathbb{Z}_2}(\bar{T}_{x_i}) = 1$ for each i and by (I3), $\gamma_{\mathbb{Z}_2}(K) \leq k < +\infty$. For (I6), Suppose now $\gamma_{\mathbb{Z}_2}(K) = n$ and let $f \in C(K; \mathbb{R}^n \backslash \{0\})$ be an odd map. Extend f to an odd function $\tilde{f}$ on X (as in the proof of (I3)). Since K is compact, there is $\delta > 0$ so that $0 \notin \tilde{f}(K^\delta)$. Hence $\gamma_{\mathbb{Z}_2}(K^\delta) \leq n = \gamma_{\mathbb{Z}_2}(K)$ and (I6) is proved.

(b) Assume h is an odd homeomorphism in $C(A, \partial \Omega)$. It is clear that $\gamma_{\mathbb{Z}_2}(A) \leq k$. If now $\gamma_{\mathbb{Z}_2}(A) = n < k$, then there is an odd function k in $C(A; \mathbb{R}^n \backslash \{0\})$. But $k \circ h^{-1}$ is then an odd mapping in $C(\partial \Omega; \mathbb{R}^n \backslash \{0\})$ contradicting the Borsuk-Ulam theorem. Hence $\gamma_{\mathbb{Z}_2}(A) = k$.

(c) Let P be the projection of X onto the k-dimensional complement Y of Z in X. Suppose $\gamma_{\mathbb{Z}_2}(A) = n$. If $\gamma_{\mathbb{Z}_2}(A \cap Z) \leq n - k - 1$, then there is an odd map $h \in C(X, \mathbb{R}^{n-k-1})$ such that h sends $A \cap Z$ into $\mathbb{R}^{n-k-1} \backslash \{0\}$. On the other hand P is an odd map in $C(X; Y)$ that sends $A \backslash Z$ to $Y \backslash \{0\}$. Since Y is linearly isomorphic to $\mathbb{R}^k$, we get that the odd and continuous map $(h, P) : X \to \mathbb{R}^{n-1}$ sends A to $\mathbb{R}^{n-1} \backslash \{0\}$ which is a contradiction.

7.4 A relaxed $\mathbb{Z}_2$-symmetric mountain pass theorem

Let X be a Banach space and assume $X = Y \oplus Z$ where Y is a subspace of dimension $k < \infty$. Let E_n be a vector subspace of X containing Y with $\dim(E_n) = n > k$. For $R > 0$, we write D_n for the ball $B_R(E_n)$ in E_n and S_n for the sphere $S_R(E_n)$. For $\rho > 0$, let $S_\rho(Z)$ be the corresponding sphere in Z. Denote now by Int_n^ρ the $S_\rho(Z)$-intersection index $\text{Ind}_{S_n}^{S_\rho(Z)}$ with boundary $S_n = S_R(E_n)$, that is

$$\text{Ind}_{S_R(E_n)}^{S_\rho(Z)}(A) = \inf\{\gamma_{\mathbb{Z}_2}(\eta(A) \cap S_\rho(Z)); \eta \in \mathcal{L}_n\}$$

where

$$\mathcal{L}_n = \{\eta \in C(X,X); \eta \text{ odd and } \eta(x) = x \text{ for } x \in S_n\}.$$

We shall need the following:

Lemma 7.21: *With the above notation, if $\rho < R$, then*

$$\text{Int}_n^\rho(\overline{B_R(E_n)}) \geq n - k.$$

Proof: Let $A = h(\overline{D_n})$ with $h \in \mathcal{L}_n$. Let $\tilde{\mathcal{O}} = \{x \in D_n; h(x) \in B_\rho\}$. Since h is odd, $0 \in \tilde{\mathcal{O}}$. Let O denote the component of $\tilde{\mathcal{O}}$ containing 0. Since D_n is bounded, $\mathcal{O}$ is a symmetric bounded neighborhood of 0 in E_n. Therefore by Lemma 7.20.b, $\gamma_{\mathbb{Z}_2}(\partial\mathcal{O}) = n$. We claim that

$$h(\partial\mathcal{O}) \subset \partial B_\rho. \tag{*}$$

Indeed, suppose $x \in \partial\mathcal{O}$ and $h(x) \in B_\rho$. If $x \in D_n$, there exists a neighborhood V of x such that $h(V) \subset B_\rho$, but then $x \notin \partial\mathcal{O}$. It follows that x is necessarily in $\partial D_n = S_n$. But if $x \in S_n$ and $h(x) \in B_\rho$, then $h(x) = x$ and $\|h(x)\| = \|x\| = R < \rho$, thus (*) holds.

Set now $W = \{x \in D_n; h(x) \in \partial B_\rho\}$. (*) implies that $W \supset \partial\mathcal{O}$. Hence by Lemma 7.20.(b), $\gamma_{\mathbb{Z}_2}(W) = n$. Since codimension$Z = k$, we get from Lemma 7.20.(c) that $\gamma_{\mathbb{Z}_2}(h(W) \cap Z) \geq n - k$. But $h(W) \subset A \cap \partial B_\rho$. This finishes the proof of the Lemma.

Now we can prove the following $\mathbb{Z}_2$-symmetric Mountain Pass Theorem.

Corollary 7.22 (the finite case): *Let φ be an even C^1-functional satisfying (PS) on $X = Y \oplus Z$ where $\dim(Y) = k < +\infty$. Assume $\varphi(0) = 0$ as well as the following conditions:*
(1) There is $\rho > 0$ and $\alpha \geq 0$ such that $\inf\varphi(S_\rho(Z)) \geq \alpha$.
(2) There exists $R > \rho$ and a subspace E of X containing Y such that $\dim(E) = n > k$ and $\sup\varphi(S_R(E)) \leq 0$.
   There exists then critical values c_i $(1 < i \leq n - k)$ for φ such that
(a) $0 \leq \alpha \leq c_1 \leq \cdots \leq c_{n-k}$.
(b) φ has at least $n - k$ distinct pairs of non-trivial critical points.
   Moreover, if $c_i = \alpha$ for some $1 \leq i \leq n - k$, we have

$$\gamma_{\mathbb{Z}_2}(K_\alpha \cap S_\rho(Z)) \geq i.$$

Proof: Follows immediately from Lemma 7.21, the fact that

$$I(\mathbb{Z}_2) = \{0\} \subset (X \setminus S_\rho(Z)) \cap \{\varphi \leq \alpha\}$$

and Corollary 7.17, F being the sphere $S_\rho(Z)$.

Corollary 7.23 (the infinite case): *Let φ be an even C^1-functional satisfying (PS) on a Banach space $X = Y \oplus Z$ with $\dim(Y) < \infty$. Assume $\varphi(0) = 0$ as well as the following conditions:*

(1) *There is $\rho > 0$ and $\alpha \geq 0$ such that $\inf \varphi(S_\rho(Z)) \geq \alpha$.*

(2) *There exists an increasing sequence $(E_n)_n$ of finite dimensional subspaces of X, all containing Y such that $\lim_n \dim(E_n) = \infty$ and for each n, $\sup \varphi(S_{R_n}(E_n)) \leq 0$ for some $R_n > \rho$.*

Then φ has an unbounded sequence of critical values.

Proof: For any vector space E_n containing Y and any $R_n > 0$, we define as above Int_n^ρ to be the $S_\rho(Z)$-intersection index $\text{Ind}_{S_R(E_n)}^{S_\rho(Z)}$ with boundary $S_R(E_n)$. Consider also for $k < j \leq \dim(E_n)$ the classes

$$\mathcal{F}_j(E_n, R_n) = \{A; \; A \text{ compact symmetric containing } S_{R_n}(E_n)$$
$$\text{and } \text{Int}_n^\rho(A) \geq j - k\}.$$

For each $k < j$, let $\widetilde{\mathcal{F}}_j = \bigcup_n \{\mathcal{F}_j(E_n, R_n); \dim(E_n) \geq j\}$. Each $\widetilde{\mathcal{F}}_j$ is a union of homotopy-stable classes with boundaries $(S_{R_n}(E_n))_n$. In view of Lemma 7.21 and the properties of the intersection index, the sequence $(\widetilde{\mathcal{F}}_j)_j$ satisfy the hypothesis of Theorem 7.13. Moreover

$$I(\mathbb{Z}_2) = \{0\} \subset (X \setminus S_\rho(Z)) \cap \{\varphi \leq \alpha\}.$$

Hence Theorem 7.13 and Remark 7.14 apply (with $F = S_\rho(Z)$) to get the result.

7.5 Multiple solutions in the presence of symmetry

Let Ω be smoothly bounded domain in $\mathbb{R}^N, N \geq 3$, and let $g : \Omega \times \mathbb{R} \to \mathbb{R}$ be a Caratheodory function with primitive $G(\cdot, t) = \int_0^t g(\cdot, s)\, ds$. Again, we are considering the following equation

$$-\Delta u = g(x, u), \quad x \in \Omega,$$
$$u = 0, \qquad x \in \partial\Omega. \tag{1}$$

Theorem 7.24: *Suppose g verifies the following conditions:*

$$g \text{ is continuous and odd: } g(x, -s) = -g(x, s), \tag{2}$$

$$|g(x, s)| \leq C(1 + |s|^{p-1}) \text{ for almost every } x \in \Omega \text{ and } s \in \mathbb{R}, \tag{3}$$

$$0 < qG(x, s) \leq g(x, s)s \text{ for almost every } x \in \Omega \text{ and } |s| \geq R_0, \tag{4}$$

where $2 \leq p < 2^* = \frac{2N}{N-2}$, $q > 2$ and R_0 is some positive constant.

Then problem (1) admits an unbounded sequence of solutions in H_0^1.

Proof: Define

$$\varphi(u) = \frac{1}{2} \int_\Omega |\nabla u|^2 dx - \int_\Omega G(x, u) dx.$$

As in Theorem 3.10, we know that φ is Fréchet differentiable on $X = H_0^1(\Omega)$, that φ verifies the Palais-Smale condition and that the assertion of the theorem is equivalent to finding an unbounded sequence of critical points for φ. So, in order to prove Theorem 7.24, we only need to show that φ verifies the rest of the hypothesis in Theorem 7.23.

First, since g is odd, φ is necessarily even. Also $\varphi(0) = 0$. Denote $0 < \lambda_1 < \lambda_2 \le \lambda_3 \le \dots$ the eigenvalues of $-\Delta$ on Ω with homogeneous Dirichlet data and let v_j be the corresponding eigenfunctions.

Claim (i): For k_0 sufficiently large, there exist $\rho > 0$ such that $\varphi(u) \ge 1$ for all $u \in Z := \mathrm{span}\{v_k; k \ge k_0\}$ with $\|u\|_{H_0^1} = \rho$.

Indeed, by condition (3), Sobolev's embedding $H_0^1(\Omega) \hookrightarrow L^{2^*}(\Omega)$ and Hölder's inequality, there exist positive constants C_1 and C_2 such that for $u \in Z$ we have

$$\varphi(u) \ge \frac{1}{2} \int_\Omega |\nabla u|^2 \, dx - C_1 \int_\Omega |u|^p \, dx - C_2$$

$$\ge \frac{1}{2} \|u\|_{H_0^1}^2 - C_1 \|u\|_2^r \|u\|_{2^*}^{p-r} - C_2$$

$$\ge \left(\frac{1}{2} - C_1 \lambda_{k_0}^{-r/2} \|u\|_{H_0^1}^{p-2} \right) \|u\|_{H_0^1}^2 - C_2$$

where $\frac{r}{2} + \frac{p-r}{2^*} = 1$. In particular, $r = N(1 - \frac{p}{2^*}) > 0$, and we may let $\rho = 2\sqrt{(C_2 + 1)}$ and choose $k_0 \in \mathbb{N}$ such that $C_1 \lambda_{k_0}^{-r/2} \rho^{p-2} \le \frac{1}{4}$ to get the claim.

Let now $Y = \mathrm{span}\{v_j; j < k_0\}$ be its orthogonal complement. It remains to prove the following:

Claim (ii): On any finite dimensional subspace $E \subset H_0^1$, there exist constants C_1, C_2, C_3 (depending on E) such that

$$\sup_{u \in \partial B_R(E)} \varphi(u) \le C_1 R^2 - C_2 R^q + C_3.$$

Indeed, condition (4) implies that for two positive constants K_1, K_2, we have

$$G(x, u) \ge K_1 \|u\|^q - K_2 \quad \text{for all } x \in \Omega \text{ and } u \in \mathbb{R}.$$

It follows that for a fixed $u \in H_0^1$, we have for any $R > 0$, that

$$\varphi(Ru) = \frac{R^2}{2} \int_\Omega |\nabla u|^2 \, dx - \int_\Omega G(x, Ru) \, dx$$

$$\le \frac{R^2}{2} \|u\|^2 - R^q K_1 \|u\|_q^q + K_2 |\Omega|.$$

This clearly implies Claim (ii).

Now Theorem 7.23 guarantees the existence of an unbounded sequence of critical values $c_k = \varphi(u_k)$ with u_k being a weak solution for (1). Since $\varphi'(u_k)u_k = 0$, we get that

$$\int_\Omega |\nabla u_k|^2 \, dx = \int_\Omega g(x, u_k)u_k \, dx \qquad (5)$$

which implies that

$$c_k = \int_\Omega \left[\frac{1}{2} g(x, u_k)u_k - G(x, u_k) \right] dx \qquad (6)$$

Since $c_k \to \infty$, we get from (5), (6) and (4), that (u_k) is unbounded in X and the proof of the Theorem is complete.

7.6 The mod 2-genus

Consider the N-dimensional real projective space

$$P^N = \{(x, -x) : x \in S^N\}.$$

The homology ring (with $\mathbb{Z}_2$ coefficients) of P^N are then

$$H^k(P^N) = \mathbb{Z}_2 \text{ if } k \le N \text{ and } H^k(P^N) = \{0\} \text{ if } k > n.$$

For any compact symmetric subset A of S^N, define the *mod 2-genus* γ_2 by

$$\gamma_2(A) = \sup\{k; (i_*)_{k-1} \ne 0\}$$

where $i : \tilde{A} = \{(x, -x) : x \in A\} \to P^N$ is the natural injection and $(i_*)_k : H_k(\tilde{A}) \to H_k(P^N)$ is the corresponding homomorphism between the k-dimensional homology groups.

The set function γ_2 enjoys properties (I1)-(I6) of an index listed above. Moreover, it satisfies the following property:

(I7) $\gamma_2(A_1 \cap A_2) \ge \gamma_2(A_1) + \gamma(A_2) - N - 1$ for all closed symmetric subsets A_1, A_2.

For the proofs we refer to [Co].

For each n, we let $\mathcal{F}_n = \{A; A \text{ closed symmetric with } \gamma_2(A) \ge n\}$. It is clearly a $\mathbb{Z}_2$-homotopy stable family. Property (I7) implies that for each k ($1 \le k \le N+1$), the families $\mathcal{F}_k$ and $\mathcal{F}_{N-k+2}$ are dual to each other. We can now state the following:

Corollary 7.25: Let φ be a C^1-functional on S^N. For each integer k ($1 \le k \le N+1$), let

$$d_k = \sup_{A \in \mathcal{F}_k} \inf_{x \in A} \varphi(x) \text{ and } c_k = \inf_{A \in \mathcal{F}_k} \max_{x \in A} \varphi(x).$$

Then:

(1) $c_k = d_{N-k+2}$ and for every min-maxing sequence $(A_n)_n$ in $\mathcal{F}_k$, every suitable max-mining sequence $(A_n^*)_n$ in $\mathcal{F}_{N-k+2}$, we have that

$$A_\infty \cap A_\infty^* \cap K_{c_k} \ne \emptyset.$$

(2) If $c_{m+k} = c_k$ for some $1 \leq k \leq m + k \leq N + 1$, then for every min-maxing sequence $(A_n)_n$ in $\mathcal{F}_{m+k}$ and every suitable max-mining sequence $(A_n^*)_n$ in $\mathcal{F}_{N-k+2}$, we have that

$$\gamma_2(A_\infty \cap A_\infty^* \cap K_{c_k}) \geq m + 1.$$

Proof: (i) Note first that since $\mathcal{F}_{N-k+2}$ is dual to $\mathcal{F}_k$, we have that $c_k \geq d_{N-k+2}$. On the other hand, if $A = \{x \in X; \varphi(x) \leq c_k - \varepsilon\}$, then necessarily, $\gamma_2(A) < k$. By $(I3)$, there exists $\delta > 0$ such that $\gamma_2(A^\delta) = \gamma_2(A) < k$. To show that the set $F = X \setminus K^\delta$ belongs to $\mathcal{F}_{N-k+2}$, note that

$$\gamma_2(F) \geq \gamma_2(X) - \gamma_2(K^\delta) \geq N - k + 2.$$

Since $F \subset \{\varphi > c_k - \varepsilon\}$, it follows that $d_{N-k+2} \geq \inf_F \varphi \geq c_k - \varepsilon$ and therefore $c_k = d_{N-k+2}$. The rest of the claim follows now from Corollary 7.11.

(ii) follows immediately from Corollary 7.11 and the fact that the γ_2-index verifies property $(I7)$.

Notes and Comments: The *duality* approach to most of the multiplicity results of this chapter is due to the author [G 1,2]. The Ljusternik and Schnirelmann type results are well known but our methods of proof are new and simple. They also cover the limiting cases for the boundary condition. For instance, the multiplicity results in the $\mathbb{Z}_2$-symmetric mountain pass theorems (Corollaries 7.22 and 7.23) are due to Rabinowitz [R 1] when the inequalities in their statements are strict. Our results show that they remain valid if we have equality. The same thing occur for Corollary 7.18 which is due to Fadell and Husseini [F-H].

The multiplicity in the degenerate cases that are not covered by the classical theorems mentioned above, is obtained from the new and surprising Corollary 7.11. In that result, the multiplicity does not come from another *inf-sup class* that gives the same critical level, but from a *dual sup-inf class*: An aesthetically appealing result that may have more theoretical ramifications in algebraic topology.

Theorem 7.24 is standard and is due to Rabinowitz [R 1], while the properties of the mod 2-genus are borrowed from Coffman [C].

8

THE PALAIS-SMALE CONDITION

AROUND A DUAL SET – EXAMPLES

In this chapter, we exhibit a few examples where the standard Palais-Smale condition is not satisfied, but the restricted ones (introduced in the previous sections) do hold. Not surprisingly, most of the examples below deal with equations involving the critical Sobolev exponent.

8.1 A positive solution for an elliptic boundary value problem involving the critical Sobolev exponent

Let Ω be a smoothly bounded domain in $\mathbb{R}^N$, $N > 2$, $2^* = \frac{2N}{N-2}$, $\lambda \in \mathbb{R}$ and consider the problem

$$\begin{cases} -\Delta u - \lambda u = u|u|^{2^*-2} & \text{in } \Omega, \\ u = 0 & \text{on } \partial\Omega. \end{cases} \quad (1)$$

Let λ_1 be the first eigenvalue of $-\Delta$ on $H_0^1(\Omega)$.

We now present the following result of Brezis-Nirenberg [B-N 1].

Theorem 8.1: *If $N \geq 4$, then for any $\lambda \in (0, \lambda_1)$, problem (1) has a strictly positive solution.*

Again, solving problem (1) is equivalent to finding critical points on $H_0^1(\Omega)$ of the energy functional

$$\varphi_\lambda(u) = \frac{1}{2} \int_\Omega (|\nabla u|^2 - \lambda|u|^2)\, dx - \frac{1}{2^*} \int_\Omega |u|^{2^*}\, dx. \quad (2)$$

However, this functional does not satisfy $(PS)_c$ for every level c. But an interesting new phenomenon emerges: the compactness is preserved in some energy range. More precisely, if we let S be the best constant appearing in the Sobolev inequality, i.e.,

$$S = \inf\{\|u\|_{H_0^1}^2;\ u \in H_0^1(\Omega), \|u\|_{2^*} = 1\} \tag{3}$$

then, the following holds:

Claim (i) : For each $\lambda > 0$, the functional φ_λ verifies $(PS)_c$ for every $c < \frac{1}{N}S^{N/2}$.

For the proof we shall need the following elementary properties of the norms in Hilbert and L^p-spaces : If $(u_m)_m$ is a sequence that converges weakly in $H_0^1(\Omega)$ and strongly in $L^2(\Omega)$ to a function $u \in H_0^1(\Omega)$, then

$$\|\nabla u_m\|_2^2 = \|\nabla(u_m - u)\|_2^2 + \|\nabla u\|_2^2 + o(1) \tag{4}$$

$$\|u_m\|_{2^*}^{2^*} = \|u_m - u\|_{2^*}^{2^*} + \|u\|_{2^*}^{2^*} + o(1) \tag{5}$$

$$\int_\Omega (u_m|u_m|^{2^*-2} - u|u|^{2^*-2})(u_m - u)\,dx = \int_\Omega |u_m - u|^{2^*}\,dx + o(1) \tag{6}$$

where $o(1) \to 0$ as $m \to \infty$.

Proof of Claim (i): Let $(u_m)_m$ in H_0^1 be such that

$$\varphi_\lambda(u_m) \to c \quad \text{and} \quad \varphi_\lambda'(u_m) \to 0.$$

To show the boundedness of (u_m) in H_0^1, note first that

$$\|u_m\|_{H_0^1}^2 = 2\varphi_\lambda(u_m) + \lambda\|u_m\|_2^2 + \frac{2}{2^*}\|u_m\|_{2^*}^{2^*}. \tag{7}$$

We also have

$$o(1)(1 + \|u_m\|_{H_0^1}) + \frac{2}{N}S^N \geq 2\varphi_\lambda(u_m) - \langle u_m, \varphi_\lambda'(u_m)\rangle$$
$$= \left(1 - \frac{2}{2^*}\right)\|u_m\|_{2^*}^{2^*}$$
$$\geq K\|u_m\|_2^{2^*}$$

where $K > 0$ and $o(1) \to 0$ as $m \to \infty$. Combining this estimate with (7), we get

$$\|u_m\|_{H_0^1}^2 \leq C + o(1)\|u_m\|_{H_0^1}$$

from which it follows that (u_m) is bounded, and hence we may assume that $u_m \to u$ weakly in $H_0^1(\Omega)$.

The Rellich-Kondrakov theorem then gives that $u_m \to u$ strongly in $L^p(\Omega)$ for all $p < 2^*$. In particular, for any $v \in C_0^\infty(\Omega)$

$$\langle v, \varphi_\lambda'(u_m)\rangle = \int_\Omega (\nabla u_m.\nabla v - \lambda u_m v - u_m|u_m|^{2^*-2}v)\,dx$$

which converges as $m \to \infty$ to

$$\int_\Omega (\nabla u.\nabla v - \lambda uv - u|u|^{2^*-2}v)\,dx = \langle v, \varphi_\lambda'(u)\rangle.$$

Hence $u \in H_0^1(\Omega)$ is a weak solution to (1). Choosing $v = u$ we have

$$0 = \langle u, \varphi_\lambda'(u) \rangle = \int_\Omega (|\nabla u|^2 - \lambda|u|^2 - |u|^{2^*}) \, dx \qquad (8)$$

and

$$\varphi_\lambda(u) = \left(\frac{1}{2} - \frac{1}{2^*}\right) \int_\Omega |u|^{2^*} \, dx = \frac{1}{N} \int_\Omega |u|^{2^*} \, dx \geq 0. \qquad (9)$$

By (4), (5) and (6) above, we have

$$\varphi_\lambda(u_m) = \varphi_\lambda(u) + \varphi_0(u_m - u) + o(1) \qquad (10)$$

and

$$\begin{aligned} o(1) &= \langle u_m - u, \varphi_\lambda'(u_m) \rangle \\ &= \langle u_m - u, \varphi_\lambda'(u_m) - \varphi_\lambda'(u) \rangle \qquad (11) \\ &= \|\nabla(u_m - u)\|_2^2 - \|u_m - u\|_{2^*}^{2^*} + o(1). \end{aligned}$$

In particular, we get from (11) that $\varphi_0(u_m - u) = \frac{1}{N}\|\nabla(u_m - u)\|_2^2 + o(1)$, while for large m, we have from (10) and (9) that

$$\begin{aligned} \varphi_0(u_m - u) &= \varphi_\lambda(u_m) - \varphi_\lambda(u) + o(1) \\ &\leq \varphi_\lambda(u_m) + o(1) \leq c' < \frac{1}{N}S^{N/2}. \end{aligned}$$

Therefore, for large m, we have

$$\|u_m - u\|_{H_0^1}^2 \leq c'' < S^{N/2}.$$

By Sobolev's inequality we finally get

$$\|u_m - u\|_{H_0^1}^2 (1 - S^{-2^*/2}\|u_m - u\|_{H_0^1}^{2^*-2}) \leq \|u_m - u\|^2 - \|u_m - u\|_{2^*}^{2^*} = o(1)$$

which shows that $u_m \to u$ strongly in $H_0^1(\Omega)$.

To set up the min-max principle, we consider the following *generalized homotopy stable family*

$$\mathcal{F}_1 = \{\gamma \in C^0([0,1]; H_0^1(\Omega)); \gamma(0) = 0, \gamma(1) \neq 0 \text{ and } \varphi_\lambda(\gamma(1)) \leq 0\}$$

whose boundary is $B = \{\varphi \leq 0\}$ and let

$$c_1 = c(\varphi_\lambda, \mathcal{F}_1) = \inf_{A \in \mathcal{F}_1} \sup_{x \in A} \varphi_\lambda(x).$$

Consider now the set

$$M = \{u \in H_0^1(\Omega) : u \neq 0 \text{ and } \langle \varphi_\lambda'(u), u \rangle = 0\}.$$

We now prove the following.

Claim (ii): If $\lambda < \lambda_1$, then M is dual to the family $\mathcal{F}_1$ and $\inf \varphi_\lambda(M) = c_1$.

Proof: First note that $B \cap M = \emptyset$ since, by (9) above we have for every $u \in M$,

$$\varphi_\lambda(u) = \left(\frac{1}{2} - \frac{1}{2^*}\right) \int_\Omega |u|^{2^*} \, dx = \frac{1}{N} \int_\Omega |u|^{2^*} \, dx > 0.$$

To prove the intersection property, fix $\gamma \in \mathcal{F}_1$ joining 0 to v where $v \neq 0$ and $\varphi(v) \leq 0$. Note that since $\lambda < \lambda_1$, we have that $\langle \varphi'_\lambda(\gamma(t)), \gamma(t) \rangle > 0$ for t close to 0. On the other hand, since $v \neq 0$, we have that $\langle \varphi'_\lambda(v), v \rangle < 2\varphi_\lambda(v) \leq 0$. It follows from the intermediate value theorem that there exists t_0 such that $\gamma(t_0) \in M$. This proves the duality and consequently,

$$c_1 \geq \inf\{\varphi_\lambda(u) : u \in H_0^1(\Omega), u \in M\}.$$

To prove the reverse inequality, set for each $u \in H_0^1$, $u \neq 0$,

$$S_\lambda(u, \Omega) = \frac{\|\nabla u\|_2^2 - \lambda\|u\|_2^2}{\|u\|_{2^*}^2}, \tag{12}$$

and consider the straight path joining 0 to u. Since $\varphi_\lambda(tu) \to -\infty$ as $t \to \infty$, we have

$$c_1 \leq \sup_{0 \leq t < \infty} \varphi_\lambda(tu)$$

$$= \sup_{0 \leq t < \infty} \left(\frac{t^2}{2}(\|\nabla u\|_2^2 - \lambda\|u\|_2^2) - \frac{t^{2^*}}{2^*}\|u\|_{2^*}^{2^*} \right)$$

$$= \frac{1}{N} S_\lambda(u, \Omega)^{N/2}.$$

On the other hand, if $u \in M$, we have

$$\frac{1}{N} S_\lambda(u, \Omega)^{N/2} = \frac{1}{N}[(\frac{\|\nabla u\|_2^2 - \lambda\|u\|_2^2}{\|u\|_{2^*}^2}) \cdot \|u\|_{2^*}^{2^*-2}]^{N/2}$$

$$= \frac{1}{N}\|u\|_{2^*}^{(2^*-2)N/2}$$

$$= \frac{1}{N}\|u\|_{2^*}^{2^*}$$

$$= \varphi_\lambda(u).$$

In other words, we have proved that

$$c_1 = \inf \varphi_\lambda(M) = \frac{1}{N} S_\lambda^{N/2}(\Omega) \tag{13}$$

where

$$S_\lambda(\Omega) = \inf\{S_\lambda(u, \Omega); u \neq 0, u \in H_0^1(\Omega)\}. \tag{14}$$

To complete the proof of the Theorem, it remains to show the following

Claim (iii): If $\lambda > 0$ and $N \geq 4$, then

$$c_1 < \frac{1}{N} S^{N/2}. \tag{15}$$

Proof: In view of (13), it is enough to prove that

$$S_\lambda(\Omega) < S.$$

To do that assume without loss of generality that $0 \in \Omega$, and consider an extremal function u_ε^*, as well as a cutoff function η equal to 1 in a neighborhood of 0. Set $u_\varepsilon = \eta u_\varepsilon^*$ and use the properties of such functions derived in Appendix A, to compute if $N \geq 5$,

$$S_\lambda(u_\varepsilon) \leq \frac{S^{N/2} - K\lambda\varepsilon + O(\varepsilon^{\frac{N-2}{2}})}{(S^{N/2} + O(\varepsilon^{n/2}))^{2/2^*}} \leq S - K\lambda\varepsilon + O(\varepsilon^{\frac{N-2}{2}}) < S$$

provided ε is sufficiently small. Similarly, if $N = 4$, we have

$$S_\lambda(u_\varepsilon) \leq S - K\lambda\varepsilon|\ln(\varepsilon)| + O(\varepsilon) < S.$$

8.2 A sign-changing solution for an elliptic boundary value problem involving the critical Sobolev exponent

We now deal with the problem of finding other solutions for the problem considered in the last section. That is

$$\begin{cases} -\Delta u - \lambda u = u|u|^{2^*-2} & \text{in } \Omega, \\ u = 0 & \text{on } \partial\Omega. \end{cases} \quad (1)$$

for $\lambda \in (0, \lambda_1)$ with λ_1 being the first eigenvalue of $-\Delta$ in $H_0^1(\Omega)$.

In Theorem 8.1, we found one solution at a level $c_1 < \frac{1}{N}S^{N/2}$, which is characterized by

$$c_1 = \min\{\varphi_\lambda(u) : u \in H_0^1(\Omega), u \in M\} \quad (2)$$

where

$$\varphi_\lambda(u) = \frac{1}{2}\int_\Omega (|\nabla u|^2 - \lambda|u|^2)\, dx - \frac{1}{2^*}\int_\Omega |u|^{2^*}\, dx \quad (3)$$

and

$$M = \{u \in H_0^1(\Omega); u \neq 0, \langle \varphi_\lambda'(u), u \rangle = 0\}. \quad (4)$$

On the other hand, the Palais-Smale condition for φ_λ may fail at level $\frac{1}{N}S^{N/2}$ and beyond (see Struwe [St1] for a discussion). Now, in order to prove the existence of more than one solution, it is reasonable to try to increase the threshold of non-compactness beyond this limit. We shall be able to do that by replacing the classical condition $(PS)_c$ by $(PS)_{F,c}$ for an appropriate subset F of M. The threshold of non-compactness can be then pushed to $c_1 + \frac{1}{N}S^{N/2}$ provided we stay close to that set F. We shall present two different examples of such a set, in order to prove the following result, first established by Cerami, Solimini and Struwe [C-S-S].

Theorem 8.2: *Suppose $N \geq 6$ and $\lambda \in (0, \lambda_1)$, then there exist at least two pairs of non-trivial solutions for (1).*

First method

Consider the set

$$F = \{u = u^+ - u^- \in H_0^1(\Omega); u^+ \in M, u^- \in M\}.$$

We have the following:

Claim (i): For $N \geq 3$, φ_λ verifies $(PS)_{F,c}$ for any $c < c_1 + \frac{1}{N}S^{N/2}$.

Proof of Claim (i): Let $(u_m)_m$ in H_0^1 be such that

$$\varphi_\lambda(u_m) \to c, \quad \varphi_\lambda'(u_m) \to 0 \tag{5}$$

and

$$\text{dist}(u_m, F) \to 0. \tag{6}$$

By (6) and (2), we have

$$\lim_{m \to \infty} \varphi_\lambda(u_m^\pm) \geq c_1. \tag{7}$$

Proceed as in the proof of Claim (i) of Theorem 8.1, to get that $u_m \to u$ weakly in $H_0^1(\Omega)$, where u is a solution of (1). Hence also $u_m^\pm \to u^\pm$ weakly in $H_0^1(\Omega)$.

Moreover, as in (10) in the proof of Theorem 8.1, we have

$$
\begin{aligned}
o(1) &= \langle \varphi_\lambda'(u_m) - \varphi_\lambda'(u), u_m^\pm - u^\pm \rangle \\
&= \|u_m - u^\pm\|^2 - \|u_m^\pm - u^\pm\|_{2^*}^{2^*} + o(1) \\
&\geq \|u_m^\pm - u^\pm\|^2 (1 - S^{-2^*/2}\|u_m^\pm - u^\pm\|^{2^*-2}) + o(1).
\end{aligned}
$$

Suppose now u_m^+ does not converge to u^+ strongly, we then obtain

$$
\begin{aligned}
\varphi_\lambda(u_m^+) &= \varphi_\lambda(u_m^+ - u^+) + \varphi_\lambda(u^+) + o(1) \\
&\geq \frac{1}{N}\|u_m^+ - u^+\|^2 + \varphi_\lambda(u^+) + o(1) \\
&\geq \frac{1}{N}S^{N/2} + o(1).
\end{aligned}
$$

This combined with (7), gives that

$$c_1 + \frac{1}{N}S^{N/2} > \lim_{m \to \infty} \varphi_\lambda(u_m) = \lim_{m \to \infty}[\varphi_\lambda(u_m^-) + \varphi_\lambda(u_m^+)] \geq c_1 + \frac{1}{N}S^{N/2},$$

a contradiction. We get the same contradiction if on the other hand, we assume that $u_m^- \not\to u^-$ strongly. The claim is proved.

To set up the min-max principle, define the function

$$f_\lambda(u) = \frac{\|u\|_{2^*}^{2^*}}{\|u\|^2 - \lambda\|u\|_2^2} \tag{8}$$

if $u \neq 0$ and $f_\lambda(0) = 0$. Note that

$$M = \{u \in H_0^1(\Omega); f_\lambda(u) = 1\}$$

and

$$F = \{u \in H_0^1(\Omega); f_\lambda(u^+) = f_\lambda(u^-) = 1\}. \tag{9}$$

Let P denote the cone of non-negative functions in $H_0^1(\Omega)$ and let $Q = [0,1] \times [0,1]$. Consider the class Γ_2 of all maps $\sigma \in C(Q, H_0^1(\Omega))$ satisfying for all $s \in [0,1]$,

(a) $\sigma(s,0) = 0$,

(b) $\sigma(0,s) \in P$,

(c) $\sigma(1,s) \in -P$,

(d) $f_\lambda(\sigma(s,1)) \geq 2$.

The class $\mathcal{F}_2 = \{\sigma(Q); \sigma \in \Gamma_2\}$ is stable under those $\mathbb{Z}_2$-equivariant homotopies that fix the boundary $B = \{0\} \cup \{f_\lambda \geq 2\}$, and which also deform the positive cone P into itself. By Remark 4.10 and since $(I - \varphi'_\lambda)(P) \subset P$, we will be able to apply Theorem 4.1 to the class $\mathcal{F}_2$ provided we show the following.

Claim (ii): F verifies conditions $(F1)$ and $(F2)$ with respect to the class $\mathcal{F}_2$.

Proof of Claim (ii): Note first that $F \cap B = \emptyset$. Indeed, $\inf \varphi_\lambda(F) \geq \inf \varphi_\lambda(M) = c_1 > 0$, while $\sup \varphi_\lambda(B) \leq 0$.

On the other hand, we have for all $\sigma \in \Gamma_2$

$$f_\lambda(\sigma(x)^+) - f_\lambda(\sigma(x)^-) \begin{cases} \geq 0 & \text{on } \{0\} \times [0,1] \\ \leq 0 & \text{on } \{1\} \times [0,1] \end{cases}$$

$$f_\lambda(\sigma(x)^+) + f_\lambda(\sigma(x)^-) - 2 \begin{cases} \geq 0 & \text{on } [0,1] \times \{1\} \\ < 0 & \text{on } [0,1] \times \{0\} \end{cases}$$

so from Miranda's theorem [Mi] we deduce that there exists $x_0 \in Q$ such that

$$f_\lambda(\sigma(x_0)^+) - f_\lambda(\sigma(x_0)^-) = 0 = f_\lambda(\sigma(x_0)^+) + f_\lambda(\sigma(x_0)^-) - 2$$

which means that $u_0 = \sigma(x_0) \in F$. Hence F is dual to $\mathcal{F}_2$ and in particular,

$$c_2 := c(\varphi_\lambda; \mathcal{F}_2) \geq \inf_{u \in F} \varphi_\lambda(u).$$

On the other hand, for $u_0 \in F$, let σ_0 be a map of Γ_2 such that

$$\sigma_0(Q) \subset \{\alpha u_0^+ + \beta u_0^- : \alpha, -\beta \in \mathbb{R}^+ \cup \{0\}\} \qquad (10)$$

As in Claim (ii) of Theorem 8.1, one can show that

$$\max_{\sigma_0(Q)} \varphi_\lambda = \varphi_\lambda(u_0). \qquad (11)$$

and condition $(F2)$ is satisfied. This completes the proof of Claim (ii).

To prove Theorem 8.2, it remains to show the following.

Claim (iii): If $N \geq 6$, then $c_1 \leq c_2 < c_1 + \frac{1}{N} S^{N/2}$.

Proof of Claim (iii): Let u_1 be the solution found in Theorem 8.1. That is

$$c_1 = \varphi_\lambda(u_1) = \inf\{\varphi_\lambda(u); f_\lambda(u) = 1\}.$$

Let u_{ε,x_0} be the extremal function corresponding to ε and $x_0 \in \Omega$. It suffices to show that

$$\sup_{\alpha,\beta \in \mathbb{R}} \varphi_\lambda(\alpha u_1 + \beta v_\varepsilon) < c_1 + S^{N/2}/N,$$

where $v_\varepsilon = \eta \cdot u_{\varepsilon,x_0}$, $\eta \in C_0^\infty(B_\rho(x_0))$, η identically 1 on $B_{\rho/2}(x_0)$, and $x_0 \in \Omega, \rho \in \mathbb{R}^+$ to be specified later.

From Appendix A, we know the following estimates:

$$\|v_\varepsilon\|^2 = S^{N/2} + O(\varepsilon^{(N-2)/2}),$$
$$\|v_\varepsilon\|_{2^*}^{2^*} = S^{N/2} + O(\varepsilon^{N/2}),$$
$$\|v_\varepsilon\|_2^2 = K_1\varepsilon + O(\varepsilon^{(N-2)/2}), K_1 > 0, \qquad (12)$$
$$\|v_\varepsilon\|_1 \le K_2\varepsilon^{(N-2)/4},$$
$$\|v_\varepsilon\|_{2^*-1}^{2^*-1} \le K_3\varepsilon^{(N-2)/4}.$$

First note that we can restrict our attention to the values where

$$\|\alpha u_1 + \beta v_\varepsilon\|_{2^*}^{2^*} < \bar{K} \text{ for some suitably large number } \bar{K}.$$

Since

$$\|\alpha u_1 + \beta v_\varepsilon\|_{2^*}^{2^*} \ge \|\beta v_\varepsilon\|_{2^*}^{2^*} + K_4\|\alpha u_1\|_{2^*}^{2^*} - K_5\beta^{2^*}\varepsilon^{N/2}$$

we get by using the second equation in (12) that if ε is small enough, then

$$\bar{K} \ge |\beta|^{2^*}[S^{N/2} + O(\varepsilon^{N/2})] + K_1|\alpha|^{2^*}\|u_1\|_{2^*}^{2^*}$$

in such a way that α and β are then bounded.

Using (12) repeatedly, we deduce

$$
\begin{aligned}
\varphi_\lambda(\alpha u_1 + \beta v_\varepsilon) \le\ & \frac{\alpha^2}{2}(\|\nabla u_1\|_2^2 - \lambda\|u_1\|_2^2) + \frac{\beta^2}{2}(\|\nabla v_\varepsilon\|_2^2 - \lambda\|v_\varepsilon\|_2^2) \\
& - K_6\{\|\beta v_\varepsilon\|_1\|\Delta(\alpha u_1)\|_{L^\infty(B_\rho(x_0))} \\
& + \|\alpha u_1\|_{L^\infty(B_\rho(x_0))}\|\beta v_\varepsilon\|_1\} \\
& - \frac{\alpha^{2^*}}{2^*}\|u_1\|_{2^*}^{2^*} - \frac{\beta^{2^*}}{2^*}\|v_\varepsilon\|_{2^*}^{2^*} \\
& + K_7\{\|\beta v_\varepsilon\|_{2^*-1}^{2^*-1}\|\alpha u_1\|_{L^\infty(B_\rho(x_0))} \\
& + \|\beta v_\varepsilon\|_1\|\alpha u_1\|_{L^\infty(B_\rho(x_0))}^{2^*-1}\} \\
\le\ & \frac{1}{N}S_\lambda(u_1,\Omega)^{N/2} + \frac{1}{N}S_\lambda(v_\varepsilon,\Omega)^{N/2} \\
& - K_6\{\|\beta v_\varepsilon\|_1\|\Delta(\alpha u_1)\|_{L^\infty(B_\rho(x_0))} \\
& + \|\alpha u_1\|_{L^\infty(B_\rho(x_0))}\|\beta v_\varepsilon\|_1\} \\
& + K_7\{\|\beta v_\varepsilon\|_{2^*-1}^{2^*-1}\|\alpha u_1\|_{L^\infty(B_\rho(x_0))} \\
& + \|\beta v_\varepsilon\|_1\|\alpha u_1\|_{L^\infty(B_\rho(x_0))}^{2^*-1}
\end{aligned}
$$

$$\leq c_1 + \frac{1}{N}S^{N/2} - K_9\varepsilon + K_8\|u_1\|_{L^\infty(B_\rho(x_0))}\varepsilon^{(N-2)/4}.$$

This implies Claim (iii) when $N \geq 7$. If $N = 6$ we deduce the desired result observing that the point x_0 and the ball $B_\rho(x_0)$ can be chosen near the boundary of Ω in such a way that $\|u\|_{L^\infty(B_\rho(x_0))}$ is so small that $K_8\|u_1\|_{L^\infty(B_\rho(x_0))} - K_9 < 0$.

To finish the proof of Theorem 8.2, it is now enough to apply Theorem 4.1, to get a critical point on F. This is clearly a sign-changing solution for (1). Hence it is different from the first one and we are done.

Second method

We now present another example of a dual set that will improve the level of compactness. The interesting feature in this candidate is that it imposes second order conditions on the almost critical sequences that are close to it, which help in proving their convergence. Define

$$\psi_\lambda(u) = \|u\|^2 - \lambda\|u\|_2^2. \tag{13}$$

We need the following.

Lemma 8.3: For every $u \neq 0$, $u \in L^{2^*}$, there exists a unique $v = v(u) \in H_0^1(\Omega)$ such that
(a) $\int_\Omega |u|^{2^*-2}v^2 = 1$ and $v \geq 0$.
(b) $\psi_\lambda(v) = \inf\{\psi_\lambda(w); w \in H, \int_\Omega |u|^{2^*-2}w^2 = 1\} = \mu_1(u)$.
   Furthermore, the map $u \to v(u)$ is continuous from L^{2^*} to H_0^1 for every $u \neq 0$.

Proof: It is clear that ψ_λ is weakly lower semi-continuous, coercive and bounded below on H_0^1. Moreover, the constraint set $C = \{w \in H : \int_\Omega |u|^{2^*-2}w^2 \, dx = 1\}$ is weakly closed in $H_0^1(\Omega)$, therefore the infimum in (b) is achieved. Any function where such infimum is achieved cannot change sign in Ω. This gives the uniqueness of $v(u)$ and therefore its continuity for non-zero u.
   Note that the couple $(v(u), \mu_1(u))$ correspond to the first eigenpair of the weighted eigenvalue problem

$$-(\Delta + \lambda)v = \mu|u|^{2^*-2}v \quad \text{in} \quad \Omega,$$
$$v = 0 \quad \text{on} \quad \partial\Omega. \tag{14}$$

If u_1 is the solution of (1) obtained in Theorem 8.1, then clearly $(v(u_1), \mu_1(u_1)) = (u_1, 1)$.
   We consider now the set

$$F' = M \cap \{u \in H_0^1; \int_\Omega |u|^{2^*-2}uv(u) = 0\}.$$

Claim (i'): For $N \geq 3$, φ_λ verifies $(PS)_{F',c}$ for any $c < c_1 + \frac{1}{N}S^{N/2}$.

Proof of Claim (i′): Let $u_k \in H_0^1$ be such that

$$\lim_k \varphi_\lambda(u_k) = c, \quad \lim_k \varphi_\lambda'(u_k) = 0 \tag{15}$$

and

$$\lim_n \text{dist}(u_k, F') = 0. \tag{16}$$

As before, (15) implies that $(u_k)_k$ is bounded and hence we can suppose it weakly convergent to $u \in H_0^1$. On the other hand, (16) implies that

$$\int_\Omega |u_k|^{2^*-2} u_k v_k = o(1) \text{ and } \int_\Omega |u_k|^{2^*-2} v_k^2 \to 1 \tag{17}$$

where $(v_k)_k$ can be supposed to be the normalized (first) eigenfunctions satisfying

$$v_k \geq 0 \text{ and } -\Delta v_k - \lambda v_k = \mu_{1,k} |u_k|^{2^*-2} v_k \text{ for some } 0 < \mu_{1,k} \leq 1. \tag{18}$$

Write now $u_k = w_k + u$ with $w_k \to 0$ weakly. Suppose $(u_k)_k$ is not norm converging. By the Blow-up technique described in Appendix C, there exist $R_k \to +\infty$, $x_k \in \Omega$ and a solution U of

$$-\Delta U = U^{2^*-1} \text{ on } \mathbf{R}^N \tag{19}$$

such that

$$\|\nabla(w_k - R_k^{\frac{N-2}{2}} U(R_k(x - x_k)))\|_2 \to 0 \tag{20}$$

and

$$\varphi_\lambda(u_k) \to \varphi_\lambda(u) + \frac{1}{N} S^{N/2}. \tag{21}$$

Since $\varphi_\lambda(u_k) \to c$ and $c < c_1 + \frac{1}{N} S^{N/2}$, it follows that $\varphi_\lambda(u) < c_1$. But u is a solution of (1), hence it belongs to M provided it is not zero. But $\inf \varphi_\lambda(M) = c_1$, hence $u = 0$ and we can therefore suppose that $u_k \to 0$ weakly and $\|\nabla(u_k - R_k^{\frac{N-2}{2}} U(R_k(x - x_k)))\|_2 \to 0$.

Let now $u_k^*(x) = (\frac{1}{R_k})^{\frac{N-2}{2}} u_k(\frac{x}{R_k} + x_k)$ for x in

$$\Omega_k := \{x \in \mathbf{R}^N; x_k + \frac{x}{R_k} \in \Omega\}.$$

We get from (20) that $u_k^* \to U$ strongly. On the other hand, if we consider $v_k^*(x) = (\frac{1}{R_k})^{\frac{N-2}{2}} v_k(\frac{x}{R_k} + x_k)$ for $x \in \Omega_k$, we get from (18) that

$$-\Delta v_k^* = \mu_{1,k} |u_k^*|^{2^*-2} v_k^* + g_k^* \tag{22}$$

with $g_k^* \to 0$ in H^{-1}.

Let now v^* be a weak limit of $(v_k^*)_k$. Combining (22) and (17) with the fact that $(u_k^*)_k$ converges strongly to U yields that for some $\mu_{1,\infty} \in [0,1]$, we have

$$-\Delta v^* = \mu_{1,\infty} |U|^{2^*-2} v^* \tag{23}$$

while

$$\int_\Omega |U|^{2^*-2}(v^*)^2 = 1, \quad \text{and} \quad \int_\Omega |U|^{2^*-2}Uv^* = 0. \tag{24}$$

This clearly implies that $v^* \neq 0$ and from (18), we also have that $v^* \geq 0$. On the other hand, by (19), we have $-\Delta U = |U|^{2^*-2}U$. Moreover, (19), (20) and (21) yield that the energy of U is strictly less than $\frac{2}{N}S^{N/2}$ which implies that U is strictly positive. It follows that U and v^* are 2 orthogonal eigenfunctions that correspond to the same simple eigenvalue. This is a contradiction from which follows that $(u_k)_k$ is norm convergent and the claim is proved.

To apply the min-max principle, we let $S_\rho = \{u \in H; \|u\| = \rho\}$ and

$$\mathcal{H} = \{h : H \to H \text{ odd homeomorphism }\}.$$

As in Chapter 7, let $\gamma_{\mathbf{Z}_2}$ denotes the Krasnoselski genus, and consider the class

$$\mathcal{F}_2' = \{A; \ A \text{ compact symmetric with } \gamma_{\mathbf{Z}_2}(h(A) \cap S_\rho) \geq 2, \ \forall h \in \mathcal{H}\}.$$

In the terminology of Chapter 7, this is the class associated with the intersection index $\gamma_{\mathbf{Z}_2}^{S_\rho}$.

Set $c_2' = \inf_{A \in \mathcal{F}_2'} \sup_A \varphi_\lambda$. We shall prove the following.

Claim (ii'): F' is dual to the class $\mathcal{F}_2'$ and $\inf \varphi_\lambda(F') = c_2'$.

Proof of Claim (ii'): First notice that the map

$$u \to \left(\frac{\|u\|^2 - \lambda\|u\|_2^2}{\|u\|_{2^*}^{2^*}}\right)^{1/(2^*-2)} u$$

defines an odd homeomorphism between any sphere S_ρ and M so that $\gamma_{\mathbf{Z}_2}(A \cap M) \geq 2$ for every $A \in \mathcal{F}_2'$. On the other hand, the map $h : A \cap M \to \mathbb{R}$ given by

$$h(u) = \int_\Omega |u|^{2^*-2}uv(u)$$

defines an odd and continuous map. Since $\gamma_{\mathbf{Z}_2}(A \cap M) \geq 2$, we get that $0 \in h(A \cap M)$ which means that $A \cap F' \neq \emptyset$ and F' is dual to $\mathcal{F}_2'$. In particular, $c_2' \geq \inf_{u \in F'} \varphi_\lambda(u)$.

To prove the reverse inequality, take $u \in F'$ and let $v(u)$ be such that $\int_\Omega |u|^{2^*-2}uv(u) = 0$. Let $w(u)$ be a minimizer for the problem:

$$\mu_2 = \inf\{\psi_\lambda(w); w \in H, \int_\Omega |u|^{2^*-2}v(u)w = 0, \int_\Omega |u|^{2^*-2}w^2 = 1\}.$$

Since $u \in F'$, we obtain

$$\mu_2 \leq \frac{\|u\|^2 - \lambda\|u\|_2^2}{\|u\|_{2^*}^{2^*}} = 1$$

Let $A =$ span $\{v(u), w(u)\}$. Clearly, $A \in \mathcal{F}_2'$, and for all non-zero elements $w \in A$ we have

$$1 \geq \mu_2 \geq \frac{\|w\|^2 - \lambda\|w\|_2^2}{\int_\Omega |u|^{2^*-2}w^2}.$$

Take $w_0 \in A$ so that $\varphi_\lambda(w_0) = \max_A \varphi_\lambda \geq c_2'$. Since A is a linear space, we derive that $w_0 \in M$. Furthermore,

$$1 \geq \frac{\|w_0\|^2 - \lambda\|w_0\|_2^2}{\int_\Omega |u|^{2^*-2}w_0^2} \geq \left(\frac{\|w_0\|_{2^*}}{\|u\|_{2^*}}\right)^{2^*-2}.$$

Consequently,

$$\varphi_\lambda(u) = \left(\frac{1}{2} - \frac{1}{2^*}\right)\|u\|_{2^*}^{2^*} \geq \left(\frac{1}{2} - \frac{1}{2^*}\right)\|w_0\|_{2^*}^{2^*} = \varphi_\lambda(w_0) \geq c_2'.$$

This completes the proof of Claim (2′).

Claim (iii′): $c_1 \leq c_2' < c_1 + \frac{1}{N}S^{N/2}$.

Proof of Claim (iii′): As in Claim (iii), let u_1 be the (first) solution found in the last section. Let u_{ε,x_0} be the extremal function corresponding to ε and $x_0 \in \Omega$. Let $v_\varepsilon = \eta \cdot u_{\varepsilon,x_0}$, $\eta \in C_0^\infty(B_\rho(x_0))$, η identically 1 on $B_{\rho/2}(x_0)$, and $x_0 \in \Omega$, $\rho \in \mathbb{R}^+$ to be specified later. Let $A_\varepsilon = $ span$\{u_1, v_\varepsilon\}$. It is clearly in the class $\mathcal{F}_2'$. The same computation as in Claim (iii) above gives that for ε small enough,

$$c_2' \leq \sup \varphi_\lambda(A_\varepsilon) < c_1 + \frac{1}{N}S^{N/2}.$$

To finish the proof of the theorem, it is now enough to apply Theorem 4.1, to get a critical point on F'. This solution for (1) is such that $\int_\Omega |u|^{2^*-2}uv(u) = 0$. It is clearly distinct from the first one.

8.3 A second solution for a non-homogeneous elliptic equation involving the critical Sobolev exponent

We reconsider the Dirichlet problem studied in section 1.3, that is

$$\begin{cases} -\Delta u = |u|^{p-2}u + f & \text{on } \Omega \\ u = 0 & \text{on } \partial\Omega. \end{cases} \tag{1}$$

where $\Omega \subset \mathbb{R}^N$ $(N \geq 3)$, is a bounded set, $H = H_0^1(\Omega)$, $f \in H^{-1}$ and $p = \frac{2N}{N-2} = 2^*$.

Recall that we are looking for the critical points of the functional

$$\varphi(u) = \frac{1}{2}\int_\Omega |\nabla u|^2 - \frac{1}{p}\int_\Omega |u|^p - \int_\Omega fu, \quad u \in H.$$

We will present a result of Tarantello [T 1] about the existence of two solutions under suitable conditions on f. The first solution was obtained in section 1.3, by a minimization procedure via Ekeland's principle on the manifold $M = \{u \in H : u \neq 0, \langle\varphi'(u), u\rangle = 0\}$. We shall now show that this solution is a local minimum for φ on H. We will then find another solution that will correspond to a saddle point of mountain-pass type for φ. To do that, we consider the following subsets of M.

$$M^+ = \{u \in M : \|\nabla u\|_2^2 - (p-1)\|u\|_p^p > 0\},$$
$$M^0 = \{u \in M : \|\nabla u\|_2^2 - (p-1)\|u\|_p^p = 0\},$$
$$M^- = \{u \in M : \|\nabla u\|_2^2 - (p-1)\|u\|_p^p < 0\}.$$

Theorem 8.4: *Suppose $f \neq 0$ and satisfies the following condition:*

$$\inf_{\|u\|_p=1} \left(K_N \|\nabla u\|_2^{\frac{N+2}{2}} - \int_\Omega fu\right) =: \mu_0 > 0 \tag{2}$$

where $K_N = \frac{4}{N-2}\left(\frac{N-2}{N+2}\right)^{\frac{N+2}{4}}$. Then

(i) *$c_1 := \inf \varphi(M)$ is finite and is achieved at a point $u_1 \in M^+$ which is a local minimum for φ on H.*

(ii) *$c_1 < c_2 := \inf \varphi(M^-)$ and c_2 is achieved at a point $u_2 \in M^-$ which is also a critical point for φ on H.*

(iii) *There exists a saddle point u_3 for φ at a level $c_3 \geq c_2$. Moreover, if $c_3 = c_2$, then M^- contains at that level, a saddle point of mountain pass type for φ.*

Recall first from Theorem 1.12, that $M^0 = \emptyset$ provided M verifies assumption (2). We shall need the following lemma, but first let us write

$$t_{\max}(u) := \left[\frac{\|\nabla u\|_2^2}{(p-1)\|u\|_p^p}\right]^{\frac{1}{p-2}}$$

for every $u \in H$, $u \neq 0$.

Lemma 8.5: *Let $f \neq 0$ satisfy (2). For every $u \in H, u \neq 0$ there exists a unique $t^+ = t^+(u) > t_{\max}(u) > 0$ such that $t^+u \in M^-$ and $\varphi(t^+u) = \max_{t \geq t_{\max}} \varphi(tu)$.*

Moreover, if $\int_\Omega fu > 0$, then there exists a unique $t^- = t^-(u)$, with $0 < t^-(u) < t_{\max}$ such that $t^-u \in M^+$ and $\varphi(t^-u) \leq \varphi(tu)$ for all $t \in [0, t^+]$.

Proof: Set $\ell(t) = t\|\nabla u\|_2^2 - t^{p-1}\|u\|_p^p$. It is clear that ℓ is concave and achieves its maximum at $t_{\max}$. Also

$$\ell(t_{\max}) = \left(\frac{1}{p-1}\right)^{\frac{p-1}{p-2}} (p-2)\frac{\|\nabla u\|_2^{2(p-1)/(p-2)}}{\|u\|_p^{p/(p-2)}} = K_N\frac{\|\nabla u\|_2^{(N+2)/2}}{\|u\|_p^{N/2}}.$$

Therefore, if $\int_\Omega fu \leq 0$, then there exists a unique $t^+ > t_{\max}$ such that $\ell(t^+) = \int_\Omega fu$ and $\ell'(t^+) < 0$. Equivalently $t^+u \in M^-$ and $\varphi(t^+u) \geq \varphi(tu)$ for all $t \geq t_{\max}$.

In case $\int_\Omega fu > 0$, we have, by assumption (2), that

$$0 < \int_\Omega fu < K_N \frac{\|\nabla u\|_2^{(N+2)/2}}{\|u\|_p^{N/2}} = \ell(t_{\max}).$$

Consequently, we have unique $0 < t^- < t_{\max} < t^+$ such that

$$\ell(t^+) = \int_\Omega fu = \ell(t^-) \quad \text{and} \quad \ell'(t^-) > 0 > \ell'(t^+).$$

Equivalently $t^+u \in M^-$ and $t^-u \in M^+$. Also $\varphi(t^+u) \geq \varphi(tu)$ for all $t \geq t^-$ and $\varphi(t^-u) \leq \varphi(tu)$ for all $t \in [0, t^+]$.

Proof of Theorem 8.4.(i): We have seen in Theorem 1.12 that $c_1 = \inf \varphi(M)$ is finite and is attained at a point $u_1 \in M$. Moreover, $c_1 < 0$ and hence $\int_\Omega fu_1 > 0$. Lemma 8.5 gives then that $u_1 \in M^+$ and

$$t^-(u_1) = 1 < \left(\frac{\|\nabla u_1\|_2^2}{(p-1)\|u_1\|_p^p} \right)^{1/(p-2)}.$$

To prove that u_1 is a local minimum for φ on H, let $\varepsilon > 0$ be sufficiently small so that for $\|w\| < \varepsilon$,

$$1 < \left(\frac{\|\nabla(u_1 - w)\|_2^2}{(p-1)\|u_1 - w\|_p^p} \right)^{1/(p-2)}. \tag{3}$$

As in the proof of Corollary 1.10, there exists a continuous function $t(w) := t_{u_1}(w) > 0$ defined on the ball $B_\varepsilon(0)$ such that $t(0) = 1$ and $t(w)(u_1 - w) \in M$ for every $w \in B_\varepsilon(0)$. By the continuity of t, we can always assume that for every $w \in B_\varepsilon(0)$,

$$t(w) < \left(\frac{\|\nabla(u_1 - w)\|_2^2}{(p-1)\|u_1 - w\|_p^p} \right)^{1/(p-2)}.$$

Hence, $t(w)(u_1 - w) \in M^+$ and for $0 < s < \left(\frac{\|\nabla(u_1 - w)\|_2^2}{(p-1)\|u_1 - w\|_p^p} \right)^{1/(p-2)}$ we have,

$$\varphi(s(u_1 - w)) \geq \varphi(t(w)(u_1 - w)) \geq \varphi(u_1).$$

From (3) we can take $s = 1$ and conclude that for every $w \in B_\varepsilon(0)$, $\varphi(u_1 - w) \geq \varphi(w)$ and hence u_1 is a local minimum.

Note that if $f \geq 0$, we get, by taking $t_1 = t^-(|u_1|) > 0$ with $t_1|u_1| \in M^+$, that necessarily $t_1 \geq 1$, and hence $\varphi(t_1|u_1|) \leq \varphi(|u_1|) \leq \varphi(u_1)$, so that we can take $u_1 \geq 0$.

In order to find another critical point, we can investigate the minimization problem

$$c_2 = \inf \varphi(M^-).$$

The same proof as in Theorem 1.12, shows that M^- verifies the hypothesis of corollary 1.10, and therefore there exists a sequence $(v_n)_n$ in M^-

such that $\varphi(v_n) \to c_1$ and $\varphi'(v_n) \to 0$. To ensure the convergence of such sequences, we need the following *restricted* Palais-Smale condition.

Claim (i): The functional φ verifies $(PS)_c$ provided $c < c_1 + \frac{1}{N}S^{N/2}$.

Proof: Assume $(u_n)_n$ is a sequence in H such that $\varphi(u_n) \to c$ and $\varphi'(u_n) \to 0$. As in the previous two examples, we can assume that the sequence $(u_n)_n$ is uniformly bounded and is weakly convergent in H_0^1 to u. Moreover, u is a solution in $H_0^1(\Omega)$ for (1). In particular $u \neq 0, u \in M$ and $\varphi(u) \geq c_1$. Write $u_n = u + v_n$ with $v_n \to 0$ weakly in H. As in (5) in the proof of Theorem 8.1, we have

$$\|u_n\|_p^p = \|u + v_n\|_p^p = \|u\|_p^p + \|v_n\|_p^p + o(1).$$

Hence, for n large, we conclude

$$c_1 + \frac{1}{N}S^{N/2} > \varphi(u + v_n)$$

$$= \varphi(u) + \frac{1}{2}\|\nabla v_n\|_2^2 - \frac{1}{p}\|v_n\|_p^p + o(1)$$

$$\geq c_1 + \frac{1}{2}\|\nabla v_n\|_2^2 - \frac{1}{p}\|v_n\|_p^p + o(1)$$

which gives

$$\frac{1}{2}\|\nabla v_n\|_2^2 - \frac{1}{p}\|v_n\|_p^p < \frac{1}{N}S^{N/2} + o(1). \qquad (4)$$

On the other hand,

$$o(1) = \langle \varphi'(u_n), u_n \rangle$$

$$= \|\nabla u\|^2 - \|u\|_p^p - \int_\Omega fu + \|\nabla v_n\|_2^2 - \|v_n\|_p^p + o(1)$$

$$= \langle \varphi'(u), u \rangle + \|\nabla v_n\|_2^2 - \|v_n\|_p^p + o(1);$$

hence, we obtain

$$\|\nabla v_n\|_2^2 - \|v_n\|_p^p = o(1). \qquad (5)$$

Assume now that $(\|v_n\|)_n$ is bounded away from zero. That is, for some constant $C > 0$ we have $\|v_n\| \geq C$ for all $n \in \mathbf{N}$. From (5), it then follows that $\|v_n\|_p^{p-2} \geq S + o(1)$ and consequently $\|v_n\|_p^p \geq S^{N/2} + o(1)$. This yields a contradiction since from (4) and (5) we have for n large enough,

$$\frac{1}{N}S^{N/2} \leq \frac{1}{N}\|v_n\|_p^p + o(1) = \frac{1}{2}\|\nabla v_n\|_2^2 - \frac{1}{p}\|v_n\|_p^p + o(1) < \frac{1}{N}S^{N/2}.$$

It follows that $u_{n_k} \to u$ strongly.

Claim (ii): There exists $v \in H$ such that M^- is dual to the family $\mathcal{F}_{u_1}^v$ of all continuous paths joining u_1 and v. Moreover, we have

$$c_1 < c_2 = \inf \varphi(M^-) \leq c_3 < c_1 + \frac{1}{N}S^{N/2} \qquad (6)$$

where $c_3 = \inf\limits_{\gamma \in \mathcal{F}^v_{u_1}} \sup\limits_{0 \le t \le 1} \varphi(\gamma(t))$.

Since $u_1 \ne 0$, there is a set of positive measure $\Sigma \subset \Omega$ such that $u_1 > 0$ on Σ (replace u_1 with $-u_1$ and f with $-f$ if necessary). As in the previous examples, we shall set $v_{\varepsilon,a}(x) = \eta_a(x).u_{\varepsilon,a}(x)$ where $u_{\varepsilon,a}$ is the extremal function corresponding to ε and $a \in \Omega$ and $\eta \in C_0^\infty(B_\varepsilon(a))$, η identically 1 on $B_{\varepsilon/2}(x_0)$. The point v is going to be of the form $v = u_1 + Rv_{\varepsilon,a}$ for suitable R, ε and $a \in \Omega$. We shall need the following estimate.

Lemma 8.6: *For every $R > 0$ and almost all $a \in \Sigma$, there exists $\varepsilon_0 = \varepsilon_0(R, a) > 0$ such that for every $0 < \varepsilon < \varepsilon_0$ and any $0 \le t \le 1$,*

$$\varphi(u_1 + tRv_{\varepsilon,a}) < c_1 + \frac{1}{N}S^{N/2}. \tag{7}$$

Proof: We have

$$\begin{aligned}
\varphi(u_1 + Rv_{\varepsilon,\alpha}) &= \int_\Omega \frac{|\nabla u_1|^2}{2} + R\int_\Omega \nabla u_1.\nabla v_{\varepsilon,a} + \frac{R^2}{2}\int_\Omega |\nabla v_{\varepsilon,a}|^2 \\
&\quad - \frac{1}{p}\int_\Omega |u_1 + Rv_{\varepsilon,a}|^p - \int_\Omega fu_1 - R\int_\Omega fv_{\varepsilon,a}.
\end{aligned} \tag{8}$$

By the estimates obtained in Appendix A, we have for all $a \in \Sigma$ and all $\varepsilon > 0$,

$$\begin{aligned}
\|u_1 + Rv_{\varepsilon,a}\|_p^p &= \|u_1\|_p^p + R^p\|v_{\varepsilon,a}\|_p^p \\
&\quad + pR\int_\Omega |u_1|^{p-2}u_1 v_{\varepsilon,a} \\
&\quad + pR^{p-1}\int_\Omega v_{\varepsilon,a}^{p-1}u_1 + o\left(\varepsilon^{\frac{N-2}{2}}\right).
\end{aligned}$$

We also have,

$$\|\nabla v_{\varepsilon,a}\|_2^2 = B + o(\varepsilon^{N-2}) \quad \text{and} \quad \|v_{\varepsilon,a}\|_p^p = A + o(\varepsilon^N) \quad \text{where}$$

$$B = \int_{\mathbf{R}^N} |\nabla v_{\varepsilon,a}(x)|^2 dx, \quad A = \int_{\mathbf{R}^N} \frac{dx}{(1 + |x|^2)^N} \quad \text{and} \quad S = \frac{B}{A^{2/p}}. \tag{9}$$

Substituting in (8) and using the fact that u_1 satisfies (1) we obtain for all $a \in \Sigma$,

$$\begin{aligned}
\varphi(u_1 + Rv_{\varepsilon,\alpha}) &= \frac{1}{2}\int_\Omega |\nabla u_1|^2 + R\int_\Omega \nabla u_1 \cdot \nabla v_\varepsilon + \frac{R^2}{2}B - \frac{1}{p}\int_\Omega |u_1|^p \\
&\quad - \frac{R^p}{p}A - R\int_\Omega |u_1|^{p-2}u_1 v_{\varepsilon,a} \\
&\quad - R^{p-1}\int_\Omega v_{\varepsilon,\alpha}^{p-1}u_1 - \int_\Omega fu_1 - R\int_\Omega fv_{\varepsilon,a} + o\left(\varepsilon^{\frac{N-2}{2}}\right) \\
&= \varphi(u_1) + \frac{R^2}{2}B - \frac{R^p}{p}A - R^{p-1}\int_\Omega v_{\varepsilon,\alpha}^{p-1}u_1 + o\left(\varepsilon^{\frac{N-2}{2}}\right).
\end{aligned}$$

Set $u_1 = 0$ outside Ω. It follows

$$\int_\Omega v_{\varepsilon,a}^{p-1} u_1 \, dx = \int_{\mathbf{R}^N} u_1(x) \eta_a(x) \frac{\varepsilon^{\frac{N+2}{2}}}{(\varepsilon^2 + |x-a|^2)^{\frac{N+2}{2}}} \, dx$$

$$= \varepsilon^{\frac{N-2}{2}} \int_{\mathbf{R}^N} u_1(x) \eta_a(x) \frac{1}{\varepsilon^N} \ell\left(\frac{x}{\varepsilon}\right) dx$$

where $\ell(x) = \dfrac{1}{(1+|x|^2)^{\frac{N+2}{2}}} \in L^1(\mathbf{R}^N)$.

Therefore, setting $D = \displaystyle\int_{\mathbf{R}^N} \frac{dx}{(1+|x|^2)^{\frac{N+2}{2}}}$, we derive for almost all $a \in \Sigma$ that

$$\int_{\mathbf{R}^N} u_1(x) \eta_a(x) \frac{1}{\varepsilon^N} \ell\left(\frac{x}{\varepsilon}\right) dx \to u_1(a) D.$$

In other words,

$$\int_\Omega v_{\varepsilon,a}^{p-1}(x) u_1(x) \, dx = \varepsilon^{\frac{N-2}{2}} u_1(a) D + o(\varepsilon^{\frac{N-2}{2}}).$$

Consequently,

$$\varphi(u_1 + R v_{\varepsilon,a}) = c_1 + \frac{R^2}{2} B - \frac{R^p}{p} A - R^{p-1} u_1(a) D \varepsilon^{\frac{N-2}{2}} + o(\varepsilon^{\frac{N-2}{2}}).$$

Define now for $s > 0$, the function

$$q(s) = \frac{s^2}{2} B - \frac{s^p}{p} A - s^{p-1} u_1(a) D \varepsilon^{\frac{N-2}{2}}$$

and let $s_\varepsilon > 0$ be a point where $q(s)$ achieves its maximum. Set $S_0 = (\frac{B}{A})^{-1/(p-2)}$ and note that since s_ε satisfies

$$s_\varepsilon B - s_\varepsilon^{p-1} A = (p-1) u_1(a) D \varepsilon^{\frac{N-2}{2}} s_\varepsilon^{p-2} \tag{10}$$

we have necessarily that $0 < s_\varepsilon < S_0$ and $s_\varepsilon \to S_0$ as $\varepsilon \to 0$.

Write $s_\varepsilon = S_0(1 - \delta_\varepsilon)$. We shall study the rate at which $\delta_\varepsilon \to 0$ as $\varepsilon \to 0$. From (10) we obtain

$$\left(\frac{B^{p-1}}{A}\right)^{1/(p-2)} (1 - \delta_\varepsilon - (1-\delta_\varepsilon)^{p-1}) = (p-1)\frac{B}{A}(1-\delta_\varepsilon)^{p-2} \varepsilon^{\frac{N-2}{2}} u_1(a) D;$$

and expanding for δ_ε we derive

$$(p-2)\left(\frac{B^{p-1}}{A}\right)^{1/(p-2)} \delta_\varepsilon = (p-1)\frac{B}{A} u_1(a) D \varepsilon^{\frac{N-2}{2}} + o\left(\varepsilon^{\frac{N-2}{2}}\right).$$

This implies

$$\varphi(u_1 + Rv_{\varepsilon,a}) \le c_1 + \frac{s_\varepsilon^2}{2}B - \frac{s_\varepsilon^p}{p}A - s_\varepsilon^{p-1}u_1(a)D\varepsilon^{\frac{N-2}{2}} + o\left(\varepsilon^{\frac{N-2}{2}}\right)$$

$$\le c_1 + \frac{S_0^p}{2}B - \frac{S_0^p}{2}A - S_0^2 B\delta_\varepsilon + S_0^p A\delta_\varepsilon$$
$$\quad - S_0^{p-1}u_1(a)D\varepsilon^{\frac{N-2}{2}} + o\left(\varepsilon^{\frac{N-2}{2}}\right)$$

$$\le c_1 + \frac{1}{N}S^{N/2} - S_0^{p-1}u_1(a)D\varepsilon^{\frac{N-2}{2}} + o\left(\varepsilon^{\frac{N-2}{2}}\right).$$

Therefore for $\varepsilon_0 = \varepsilon_0(R, a) > 0$ sufficiently small we conclude that for all $0 < \varepsilon < \varepsilon_0$,

$$\varphi(u_1 + Rv_{\varepsilon,a}) < c_1 + \frac{1}{N}S^{N/2}. \tag{11}$$

Back to the proof of Claim (ii), we first notice that under assumption (2) on f, the manifold M^- disconnects H in exactly two connected components U_1 and U_2. Indeed, by Lemma 8.5, we can find for every $u \in H$ with $\|u\| = 1$, a unique $t^+(u) > 0$ such that

$$t^+(u)u \in M^- \quad \text{and} \quad \varphi(t^+(u)u) = \max_{t \ge t_{\max}} \varphi(tu).$$

The uniqueness of $t^+(u)$ and its extremal property give that $t^+(u)$ is a continuous function of u on $H \setminus \{0\}$. It follows that the function $k(u) = \frac{1}{\|u\|}t^+(\frac{u}{\|u\|})$ is continuous on $H \setminus \{0\}$. Set

$$U_1 = \{u \in H \setminus \{0\}; k(u) > 1\} \cup \{0\} \quad \text{and} \quad U_2 = \{u : k(u) < 1\}.$$

Clearly $M^- = \{u : k(u) = 1\}$, $H \setminus M^- = U_1 \cup U_2$ and $u_0 \in M^+ \subset U_1$.

It remains now to find $v = u_1 + w$ such that

$$v \in U_2 \quad \text{and} \quad \sup_{0 \le t \le 1} \varphi(u_1 + tw) < c_1 + \frac{1}{N}S^{N/2} \tag{12}$$

since then

$$c_2 = \inf \varphi(M^-) \le c_3 = c(\varphi, \Gamma_{u_1}^v)$$
$$\le \sup\{\varphi(u_1 + tw); 0 \le t \le 1\}$$
$$< c_1 + \frac{1}{N}S^{N/2}$$

and Claim (ii) will be proved.

To obtain (12), we can find, modulo an easy computation, a constant $C > 0$ such that $0 < t^+(u) < C$ for all u with $\|u\| = 1$. Set $R_0 = (\frac{1}{B}|C^2 - \|u_1\|^2|)^{1/2} + 1$ and fix $a \in \Sigma$. Find $\varepsilon_0 = \varepsilon(R_0, a) > 0$, such that the estimate in Lemma 8.6 holds for all $0 < \varepsilon < \varepsilon_0$. We claim that for $\varepsilon > 0$ small

$$v_\varepsilon := u_1 + R_0 \eta_a u_{\varepsilon,a} \in U_2. \tag{13}$$

Indeed
$$\|\nabla v_\varepsilon\|_2^2 = \|\nabla(u_1 + R_0\eta_a u_{\varepsilon,a})\|_2^2$$
$$= \|u_1\|_2^2 + R_0^2 B + O(1)$$
$$> C^2 \geq \left(t^+\left(\frac{v_\varepsilon}{\|v_\varepsilon\|}\right)\right)^2$$
for $\varepsilon > 0$ small enough. Hence $k(v_\varepsilon) < 1$ and $v_\varepsilon \in U_2$.

To finish the prove of Theorem 8.4, we first note that if u is a local minimum or a limit of local minima for φ then necessarily
$$\|\nabla u\|_2^2 - (p-1)\|u\|_p^p \geq 0,$$
and hence u cannot be in M^-. So we distinguish two cases: either $c_2 = c_3$, which means that Theorem 6.6 applies and we get, in view of the above observation, that M^- contains a saddle point u_2 of mountain-pass type. Otherwise, $c_3 > c_2$ and Corollary 6.4 yields a saddle point u_3 for φ at the level c_3.

Notes and Comments: Motivated by the work of T. Aubin on the Yamabe problem, Brezis-Nirenberg ([B-N 1]), resurrected the interest in non-linear elliptic equations that correspond to the critical Sobolev exponent. That case was largely ignored because of a dramatic counterexample due to Pohozaev, (see [St] for a discussion). In that same paper [B-N 1], they established, among other things, the existence of a strictly positive solution, while emphasizing the fact that the Palais-Smale condition may only hold up to a certain level. The existence of a second sign-changing solution is due to Cerami-Solimini-Struwe [C-S-S]. The method of Tarantello [T 2] is more recent and, we believe, is more interesting since it exploits the second order properties of the functional. We will come back to this topic in chapter 11. The non-homogeneous case (Theorem 8.4) is very recent and is also due to Tarantello [T 1] where she proves the same results under weaker conditions. We believe that these three examples convincingly illustrate the relevance of the strong form of the min-max principle in conjunction with the Palais-Smale condition around dual sets.

9

MORSE INDICES OF MIN-MAX CRITICAL POINTS —

THE NON-DEGENERATE CASE

In this chapter, we try to relate the *topological* properties of the homotopy-stable class $\mathcal{F}$ to the *Morse indices* of those critical points obtained by min-maxing over $\mathcal{F}$ and which are located on an a priori given dual set. We shall be able to find one-sided relations between the *Morse index* and the *homotopic* (resp. *cohomotopic*) dimension of the class, while for *homological* and *cohomological* families, two-sided estimates are available.

We then give three examples of homogeneous non-linear eigenvalue problems, where we use the extra information on the Morse index of the variationally obtained solution, to be able to show that it is not the trivial one.

9.1 Homotopic, cohomotopic and homological families

To avoid repetition and to cover several examples simultaneously we are led to introduce the notions that will follow. We shall always assume that the group G which acts on the manifold under study X, is also acting on $\mathbb{R}^k$ and on the standard k-dimensional Euclidian sphere S^k.

If D and E are two topological spaces on which G acts continuously, we shall denote by $C_G(D; E)$ the set of all continuous and G-equivariant maps from D into E. The G-invariant sets will sometimes be called G-sets. In the sequel, B will be a fixed closed G-subset of the manifold X. We introduce the following notions.

Definition 9.1: A family $\mathcal{F}$ of G-subsets of X is said to be G-*homotopic of dimension n with boundary B* if there exists a compact G-subset D

of $\mathbb{R}^n$, containing a closed subset D_0, and a continuous G-invariant function σ from D_0 onto B such that

$$\mathcal{F} = \{A \subset X;\, A = f(D) \text{ for some } f \in C_G(D;X) \text{ with } f = \sigma \text{ on } D_0\}.$$

Dually, we can introduce the *cohomotopic classes*. For that, fix a G-equivariant continuous map $\sigma^* : B \to S^k$ and for any closed G-subset A of X containing B, set

$$\gamma_G(A; B, \sigma^*) = \inf\{n; \exists f \in C_G(A; S^n) \text{ with } f = \sigma^* \text{ on } B\}.$$

Definition 9.2: A family $\mathcal{F}$ of subsets of X is said to be G-*cohomotopic of dimension n with boundary B* if there exists a continuous and G-equivariant $\sigma^* : B \to S^n$ such that

$$\mathcal{F} = \{A;\, A \text{ compact } G\text{-subset of } X,\, A \supset B \text{ and } \gamma_G(A; B, \sigma^*) \geq n\}.$$

Definition 9.3: A family $\mathcal{F}$ of G-subsets of X is said to be *a homological family of dimension n with boundary B* if for some non-trivial class α in the n-dimensional relative homology group $H_n(X, B)$ we have that

$$\mathcal{F} =: \mathcal{F}(\alpha) = \{A;\, A \text{ compact } G\text{-subset of } X,\, A \supset B \text{ and } \alpha \in \operatorname{Im}(i_*^A)\}$$

where i_*^A is the homomorphism $i_*^A : H_n(A, B) \to H_n(X, B)$ induced by the *immersion* $i : A \to X$.

Suppose now that F is a closed subset of X that is disjoint from B. It is readily seen that F is dual to $\mathcal{F}(\alpha)$ if and only if $\alpha \notin Im(i_*)$ where $i_* : H_n(X \setminus F, B) \to H_n(X, B)$.

Dually, we can say that $\mathcal{F}$ is *a cohomological family of dimension n* if for some non-trivial class β in the n-dimensional cohomology group $H^n(X)$ we have that

$$\mathcal{F} = \mathcal{F}(\beta) = \{A;\, A \text{ compact } G\text{-subset of } X \text{ and } \beta \in \operatorname{Ker}(i_{X \setminus A}^*)\}$$

where $i_{X \setminus A}^* : H^n(X) \to H^n(X \setminus A)$ is the *restriction map*.

We shall only use singular homology with rational or real coefficients. If φ is G-invariant we shall use equivariant homology in a trivial fashion since we will always assume that the action of G is free and hence everything will be reduced to the quotient space X/G.

We now state the two main results of this chapter. We shall write $m(x)$ for the *Morse index* of the critical point x (see the definitions below after the statements of the results).

Theorem 9.4: *Let G be a compact Lie group acting freely and differentiably on a complete C^2-Riemannian manifold X. Let φ be a G-invariant C^2-functional on X and consider a G-homotopic family $\mathcal{F}$ (resp. a G-cohomotopic family $\bar{\mathcal{F}}$) of dimension n with closed boundary B. Let $\mathcal{F}^*$*

(resp. $\bar{\mathcal{F}}^*$) be a family dual to $\mathcal{F}$ (resp. $\bar{\mathcal{F}}$) such that

$$c := \sup_{F \in \mathcal{F}^*} \inf_{x \in F} \varphi(x) = \inf_{A \in \mathcal{F}} \max_{x \in A} \varphi(x)$$

$$(\text{resp. } \bar{c} := \sup_{F \in \bar{\mathcal{F}}^*} \inf_{x \in F} \varphi(x) = \inf_{A \in \bar{\mathcal{F}}} \max_{x \in A} \varphi(x)).$$

Assume that φ verifies $(PS)_c$ (resp. $(PS)_{\bar{c}}$) along a min-maxing sequence $(A_k)_k$ in $\mathcal{F}$ (resp. $(\bar{A}_k)_k$ in $\bar{\mathcal{F}}$), and a suitable max-mining sequence $(F_k)_k$ in $\mathcal{F}^*$ (resp. $(\bar{F}_k)_k$ in $\bar{\mathcal{F}}^*$) and that $K_c \cap F_\infty \cap A_\infty$ (resp. $K_{\bar{c}} \cap \bar{F}_\infty \cap \bar{A}_\infty$) consists of non-degenerate critical orbits. Then the following hold:

(1) There exists x_1 in $K_c \cap F_\infty \cap A_\infty$ with $m(x_1) \leq n$.

(2) There exists x_2 in $K_{\bar{c}} \cap \bar{F}_\infty \cap \bar{A}_\infty$ with $m(x_2) \geq n$.

(3) If $\mathcal{F} \subset \bar{\mathcal{F}}$ and $c = \bar{c}$, then there exists x_3 in $K_c \cap \bar{F}_\infty \cap A_\infty$ with $m(x_3) = n$.

The following deal with the homological case.

Theorem 9.5: Let G be a compact Lie group acting freely and differentiably on a complete C^2-Riemannian manifold X and let φ be a G-invariant C^2-functional on X. Consider a G-homological (or cohomological) family $\tilde{\mathcal{F}}$ of dimension n with boundary B such that $c(\varphi, \tilde{\mathcal{F}}) =: c$ is finite. let F be a closed G-invariant subset of X satisfying (F1) and (F2) with respect to $\tilde{\mathcal{F}}$ and suppose that φ verifies $(PS)_{F,c}$ along a min-maxing sequence $(A_n)_n$ in $\tilde{\mathcal{F}}$.

If $K_c \cap F \cap A_\infty$ consists of non-degenerate critical points then there exists x in $F \cap K_c \cap A_\infty$ such that $m(x) = n$.

If we suppose that $\sup \varphi(B) < c$, then the above theorems apply to the dual set $F = \{\varphi \geq c\}$ and we get the following.

Corollary 9.6: Let G be a compact Lie group acting freely and differentiably on a complete C^2-Riemannian manifold X. Let φ be a G-invariant C^2-functional on X and consider a G-homotopic family $\mathcal{F}$ (resp. a G-cohomotopic family $\bar{\mathcal{F}}$) (resp. a G-homological family $\tilde{\mathcal{F}}$) of dimension n with closed boundary B. Set $c = c(\varphi, \mathcal{F})$ (resp. $\bar{c} = c(\varphi, \bar{\mathcal{F}})$) (resp. $\tilde{c} = c(\varphi, \tilde{\mathcal{F}})$) and assume that $\sup \varphi(B) < c$ (resp. $\sup \varphi(B) < \bar{c}$) (resp. $\sup \varphi(B) < \tilde{c}$). If φ verifies $(PS)_c$ (resp. $(PS)_{\bar{c}}$) (resp. $(PS)_{\tilde{c}}$) along a min-maxing sequence $(A_k)_k$ and if the set $K_c \cap A_\infty$ (resp. $K_{\bar{c}} \cap A_\infty$) (resp. $K_{\tilde{c}} \cap A_\infty$) consists of non-degenerate critical orbits, then the following hold:

(i) There exists $x \in K_c \cap A_\infty$ with $m(x) \leq n$ (resp. $x \in K_{\bar{c}} \cap A_\infty$ with $m(x) \geq n$) (resp. $x \in K_{\tilde{c}} \cap A_\infty$ with $m(x) = n$).

(ii) If $\mathcal{F} \subset \bar{\mathcal{F}}$ and $c = \bar{c}$, then there exists x_3 in $K_c \cap A_\infty$ with $m(x_3) = n$.

9.2 The Morse lemma

To establish these Theorems, we start by recalling some basic concepts of Morse theory. The following lemma is standard.

Lemma 9.7: *Assume φ is a C^2-functional on a complete C^2-Riemanian manifold X modelled on a Hilbert space E. If v_0 is a non-degenerate critical point for φ (i.e. if $d^2\varphi(v_0)$ is invertible), then there exists a Lipschitz homeomorphism H from a neighborhood W of 0 in E onto a neighborhood M of v_0 with $H(0) = v_0$ in such a way that*

$$\varphi(H(z)) = \varphi(v_0) + \|z_+\|^2 - \|z_-\|^2$$

where $z \to (z_-, z_+)$ corresponds to the decomposition of E into the positive and negative spaces E_+ and E_- associated with the operator $d^2\varphi(v_0)$.

The Morse index of v_0 will be the dimension of E_-.

Proof: Assume without loss of generality that $v_0 = 0$ and that X itself is a Hilbert space. Write φ'' for $d^2\varphi$. We need to prove the existence of a Lipschitz homeomorphism H from a neighborhood W of 0 into a neighborhood M of 0 satisfying $H(0) = 0$ and for every $z \in X$

$$\varphi(H(z)) = \varphi(0) + \frac{1}{2}\langle\varphi''(0)z, z\rangle. \tag{1}$$

For that define, near $[0, 1] \times \{0\}$, the function

$$F(t, z) = (1 - t)(\varphi(0) + \frac{1}{2}\langle\varphi''(0)z, z\rangle) + t\varphi(z)$$

and the vector field

$$f(t, z) = \begin{cases} -F_t(t, z)\|F_z(t, z)\|^{-2}F_z(t, z) & if \quad z \neq 0 \\ 0 & if \quad z = 0. \end{cases}$$

We need to show that $f(t, .)$ is Lipschitz on a suitable neighborhood of 0. For that, define the function

$$\psi(z) = \varphi(z) - \varphi(0) - \frac{1}{2}\langle\varphi''(0)z, z\rangle$$

and note that $\psi(0) = 0$, $\psi'(0) = 0$ and $\psi''(0) = 0$. Consequently,

$$\psi(z) = \int_0^1 (1 - s)\langle\psi''(sz)z, z\rangle \, ds \quad \text{and} \quad \psi'(z) = \int_0^1 \psi''(sz)z \, ds.$$

Thus, for each $\varepsilon > 0$, there exists $\delta(\varepsilon) > 0$ such that for $\|z\| \leq \delta(\varepsilon)$,

$$|\psi(z)| \leq \varepsilon\|z\|^2 \quad \text{and} \quad \|\psi'(z)\| \leq \varepsilon\|z\|. \tag{2}$$

Since 0 is non-degenerate, there exists $K > 0$ such that for all $z \in X$,

$$K^{-1}\|z\| \leq \|\varphi''(0)z\| \leq K\|z\|. \tag{3}$$

It follows that for $z \neq 0$, we have

$$f(t, z) = -\psi(z)\|\varphi''(0)z + t\psi'(z)\|^{-2}(\varphi''(0)z + t\psi'(z)).$$

Let $\varepsilon = \frac{1}{2K}$, and use (2) and (3) to obtain for $\|z\| \leq \delta(\varepsilon)$,

$$|f(t,z)| \leq 2K(K+\varepsilon)\varepsilon\|z\|. \tag{4}$$

Since $f(t,0) = 0$, f is continuous. Let $\rho \in (0, \delta(\varepsilon))$ be such that for all $\|z\| \leq \rho$,

$$\|\psi''(z)\| \leq 1. \tag{5}$$

Using (2), (3) and (5) it is easy to find $K_1 > 0$ such that $\|f_z(t,z)\| \leq K_1$ for $\|z\| \leq \rho$ and $z \neq 0$. This coupled with the mean value theorem yields some constant $C > 0$ such that

$$\|f(t,z_1) - f(t,z_2)\| \leq C\|z_1 - z_2\|$$

for all z_1, z_2 in B_ρ.

It follows that the Cauchy problem

$$\begin{cases} \dot{\eta} & = f(t,\eta) \\ \eta(0) & = z \end{cases}$$

has a continuous solution η for z in some open neighborhood W of 0. Note that

$$\frac{d}{dt}F(t,\eta(t)) = F_t(t,\eta(t)) + \langle F_z(t,\eta(t)), \dot{\eta}(t) \rangle = 0$$

and in particular

$$\varphi(0) + \frac{1}{2}\langle \varphi''(0)z, z \rangle = F(0,z) = F(1,\eta(1,z)) = \varphi(\eta(1,z))$$

which means that the homeomorphism $H(z) = \eta(1,z)$ verifies the claim of the lemma.

9.3 The transformation of Lazer-Solimini

In the sequel, we shall make consistent use of the following transformation. Assume v_0 is a non-degenerate critical point for φ and let H be the *change of variables map* associated to v_0 by the Morse Lemma. Choose $r_1 > 0$ and $r_2 > 0$ small enough so that if B_- (resp. B_+) denotes the closed ball in E_- (resp. E_+) of radius r_1 (resp. r_2) centered at 0, then $2B_- + B_+$ is contained in the domain of H. Assume $4r_1^2 < r_2^2$, $0 < \varepsilon < r_2^2 - 4r_1^2$ and denote $C' = H(2B_- + B_+)$ and $C = H(B_- + B_+)$. Let α be a Lipschitz function from $\mathbb{R}$ to $[0,1]$ so that $\alpha = 0$ on $(-\infty, 0]$ and $\alpha = 1$ on $[1, +\infty)$. Let $\eta : E \to E$ be defined by

$$\eta(z_- + z_+) = z_- + \alpha\left(\frac{\|z_-\|}{r_1} - 1\right)z_+$$

and consider the following transformation $K : X \to X$

$$K(x) = \begin{cases} x & \text{on } X \setminus C' \\ H \circ \eta \circ H^{-1}(x) & \text{on } C' \end{cases}$$

We summarize the properties of this function in the following lemma whose easy proof is left to the reader.

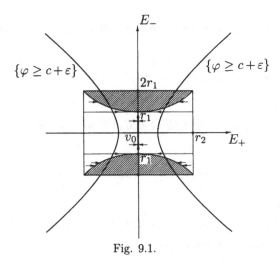

Fig. 9.1.

Lemma 9.8 (Fig 9.1): *For any non-degenerate critical point v_0, we can associate by the above procedure, a family $\{E_-(v_0), E_+(v_0), B_-(v_0), B_+(v_0), \varepsilon(v_0), H(v_0), C'(v_0), K(v_0)\}$ such that the following holds:*

(i) K is continuous on $X \setminus H(2B_- + \partial B_+) \supset \{x \in X; \varphi(x) \le \varphi(v_0) + \varepsilon\}$.

(ii) $\varphi \circ K \le \varphi$ on X.

(iii) $K(X \setminus \overset{\circ}{C}) \subset X \setminus \overset{\circ}{C}$.

(iv) $K(\partial C \cap \{\varphi \le \varphi(v_0) + \varepsilon\}) = H(\partial B_-)$.

(v) $K(C) = H(B_-)$.

In the sequel, we shall say that $C = C(v_0)$ is a *Morse neighborhood for v_0* with the implicit implication that it is associated to a family $\{E_-(v_0), E_+(v_0), B_-(v_0), B_+(v_0), \varepsilon(v_0), H(v_0), C'(v_0), K(v_0)\}$ by the above procedure.

If φ is G-invariant and the action of G is free on X, we shall just consider the Morse theory of the function $\tilde{\varphi}$ induced unambiguously by φ on the manifold X/G.

For any $d \in \mathbb{R}$ and any family of sets $\mathcal{F}$, we shall denote by $\mathcal{F}_d$ the subfamily

$$\mathcal{F}_d = \{A \in \mathcal{F}; \ \sup \varphi(A) \le d\}.$$

We now prove the following.

Lemma 9.9: *Under the hypothesis of Theorem 9.4, assume $u_1, \ldots, u_s$ (resp. $v_1, \ldots, v_r$) are non-degenerate critical points for φ on the set $K_c \cap A_\infty \cap F_\infty$ (resp. $K_{\bar{c}} \cap \bar{A}_\infty \cap \bar{F}_\infty$). Let $C_1, \ldots, C_s$ (resp. $C_1^*, \ldots, C_r^*$) be their corresponding Morse neighborhoods while $K_1, \ldots, K_s$ (resp. $K_1^*, \ldots, K_r^*$) are their associated transformations.*

(1) If $\min\limits_{1\leq i\leq s} m(u_i) > n$, then there is $\delta > 0$ so that for k large enough, the sets

$$F'_k := (K_s \circ K_{s-1} \circ \cdots \circ K_1)^{-1}(F_k) \cap \{\varphi \geq \inf_{F_k} \varphi\} \setminus \bigcup_{i=1}^{s} \overset{\circ}{C}_i$$

are dual to $\mathcal{F}_{c+\delta}$.

(2) If $\max\limits_{1\leq i\leq r} m(v_i) < n$, then there is $\delta > 0$ so that for k large enough, the sets

$$\bar{F}'_k := (K^*_r \circ \cdots \circ K^*_1)^{-1}(\bar{F}_k) \cap \{\varphi \geq \inf_{\bar{F}_k} \varphi\} \setminus \bigcup_{i=1}^{r} \overset{\circ}{C}^*_i$$

are dual to $\bar{\mathcal{F}}_{\bar{c}+\delta}$.

(3) If $\mathcal{F} \subset \bar{\mathcal{F}}$, $c = \bar{c}$, $\min\limits_{1\leq i\leq s} m(u_i) > n$ and $\max\limits_{1\leq i\leq r} m(v_i) < n$, then there is $\delta > 0$ so that for k large enough, the sets

$$\bar{F}''_k = (K^*_r \circ \cdots K^*_1 \circ K_s \cdots \circ K_1)^{-1}(\bar{F}_k) \cap \{\varphi \geq \inf_{\bar{F}_k} \varphi\} \setminus \bigcup_{i=1}^{r} \overset{\circ}{C}^*_i \cup \bigcup_{j=1}^{s} \overset{\circ}{C}_j$$

are dual to $\mathcal{F}_{c+\delta}$.

Proof: (1) Assume first, for simplicity, that we have only one critical point u. Let $E_-, E_+, B_-, B_+, \varepsilon, H, C'$ and K be the family associated to u. Recall that $C' = H(2B_- + B_+)$ which is a neighborhood of u. Moreover, since $u \in F_\infty$ and $F_k \cap B = \emptyset$ for each k, we can assume that C' is disjoint from B.

Let $A \in \mathcal{F}$ be such that $\sup \varphi(A) \leq c + \varepsilon$. That is, there exists a continuous map $f : D \to X$ such that $A = f(D)$ and $f = \sigma$ on D_0.

Let $L = f^{-1}(\overset{\circ}{C})$. By property (iv) of K we have that $K \circ f$ maps ∂L to $H(\partial B_-)$ which is homeomorphic to ∂B_-. Since $n < m(u) = \dim(E_-)$ and D is a subset of $\mathbb{R}^n$, we can extend $K \circ f$ to a continuous map $g : L \to H(\partial B_-)$ (Appendix D.9). Consider now the continuous map on D defined by

$$f_1(x) = \begin{cases} K \circ f & \text{if } x \in D \setminus L \\ g(x) & \text{if } x \in L. \end{cases}$$

Since $B \cap C' = \emptyset$ and K is the identity on $X \setminus C'$ we have that $f_1 = \sigma$ on $D_0 \subset D \setminus f^{-1}(C')$ which means that $f_1(D) \in \mathcal{F}$. It follows that $f_1(D) \cap F_k \neq \emptyset$ for each n. On the other hand,

$$\sup \varphi(f_1(L)) = \sup \varphi(g(L)) \leq \sup \varphi(H(\partial B_-)) = c - \tau$$

for some $\tau > 0$. Since $\inf \varphi(F_k) \geq c - \rho_k$ with $\rho_k \to 0$, we have necessarily that $F_k \cap (K \circ f(D \setminus L)) = F_k \cap K(A \setminus \overset{\circ}{C}) \neq \emptyset$ as long as $\rho_k < \tau$. For such a k, this implies that $A \cap (K^{-1}(F_k) \cap \{\varphi \geq \inf \varphi(F_k)\} \setminus \overset{\circ}{C}) \neq \emptyset$ since $\varphi(K(x)) \leq \varphi(x)$ on $X \setminus \overset{\circ}{C}$.

For the general case, let E^i_-, E^i_+, B^i_-, B^i_+, ε_i, H_i, C'_i and K_i be the family associated to each u_i. Since the critical points $(u_i)^s_{i=1}$ are isolated and belong to F_∞, we can assume that B and each C'_i $(1 \le i \le s)$ are mutually disjoint. Let $\delta = \min_{1 \le i \le s} \varepsilon_i$. Starting with $f = f_0$, one can construct as above, by induction, a sequence of continuous functions $f_i : D \to X$ $(1 \le i \le s)$ such that:

(a) $f_i = \sigma$ on D_0,

(b) $f_i = K_i \circ f_{i-1}$ on $D \setminus f_{i-1}^{-1}(\overset{\circ}{C}_i)$,

(c) f_i maps $f_{i-1}^{-1}(\overset{\circ}{C}_i)$ to $H_i(\partial E^i_-)$.

It follows that $f_s(D) \cap E_k \ne \emptyset$. As before, since $\inf \varphi(F_k) \to c$ as $k \to \infty$, $\sup \varphi(H(\partial E^s_-)) < c$ and $\varphi(K_s(x)) \le \varphi(x)$ on $X \setminus \overset{\circ}{C}_s$, we get for k large enough that

$$K_s(f_{s-1}(D) \setminus \overset{\circ}{C}_s) \cap F_k \ne \emptyset$$

and

$$f_{s-1}(D) \cap (K_s^{-1}(F_k) \setminus \overset{\circ}{C}_s) = f_{s-1}(D) \cap \{\varphi \ge \inf_{F_k} \varphi\} \cap (K_s^{-1}(F_k) \setminus \overset{\circ}{C}_s) \ne \emptyset$$

To get the conclusion of the lemma it is enough to proceed by backward induction in the same fashion as above by using in addition the fact that $K_{i-1}^{-1}(K_i^{-1}(L) \setminus \overset{\circ}{C}_i) \subset (K_i \circ K_{i-1})^{-1}(L) \setminus \overset{\circ}{C}_i$ for each L since K_{i-1} is the identity on $\overset{\circ}{C}_i$.

(2) Similarly, let E_-, E_+, B_-, B_+, ε^*, H, C'^* and K be the family associated to the Morse neighborhood of the critical point v in $\bar{F}_\infty$ at the level $\bar{c}$. We shall first prove that for any A in $\bar{\mathcal{F}}$ with $\sup \varphi(A) \le \bar{c} + \varepsilon^*$ we have that $K(A) \setminus \overset{\circ}{C}^* \in \bar{\mathcal{F}}$. Suppose not and assume without loss of generality that $C'^* \cap B = \emptyset$ and hence $B \subset K(A) \setminus \overset{\circ}{C}^*$. There exists then a continuous function $f : K(A) \setminus \overset{\circ}{C}^* \to S^{n-1}$ which is equal to σ on B. Consider now the restriction h of $f \circ H$ to $H^{-1}(K(A) \cap \partial C^*)$. Since $H^{-1}(K(A) \cap C^*) \subset E_-$ and $\dim(E_-) < n$, there exists a continuous extension g for h from $H^{-1}(K(A) \cap C^*)$ to S^{n-1} (Appendix D.9). The function f_1 defined by

$$f_1(x) = \begin{cases} f(x) & \text{if } x \in K(A) \setminus \overset{\circ}{C}^* \\ g \circ H^{-1}(x) & \text{if } x \in K(A) \cap C^* \end{cases}$$

is clearly continuous from $K(A)$ into S^{n-1} and is equal to σ on B. It follows that $\gamma_G(A, B, \sigma) \le \gamma_G(K(A), B, \sigma) \le n - 1$ which is a contradiction. Hence $K(A) \setminus \overset{\circ}{C}^* \in \bar{\mathcal{F}}$ from which follows that $[K(A) \setminus \overset{\circ}{C}^*] \cap \bar{F}_k \ne \emptyset$. This implies that for each k,

$$A \cap [K^{-1}(\bar{F}_k) \cap \{\varphi \ge \bar{c}\} \setminus \overset{\circ}{C}^*] \ne \emptyset.$$

The general case follows by an immediate induction and is left to the interested reader.

(3) Let $\delta = \min\{\varepsilon_1, \ldots, \varepsilon_s, \varepsilon_1^*, \ldots, \varepsilon_r^*\}$ where $(\varepsilon_i)_{i=1}^s$ (resp. $(\varepsilon_i^*)_{i=1}^r$) are associated with $(u_i)_{i=1}^s$ (resp. $(v_i)_{i=1}^r$). Since $\min\limits_{1 \leq i \leq s} m(u_i) > n$, we can associate for every $A \in \mathcal{F}$ with $\sup \varphi(A) \leq c + \delta$, the maps $f_1, \ldots, f_s$ as in part (1). Since $\mathcal{F} \subset \bar{\mathcal{F}}$, we get that $f_s(A) \in \bar{\mathcal{F}}$ and hence by part (2) we have, since $c = \bar{c}$ and $\max\limits_{1 \leq i \leq r} m(u_i) < n$, that

$$f_s(A) \cap [(K_r^* \circ \cdots \circ K_1^*)^{-1}(F_k) \cap \{\varphi \geq \inf \varphi(\bar{F}_k)\} \setminus \bigcup_{i=1}^r \overset{\circ}{C}_i^*] \neq \emptyset$$

for k large enough. Since now $\{C_1, \ldots, C_s, C_1^*, \ldots, C_r^*\}$ are mutually disjoint and since K_j (resp. K_j^*) is the identity on $\bigcup\limits_{i \neq j} \overset{\circ}{C}_i \cup \bigcup\limits_{i^*} \overset{\circ}{C}_i^*$ (resp. $(\bigcup\limits_i \overset{\circ}{C}_i) \cup \bigcup\limits_{i \neq j} \overset{\circ}{C}_i^*$), one can easily verify claim (3) of the lemma.

Proof of Theorem 9.4: (1) $K_c \cap F_\infty \cap A_\infty$ consists necessarily of a finite number of isolated critical orbits $\{u_1, \ldots, u_s\}$. Let $(C_i', C_i, K_i)_{i=1}^s$ be the families associated to these orbits in Lemma 9.8. Since they are isolated critical points on the level c, we can assume that there are no critical points in $\bigcup\limits_{i=1}^s (C_i' \setminus C_i)$ on the level c.

Suppose now $\min\limits_{1 \leq i \leq s} m(u_i) > n$. By Lemma 9.9, there exists $\delta > 0$ such that for k large enough, the sequence

$$F_k' = (K_s \circ K_{s-1} \circ \cdots \circ K_1)^{-1}(F_k) \cap \{\varphi \geq \inf \varphi(F_k)\} \setminus \bigcup_{i=1}^s \overset{\circ}{C}_i$$

is dual to $\mathcal{F}_{c+\delta}$. Since $(F_k)_k$ is a suitable max-mining for φ on $\mathcal{F}$, the same obviously holds for $(F_k')_k$. Moreover, by Lemma 9.8, we have that $F_k' \subset F_k \cup (\cup_{i=1}^s C_i' \setminus \overset{\circ}{C}_i)$. By Lemma 10.12 of the next chapter, the neighborhood $C' = \cup_{i=1}^s C_i'$ can be chosen in such a way that φ' is a proper map on C'. It follows that φ verifies $(PS)_c$ along the min-maxing sequence $(A_k)_k$ in $\mathcal{F}$ and the max-mining sequence $(F_k')_k$ in $\mathcal{F}_{c+\delta}^*$. Theorem 4.1 applies and we get that $F_\infty' \cap K_c \cap A_\infty \neq \emptyset$. But since there are no critical points at level c in $\bigcup\limits_{i=1}^s C_i' \setminus \overset{\circ}{C}_i$, we have a contradiction.

Assertions (2) and (3) can be proved in an identical manner.

We now deal with the homological case.

Lemma 9.10: *Under the hypothesis of Theorem 9.5, assume $u_1, \dots u_s$ are non-degenerate critical orbits for φ at the level c on the set F. Let $C_1, \dots, C_s$ be their corresponding Morse neighborhoods. If $m(u_i) \neq n$ for all $1 \leq i \leq s$, then there is $\delta > 0$ such that if we set $F' = F \cup (L_c \cap (\cup_{i=1}^{s} C_i'))$, then the canonical map from $H_n(\varphi^{c+\delta} \setminus F', B)$ to*

$$H_n\left((\varphi^{c+\delta} \setminus F') \cup \bigcup_{i=1}^{s} \overset{\circ}{C}_i, B\right) \text{ is surjective.}$$

Proof: For simplicity, we shall deal with the case of one critical point u. The rest will follow by an immediate induction. Let E_-, E_+, B_-, B_+, ε, H, C' and K be the Morse family associated to u. Consider the following Mayer-Vietoris exact sequence and its image by the homomorphism K_* induced by K.

$$H_n(\varphi^{c+\delta} \setminus F') \oplus H_n(\overset{\circ}{C}) \to H_n((\varphi^{c+\delta} \setminus F') \cup \overset{\circ}{C})) \to H_{n-1}((\varphi^{c+\delta} \setminus F') \cap \overset{\circ}{C}).$$

Note that $K(\overset{\circ}{C}) = H(B_-)$ and since $\dim(E_-) \neq n$ we have that $H_n(K(\overset{\circ}{C})) = \{0\}$. On the other hand, since $u \in F$ and $\inf \varphi(F) \geq c$, we have

$$K((\varphi^{c+\delta} \setminus F') \cap \overset{\circ}{C}) = H(B_- \setminus \{u\}).$$

But $H(\partial B_-)$ is a deformation retract of $H(B_- \setminus \{u\})$, hence

$$H_{n-1}(K((\varphi^{c+\delta} \setminus F') \cap \overset{\circ}{C})) \simeq H_{n-1}(H(\partial B_-)) = \{0\}$$

since again $\dim(E_-) \neq n$.

Finally, we get that the map

$$H_n(K(\varphi^{c+\delta} \setminus F'), B) \to H_n(K((\varphi^{c+\delta} \setminus F') \cup \overset{\circ}{C}), B)$$

is surjective. Note now that K maps $X \setminus F'$ into itself and since K is homotopic to the identity in $\varphi^{c+\delta}$, we get the claim of the lemma.

Proof of Theorem 9.5: $K_c \cap F \cap A_\infty$ consists necessarily of a finite number of isolated non-degenerate critical orbits $\{u_1, \dots, u_s\}$. let $(C_i)_{i=1}^{s}$ be the Morse neighborhoods associated to these points in Lemma 9.8. Since $\tilde{\mathcal{F}}$ is a homological family of dimension n that is dual to F, it follows that $\tilde{\mathcal{F}} = \mathcal{F}(\alpha)$ for some α which is not in the range of $i_* : H_n(\varphi^{c+\delta} \setminus F, B) \to H_n(\varphi^{c+\delta}, B)$. Since the map

$$H_n(\varphi^{c+\delta} \setminus F', B) \to H_n((\varphi^{c+\delta} \setminus F') \cup \bigcup_{i=1}^{s} \overset{\circ}{C}_i), B)$$

is surjective by Lemma 9.10, we get that $F'' = F' \setminus \bigcup_{i=1}^{s} \overset{\circ}{C}_i$ is dual to the class $\tilde{\mathcal{F}}$. By Theorem 4.1, $F'' \cap K_c \cap A_\infty \neq \emptyset$ which is a contradiction, since there are no critical points at level c in $\bigcup_{i=1}^{s}(C'_i \setminus \overset{\circ}{C}_i)$.

Remark 9.11: Note that the results in Theorems 9.4 and 9.5 remain valid if the min-max is taken over a union of homotopic (resp. cohomotopic) (resp. homological) families of the same dimension but possibly with different boundaries, provided these boundaries are all disjoint from the set F. This follows from the above proof and Remark 4.11.

9.4 Morse indices in the saddle point theorem

Let $X = Y \oplus Z$ with $\dim(Y) = n$ and consider the following class
$$\mathcal{F} = \{A; \exists h : B_Y \to X \text{ continuous, } h(x) = x \text{ on } S_Y \text{ and } A = h(B_Y)\}.$$
It is clear that $\mathcal{F}$ is a homotopic class of dimension n with boundary S_Y. Let now
$$\bar{\mathcal{F}} = \{A; A \text{ compact, } A \supset S_Y \text{ and } 0 \in f(A) \text{ whenever } f \in C(A;Y)$$
$$\text{and } f(x) = x \text{ on } S_Y\}.$$

$\bar{\mathcal{F}}$ is clearly a cohomotopic class of dimension n and boundary S_Y. Note also that $\mathcal{F} \subset \bar{\mathcal{F}}$.

Regard now $\sigma = [S_Y]$ as the generator of the homology $H_{n-1}(S_Y, \emptyset)$ and let $\beta \in H_n(X, S_Y)$ such that $\partial \beta = \sigma$. Consider $\tilde{\mathcal{F}} = \tilde{\mathcal{F}}(\beta)$ to be the corresponding homological family. Since $\sigma \neq 0$ in $H_{n-1}(X \setminus Z)$, it follows that Z is dual to the class $\tilde{\mathcal{F}}$.

Corollary 9.12: *Let φ be a C^2-functional on the Hilbert space X such that*
$$\alpha := \inf \varphi(Z) \geq 0 \geq \sup \varphi(S_Y).$$
Let $c = c(\varphi, \mathcal{F})$, $\bar{c} = c(\varphi, \bar{\mathcal{F}})$ and $\tilde{c} = c(\varphi, \tilde{\mathcal{F}})$. Assume that φ verifies (PS) and that $d^2\varphi$ is non-degenerate on the critical set. The following hold:

  If $0 < \bar{c}$ then,

(1) there exists x_1 in K_c with $m(x_1) \leq n$;

(2) there exists x_2 in $K_{\bar{c}}$ with $m(x_2) \geq n$;

(3) there exists x_3 in $K_{\bar{c}}$ with $m(x_3) = n$.

(4) If $c = \bar{c}$, there exists x_4 in K_c with $m(x_4) = n$.

  Moreover,

(5) if $\bar{c} = \alpha \geq 0$, there exists x_5 in $K_{\bar{c}} \cap Z$ with $m(x_5) = n$.

(6) if $\tilde{c} = \alpha \geq 0$, there exists x_6 in $K_{\tilde{c}} \cap Z$ with $m(x_6) = n$.

Proof: Clearly $c \geq \bar{c} \geq \alpha \geq 0$ and if $\bar{c} > 0$, then (1), (2), (3) and (4) follow immediately from Corollary 9.6, while (6) is a consequence of

Theorem 9.5. Suppose now $\bar{c} = \alpha$, then by Theorem 9.4, there exists x_5 in $K_{\bar{c}} \cap Z$ with $n \leq m(x_5)$. It remains to notice that since $\varphi \geq \alpha = \bar{c} = \varphi(x_5)$ on Z, we have that $m(x_5) \leq \operatorname{codim}(Z) = n$.

9.5 An asymptotically linear non-resonant boundary value problem

Let $\Omega \subset \mathbb{R}^N$ be a smoothly bounded domain. We consider the following problem

$$-\Delta u - \lambda u = g(u) \quad \text{in} \quad \Omega,$$
$$u = 0 \quad \text{on} \quad \partial\Omega. \tag{1}$$

Again, we denote by $(\lambda_i)_i$ the eigenvalues of $-\Delta$ on $H_0^1(\Omega)$. We shall now prove that a non-trivial solution for (1) exists, provided some eigenvalue lies between λ and $\lambda + g'(0)$. More precisely, we have the following.

Theorem 9.13: *Let g be a bounded real C^1-function such that $g(0) = 0$. Assume that λ verifies*

$$\lambda_n < \lambda < \lambda_{n+1} \tag{2}$$

and that

$$\text{either } g'(0) < \lambda_n - \lambda \quad \text{or} \quad g'(0) > \lambda_{n+1} - \lambda. \tag{3}$$

Then, problem (1) has at least one non-trivial solution.

Proof: As usual, the solutions of (1) correspond to the critical points of the C^2 functional

$$\varphi(u) = \frac{1}{2} \int_\Omega |\nabla u|^2 - \lambda |u|^2)\, dx - \int_\Omega G(u)\, dx$$

on $X = H_0^1$, where G denotes a primitive of g.

Note that the Hessian of φ at 0 is given by the operator

$$d^2\varphi(0)v = -\Delta v - \lambda v - g'(0)v$$

which implies that the Morse index $m(0)$ of φ at the critical point 0 is given by the largest integer i such that $\lambda_i < \lambda + g'(0)$. In view of hypothesis (3), it follows that in order to find a non-trivial solution, it is enough to find a critical point u of φ with Morse index equal to n. To do that, let us first prove

Claim 1: φ verifies $(PS)_c$ for any $c \in \mathbb{R}$.

Proof of Claim 1: Note that by Lemma 3.11, the functional φ verifies $(PS)_{F,c}$ for any c and any bounded set F, so that if $(u_n)_n$ is a sequence in H_0^1 is such that $\varphi(u_n)$ is bounded and $\varphi'(u_n) \to 0$, it only remains to show that it is necessarily bounded.

Suppose not and note that

$$\lim_n(-\Delta u_n - \lambda u_n - g(u_n)) = 0 \quad \text{in } H^{-1}. \tag{4}$$

Hence, if we set $v_n = \frac{u_n}{\|u_n\|}$, we get from (4) that

$$\lim_n(-\Delta v_n - \lambda v_n - \frac{g(u_n)}{\|u_n\|}) = 0 \quad \text{in } H^{-1}. \tag{5}$$

Since clearly $\frac{g(u_n)}{\|u_n\|} = 0$ in H^{-1}, it follows that any weak cluster point $\bar{v}$ of $(v_n)_n$, verifies $-\Delta \bar{v} - \lambda \bar{v} = 0$. But $\bar{v}$ cannot be 0, since otherwise the right-hand side of the following identity

$$\frac{\varphi(u_n)}{\|u_n\|^2} = \frac{1}{2} - \frac{1}{2}\lambda\frac{\|u_n\|_2^2}{\|u_n\|^2} - \int_\Omega \frac{G(u_n)}{\|u_n\|^2}\,dx,$$

will go to $1/2$, while the left-side goes to 0.

But the fact that $\bar{v} \neq 0$ contradicts the fact that λ is not in the spectrum of $-\Delta$. Hence $(u_n)_n$ is necessarily bounded in H_0^1.

To set up the saddle point theorem, let Y be the subspace of H_0^1 spanned by the eigenfunctions corresponding to the eigenvalues $(\lambda_i)_{i \leq n}$ and let Z be its linear suplement. We need to show the following.

Claim 2: For R large enough, we have

$$\sup \varphi(S_Y(R)) < \inf \varphi(Z).$$

Indeed, if $u \in Z$,

$$\varphi(u) \geq \frac{1}{2}(1 - \frac{\lambda}{\lambda_{n+1}})\int_\Omega |\nabla u|^2\,dx - \int_\Omega G(u)\,dx.$$

Since g is bounded, the last term grows linearly, while the first one is quadratic with a positive coefficient. Hence φ is bounded below on Z.

On the other hand, if $u \in Y$ and $\|u\| = R$, then

$$\varphi(u) \leq \frac{1}{2}(1 - \frac{\lambda}{\lambda_n})R^2 - \int_\Omega G(u)\,dx.$$

This clearly implies that $\lim_{R\to\infty} \sup \varphi(S_R(Y)) = -\infty$ and therefore Claim 2.

The conclusion of the Theorem is now an immediate application of Corollary 9.12 above.

9.6 An asymptotically linear resonant problem of Landesman-Lazer type

We consider again the same problem

$$-\Delta u - \lambda u = g(u) \quad \text{in } \Omega,$$
$$u = 0 \quad \text{on } \partial\Omega. \tag{1}$$

But now λ is allowed to be one of the eigenvalues $(\lambda_i)_i$ of $-\Delta$ on $H_0^1(\Omega)$. We shall prove the following.

Theorem 9.14: *Let g be a bounded real C^1-function with $g(0) = 0$ and such that the limits $\alpha = \lim_{s \to +\infty} g(s)$ and $\beta = \lim_{s \to -\infty} g(s)$ are finite. Assume that λ verifies for some integers n, k,*

$$\lambda_n < \lambda_{n+1} = \lambda = \lambda_{n+k} < \lambda_{n+k+1} \tag{2}$$

and that

$$\beta \int_\Omega e^+ \, dx - \alpha \int_\Omega e^- \, dx > 0 \tag{3}$$

for every eigenfunction e corresponding to the eigenvalue λ. Assume also that

$$\text{either } g'(0) < \lambda_n - \lambda \text{ or } g'(0) > \lambda_{n+1} - \lambda. \tag{4}$$

Then problem (1) has at least one non-trivial solution.

Proof: As usual, the solutions of (1) correspond to the critical points of the C^2-functional

$$\varphi(u) = \frac{1}{2} \int_\Omega (|\nabla u|^2 - \lambda |u|^2) \, dx - \int_\Omega G(u) \, dx$$

on $X = H_0^1$, where G denotes a primitive of g. As before, and in view of hypothesis (4), we need to find a critical point u of φ with Morse index equal to n. To do that, let us first prove

Claim 1: φ verifies $(PS)_c$ for any $c \in \mathbb{R}$.

Proof of Claim 1: Again by Lemma 3.11, the functional φ verifies $(PS)_{F,c}$ for any c and any bounded set F, so that it remains to show that $(u_n)_n$ is bounded in H_0^1 whenever it verifies $\varphi'(u_n) \to 0$ and $\varphi(u_n)$ bounded.

Assume $(u_n)_n$ is not bounded and that

$$\lim_n (-\Delta u_n - \lambda u_n - g(u_n)) = 0 \text{ in } H^{-1}. \tag{5}$$

As before, we may assume that $(u_n/\|u_n\|)_n$ is weakly convergent in H_0^1 and hence in L^2 to e. Also, as in the last example, one can easily show that $e \neq 0$ a.e. We then conclude that $u_n(x) \to +\infty$ a.e on $\{e > 0\}$ and $u_n(x) \to -\infty$ a.e on $\{e < 0\}$. It follows that $g(u_n) \to \tilde{g}$ a.e and hence in L^{2^*}, where

$$\tilde{g} = \alpha \chi_{\{e>0\}} + \beta \chi_{\{e<0\}}.$$

Use now (5) and an integration by parts to get that the following expression

$$\int_\Omega (-\Delta u_n - \lambda u_n) e - \int_\Omega g(u_n) e = \int_\Omega u_n (-\Delta e - \lambda e) - \int_\Omega g(u_n) e$$

$$= - \int_\Omega g(u_n) e$$

goes to zero.

It follows that

$$\beta \int_\Omega e^+ \, dx - \alpha \int_\Omega e^- \, dx = 0$$

which clearly contradicts (3). Hence $(u_n)_n$ is bounded in H_0^1 and the (PS) condition follows.

To set up the saddle point theorem, let Y be the subspace of H_0^1 spanned by the eigenfunctions corresponding to the eigenvalues $(\lambda_i)_{i \leq n}$ and let Z be its linear suplement. We need to show the following.

Claim 2: For R large enough, we have

$$\sup \varphi(S_Y(R)) < \inf \varphi(Z).$$

Indeed, as in the preceeding example, we have if $u \in Y$ and $\|u\| = R$, that

$$\varphi(u) \leq \frac{1}{2}(1 - \frac{\lambda}{\lambda_n})R^2 - \int_\Omega G(u) \, dx.$$

This implies that $\lim_{R \to \infty} \sup \varphi(S_R(Y)) = -\infty$.

On the other hand, we need to show that φ is bounded below on Z. For that, we let P be the orthogonal projection on the finite dimensional eigenspace W corresponding to λ and we note that if $u \in Z$, then

$$\varphi(u) \geq \frac{1}{2}(1 - \frac{\lambda}{\lambda_{n+k+1}}) \int_\Omega |\nabla v|^2 \, dx - \int_\Omega G(u) \, dx, \qquad (7)$$

where $v = (I - P)u$ and $w = Pu$.

Suppose now $(u_n)_n$ is a sequence in Z such that $\varphi(u_n) \to -\infty$. Clearly $\|w_n\| \to +\infty$ since otherwise, $\|v_n\| \to \infty$ and by (7) and the fact that g is bounded, $\varphi(u_n) \to +\infty$.

As in the proof of Claim 1, let e be a weak limit for $(w_n/\|w_n\|)_n$ and note that

$$\frac{G(u_n)}{\|w_n\|} \to \tilde{g}e \text{ in } L^1. \qquad (8)$$

So from (7), we obtain

$$\varphi(u_n) \geq \frac{1}{2}(1 - \frac{\lambda}{\lambda_{n+k+1}}) \int_\Omega |\nabla v_n|^2 \, dx - \|w_n\| \int_\Omega \frac{G(u_n)}{\|w_n\|} \, dx \qquad (9)$$

$$\geq -\|w_n\|\alpha_n$$

where, by (8) the coefficients

$$\alpha_n = \int_\Omega \frac{G(u_n)}{\|w_n\|} \, dx \to \alpha \int_\Omega e^+ \, dx - \beta \int_\Omega e^- \, dx.$$

Since $-e$ is also an eigenfunction corresponding to λ, it follows from hypothesis (3) that

$$\lim_n \alpha_n = \alpha \int_\Omega e^+ \, dx - \beta \int_\Omega e^- \, dx = -\beta \int_\Omega (-e)^+ \, dx + \alpha \int_\Omega (-e)^- \, dx < 0.$$

Hence, from (9), we get that $\varphi(u_n) \to +\infty$. A contradiction that completes the proof of Claim 2.

The conclusion of Theorem 9.14 is now an immediate application of Corollary 9.11 above.

Remark 9.15: By using a slight variation of the above argument, one can get a similar existence result if the inequality in the Landesman-Lazer condition (3) is reversed. However, in this case, the Morse index of the solution u is then equal to $n+k$. This implies the following result.

Theorem 9.16: *Let g be a bounded real C^1-function with $g(0) = 0$ and such that the limits $\alpha = \lim_{s\to+\infty} g(s)$ and $\beta = \lim_{s\to-\infty} g(s)$ are finite. Assume that λ verifies for some integers n, k,*

$$\lambda_n < \lambda_{n+1} = \lambda = \lambda_{n+k} < \lambda_{n+k+1}, \tag{2'}$$

and that

$$\beta \int_\Omega e^+ \, dx - \alpha \int_\Omega e^- \, dx < 0 \tag{3'}$$

for every eigenfunction e corresponding to the eigenvalue λ. Assume also that

$$\text{either } g'(0) < \lambda_{n+k} - \lambda \text{ or } g'(0) > \lambda_{n+k+1} - \lambda. \tag{4'}$$

Then problem (1) has at least one non-trivial solution.

9.7 An asymptotically linear strongly resonant elliptic problem

We reconsider now the homogeneous counterpart of the problem studied in section 3.5:

$$\begin{aligned} -\Delta u - \lambda u = g(u) \text{ in } \Omega, \\ u = 0 \quad \text{on } \partial\Omega, \end{aligned} \tag{1}$$

where λ is again one of the eigenvalues $(\lambda_i)_i$ of $-\Delta$ on $H_0^1(\Omega)$. However, the function g can satisfy the following *strong resonance condition*:

$$\lim_{|s|\to\infty} g(s) = 0 \text{ and } \int_{-\infty}^{+\infty} G(s) \, ds = 0 \tag{2}$$

or, in case g oscillates at infinity, we will assume that

$$g \text{ has mean zero and is } T\text{-periodic for some } T > 0. \tag{3}$$

We can prove the following.

Theorem 9.17: *Assume that λ verifies for some integers n, k,*

$$\lambda_n < \lambda_{n+1} = \lambda = \lambda_{n+k} < \lambda_{n+k+1}, \tag{4}$$

and that

$$\text{either } g'(0) < \lambda_n - \lambda \ \text{ or } \ g'(0) > \lambda_{n+k+1} - \lambda. \tag{5}$$

If g verifies either (2) or (3), then problem (1) has at least one non-trivial solution.

Proof: Note that the existence of a solution of (1) was established in section 3.5, via the saddle point Theorem. However, two cases had to be distinguished. In the first, the (PS) condition holds and a critical point was found via an n-dimensional saddle point Theorem. In the second case, only the (PS) condition along a min-maxing sequence of sets holds, and a critical point was found via an $(n + k)$-dimensional saddle point theorem. By Corollary 9.12, we can find such a critical point with Morse index either equal to n or to $n+k$. Hypothesis (5) now insures that such a solution cannot be the trivial one.

Notes and Comments: It has been established by many authors and in various examples that one can find one-sided relations between the Morse index and the *homotopic* (resp. *cohomotopic*) dimension of a given class (See Hofer [H 1], Ekeland-Hofer [E-H 1], Lazer-Solimini [L-S], Bahri [Ba 1], Bahri-Lions [Ba-L]), while for *homological* and *cohomological* families, two-sided estimates are available (Viterbo [V], Ekeland-Hofer [E-H 2]).

In this chapter, we follow the approach in Ghoussoub [G 1] which combines the new min-max principle with the ideas of Lazer-Solimini [L-S]. The novelty here is that one manages to find such points by only assuming the Palais-Smale condition to hold along one min-maxing sequence in the class $\mathcal{F}$ and a suitable max-mining sequence in $\mathcal{F}^*$. Moreover, we can find critical points with the right Morse index on any -a priori given- dual set F. The interest of such results will become more apparent in the next chapter.

The result concerning the existence of a nontrivial solution for the case of nonresonance problems, is due to Amann and Zehnder [A-Z]. For the resonance cases, we have followed the exposition of Solimini [So 3] where one can find more details about the earlier work of other authors on these problems.

10

MORSE INDICES OF MIN-MAX CRITICAL POINTS —

THE DEGENERATE CASE

In this chapter, we shall refine yet another perturbation technique developed by Marino-Prodi, in order to reduce the general problem to the non-degenerate situation where the results of the previous chapter can be applied.

The proof consists of perturbing the function φ in a neighborhood N of the critical set to obtain a non-degenerate function ψ on which the results of Chapter 9 may apply. Since we want ψ to have a critical point of the right index in that neighborhood, we need to min-max it on what we shall call the *homotopic* (resp. *cohomotopic*, resp. *homological*) *restriction* of the original family $\mathcal{F}$ to the neighborhood N.

Two extra complications are to be overcome during the application of the standard Marino-Prodi method. First, we need to find the right critical point to be close to - a priori given - dual sets. Secondly, we are only assuming the Palais-Smale condition to hold along a fixed min-maxing sequence of sets in $\mathcal{F}$ and a given max-mining sequence of dual sets. In total, three perturbations of the function coupled with an appropriate change in the homotopy-stable class will be necessary.

10.1 Multiplicity results for critical points with a given Morse index

In the sequel, $K_c(\varphi)$ will denote the set of critical points of the function φ at the level c. When no confusion is possible, we will drop φ from the above notation.

Definition 10.1: A compact subset L of K_c is said to be *an isolated* *critical set for φ in K_c* if it has a neighborhood in which φ has no critical points at the level c other than the ones that are already in L.

We shall deal with the following kind of degenerate critical points. Assume $v_0 \in K_c$ is such that $d^2\varphi(v_0)$ is a Fredholm operator on the associated Hilbert space E, then $E = E_0 \oplus E_1$ where E_0 and E_1 denote the kernel and range of $d^2\varphi(v_0)$ respectively. Moreover, $\dim(E_0) < +\infty$ and $d^2\varphi(v_0)$ is invertible on E_1. From basic spectral theory, it follows that $E_1 = E_+ \oplus E_-$ where E_+ and E_- are the positive and negative spaces associated with $d^2\varphi(v_0)$ restricted to E_1. The *augmented Morse index of v_0* will be

$$m^*(v_0) = \dim(E_0) + \dim(E_-).$$

We start by stating the main results of this chapter.

Theorem 10.2: *Let G be a compact Lie group acting freely and differentiably on a complete C^2-Riemannian manifold X and let φ be a G-invariant C^2-functional on X. Consider a G-homotopic family $\mathcal{F}$ (resp. a G-cohomotopic family $\bar{\mathcal{F}}$) of dimension n with closed boundary B and let $\mathcal{F}^*$ (resp. $\bar{\mathcal{F}}^*$) be a family dual to $\mathcal{F}$ (resp. $\bar{\mathcal{F}}$) such that*

$$c := \inf_{A \in \mathcal{F}} \max_{x \in A} \varphi(x) = \sup_{F \in \mathcal{F}^*} \inf_{x \in F} \varphi(x)$$

$$\left(\text{resp. } \bar{c} := \inf_{A \in \bar{\mathcal{F}}} \max_{x \in A} \varphi(x)) = \sup_{F \in \bar{\mathcal{F}}^*} \inf_{x \in F} \varphi(x)\right).$$

Assume that φ verifies $(PS)_c$ (resp. $(PS)_{\bar{c}}$) along a min-maxing sequence (A_k) in $\mathcal{F}$ (resp. in $\bar{\mathcal{F}}$) and a suitable max-mining sequence $(F_k)_k$ in $\mathcal{F}^*$ (resp. in $\bar{\mathcal{F}}^*$), and that $K_c \cap L$ (resp. $K_{\bar{c}} \cap L$) where L is either A_∞ or F_∞, is an isolated compact subset in K_c (resp. $K_{\bar{c}}$) on which $d^2\varphi$ is Fredholm. Then, the following hold:
(1) There exists x_1 in $K_c \cap F_\infty \cap A_\infty$ with $m(x_1) \leq n$.
(2) There exists x_2 in $K_{\bar{c}} \cap F_\infty \cap A_\infty$ with $m^*(x_2) \geq n$.
(3) If $\mathcal{F} \subset \bar{\mathcal{F}}$ and $c = \bar{c}$, then there exists x_3 in $K_c \cap F_\infty \cap A_\infty$ with $m(x_3) \leq n \leq m^*(x_3)$.

We shall also prove the following multiplicity result. First, let us write

$$K_c^-(n) = \{x \in K_c; m(x) \leq n\}$$

$$K_c^+(n) = \{x \in K_c; n \leq m^*(x)\}$$

and

$$K_c(n) = K_c^-(n) \cap K_c^+(n).$$

Theorem 10.3: *Under the hypothesis of Theorem 10.2, assume that Ind_G is an index associated to the group G.*

(1) If $K_c^-(n) \cap F_\infty$ is isolated in $K_c \cap F_\infty$, then
$$\text{Ind}_G(K_c^-(n) \cap F_\infty \cap A_\infty) \geq \inf\{\text{Ind}_G(A \cap F); A \in \mathcal{F}, F \in \mathcal{F}^*\}.$$
(2) If $K_{\bar{c}}^+(n) \cap F_\infty$ is isolated in $K_c \cap F_\infty$, then
$$\text{Ind}_G(K_{\bar{c}}^+(n) \cap F_\infty \cap A_\infty) \geq \inf\{\text{Ind}_G(A \cap F); A \in \bar{\mathcal{F}}, F \in \bar{\mathcal{F}}^*\}.$$
(3) If $\mathcal{F} \subset \bar{\mathcal{F}}$, $c = \bar{c}$ and if $K_c(n) \cap F_\infty$ is isolated in $K_c \cap F_\infty$ then
$$\text{Ind}_G(K_c(n) \cap F_\infty \cap A_\infty) \geq \inf\{\text{Ind}_G(A \cap F); A \in \mathcal{F}, F \in \bar{\mathcal{F}}^*\}.$$

The following deal with the homological case.

Theorem 10.4: *Let G be a compact Lie group acting freely and differentiably on a complete C^2-Riemannian manifold X and let φ be a G-invariant C^2-functional on X. Consider a G-homological (or cohomological) family $\tilde{\mathcal{F}}$ of dimension n with boundary B such that $c(\varphi, \tilde{\mathcal{F}}) =: c$ is finite. Let F be a closed G-invariant subset of X satisfying (F1) and (F2) with respect to $\tilde{\mathcal{F}}$ and suppose that φ verifies $(PS)_{F,c}$ along a min-maxing sequence $(A_n)_n$ in $\tilde{\mathcal{F}}$. If $K_c \cap F$ or $K_c \cap A_\infty$ is a compact isolated subset of K_c on which $d^2\varphi$ is Fredholm, then there exists x in $F \cap K_c \cap A_\infty$ such that $m(x) \leq n \leq m^*(x)$.*

Moreover, if Ind_G is an index associated to the group G and if $K_c(n) \cap F$ is isolated in $K_c \cap F$, then:
$$\text{Ind}_G(K_c(n) \cap F \cap A_\infty) \geq \inf\{\text{Ind}_G(A \cap F); A \in \tilde{\mathcal{F}}\}.$$

The hypothesis of *isolation* imposed above is not usually desirable since it is not easily verifiable in practice. We can do without it, provided we do not insist on the critical points to be in both A_∞ and F_∞. For instance, if we suppose that $\sup \varphi(B) < c$, then the above theorem applies to the dual set $F = \{\varphi \geq c\}$ since $K_c \cap F = K_c$ is obviously isolated in itself and we get the following:

Corollary 10.5: *Let G be a compact Lie group acting freely and differentiably on a complete C^2-Riemannian manifold X. Let φ be a G-invariant C^2-functional on X and consider a G-homotopic family $\mathcal{F}$ (resp. a G-cohomotopic family $\bar{\mathcal{F}}$) (resp. a G-homological family $\tilde{\mathcal{F}}$) of dimension n with closed boundary B. Set $c = c(\varphi, \mathcal{F})$ (resp. $\bar{c} = c(\varphi, \bar{\mathcal{F}})$) (resp. $\tilde{c} = c(\varphi, \tilde{\mathcal{F}})$) and assume that $\sup \varphi(B) < c$ (resp. $\sup \varphi(B) < \bar{c}$) (resp. $\sup \varphi(B) < \tilde{c}$). If φ verifies $(PS)_c$ (resp. $(PS)_{\bar{c}}$) (resp. $(PS)_{\tilde{c}}$) along a min-maxing sequence $(A_n)_n$ and if $d^2\varphi$ is Fredholm on K_c (resp. $K_{\bar{c}}$) (resp. $K_{\tilde{c}}$). Then the following hold:*
(1) *There exists $x \in K_c \cap A_\infty$ with $m(x) \leq n$ (resp. $x \in K_{\bar{c}} \cap A_\infty$ with $m^*(x) \geq n$) (resp. $x \in K_{\tilde{c}} \cap A_\infty$ with $m(x) \leq n \leq m^*(x)$).*
(2) *If $\mathcal{F} \subset \bar{\mathcal{F}}$ and $c = \bar{c}$, then there exists x_3 in $K_c \cap A_\infty$ such that $m(x_3) \leq n \leq m^*(x_3)$.*

Moreover, if Ind_G is an index associated to the group G, then:

(3) If $K_c^-(n)$ is isolated in K_c, we have :
$$\mathrm{Ind}_G(K_c^-(n) \cap A_\infty) \geq \inf\{\mathrm{Ind}_G(A \cap \{\varphi \geq c\}); A \in \mathcal{F}\}.$$

(4) If $K_{\bar{c}}^+(n)$ is isolated in $K_{\bar{c}}$, we have :
$$\mathrm{Ind}_G(K_{\bar{c}}^+(n) \cap A_\infty) \geq \inf\{\mathrm{Ind}_G(A \cap \{\varphi \geq \bar{c}\}); A \in \mathcal{F}\}.$$

(5) If $K_{\bar{c}}^-(n)$ is isolated in K_c, we have :
$$\mathrm{Ind}_G(K_{\bar{c}}(n) \cap A_\infty) \geq \inf\{\mathrm{Ind}_G(A \cap \{\varphi \geq \bar{c}\}); A \in \mathcal{F}\}.$$

We can also locate the right critical points on F_∞ without assuming the *isolation* hypothesis. Unfortunately, it is not an immediate consequence of Theorem 10.2 and another perturbation is needed to prove the following result.

Theorem 10.6:   *Let G be a compact Lie group acting freely and differentiably on a complete C^2-Riemannian manifold X. Let φ be a G-invariant C^2-functional on X and consider a G-homotopic family $\mathcal{F}$ (resp. a G-cohomotopic family $\bar{\mathcal{F}}$) of dimension n with closed boundary B. Let $\mathcal{F}^*$ (resp. $\bar{\mathcal{F}}^*$) be a family dual to $\mathcal{F}$ (resp. $\bar{\mathcal{F}}$) such that*
$$c := \inf_{A \in \mathcal{F}} \max_{x \in A} \varphi(x) = \sup_{F \in \mathcal{F}^*} \inf_{x \in F} \varphi(x)$$
$$(\text{resp. } \bar{c} := \inf_{A \in \bar{\mathcal{F}}} \max_{x \in A} \varphi(x)) = \sup_{F \in \bar{\mathcal{F}}^*} \inf_{x \in F} \varphi(x)).$$

If φ verifies $(PS)_c$ (resp. $(PS)_{\bar{c}}$) along a suitable max-mining sequence $(F_k)_k$ in $\mathcal{F}^*$ (resp. in $\bar{\mathcal{F}}^*$) and if $d^2\varphi$ is Fredholm on $K_c \cap F_\infty$ (resp. $K_{\bar{c}} \cap F_\infty$), then the following hold:

(1) *There exists $x \in K_c \cap F_\infty$ with $m(x) \leq n$ (resp. $x \in K_{\bar{c}} \cap F_\infty$ with $m^*(x) \geq n$).*

(2) *If $\mathcal{F} \subset \bar{\mathcal{F}}$ and $c = \bar{c}$, then there exists x_3 in $K_c \cap F_\infty$ such that $m(x_3) \leq n \leq m^*(x_3)$.*

Moreover, if Ind_G is an index associated to the group G, we obtain:

(3) $\mathrm{Ind}_G(K_c^-(n) \cap F_\infty) \geq \inf\{\mathrm{Ind}_G(A \cap F); A \in \mathcal{F}, F \in \mathcal{F}^*\}.$

(4) $\mathrm{Ind}_G(K_{\bar{c}}^+(n) \cap F_\infty) \geq \inf\{\mathrm{Ind}_G(A \cap F); A \in \mathcal{F}, F \in \bar{\mathcal{F}}^*\}.$

Remark 10.7: (i) Note that if $\sup \varphi(B) < c$ (resp. $\sup \varphi(B) < \bar{c}$), the above theorem applied to the dual set $F = \{\varphi \geq c\}$ yields that
$$\mathrm{Ind}_G(K_c^-(n)) \geq \inf\{\mathrm{Ind}_G(A \cap \{\varphi \geq c\}); A \in \mathcal{F}\}$$
and
$$\mathrm{Ind}_G(K_{\bar{c}}^+(n)) \geq \inf\{\mathrm{Ind}_G(A \cap \{\varphi \geq \bar{c}\}); A \in \mathcal{F}\}$$
provide that φ verifies $(PS)_c$ and that $d^2\varphi$ is Fredholm on K_c.

(ii) In the homological case, one can prove the same kind of result as in

Theorem 10.6, provided F_∞ is replaced by a single dual set F satisfying (F2).

From the above, we shall deduce the following useful result:

Theorem 10.8: *Let G, X and φ be as in Theorem 10.2 and let $(\mathcal{F}_j)_{j=1}^N$ be a decreasing sequence of G-homotopic (resp. G-cohomotopic) (resp. G-homological) families of dimensions $(n_j)_{j=1}^N$, with boundaries $(B_j)_{j=1}^N$ and satisfying the following excision property with respect to an index Ind_G:*

(E) For every $1 \le j \le j + p \le N$, any A in $\mathcal{F}_{j+p}$ and any U open and invariant such that $\bar{U} \cap B_j = \emptyset$ and $\mathrm{Ind}_G(\bar{U}) \le p$, we have $A \setminus U \in \mathcal{F}_j$.

Let F be a closed invariant set that is dual to each $\mathcal{F}_j$ and such that $\sup \varphi(B_j) \le d := \inf \varphi(F)$ for $1 \le j \le N$. Set $c_j = c(\varphi, \mathcal{F}_j)$ and assume that for every $1 \le j \le N$, φ verifies $(PS)_{c_j}$ and $d^2\varphi$ is Fredholm on K_{c_j}. Then, the following hold:

(1) If $M = \sup\{k \ge 0; c_k = d\}$, then

$$\mathrm{Ind}_G(K_{c_M}^-(d_M) \cap F) \ge M,$$

$$(\text{resp. } \mathrm{Ind}_G(K_{c_M}^+(d_M) \cap F) \ge M),$$

$$(\text{resp. } \mathrm{Ind}_G(K_{c_M}(d_M) \cap F) \ge M).$$

(2) If $c_j = c_{j+p}$ for $M < j \le j + p \le N$, then

$$\mathrm{Ind}_G(K_{c_j}^-(n_{j+p})) \ge p + 1,$$

$$(\text{resp. } \mathrm{Ind}_G(K_{c_j}^+(n_{j+p})) \ge p + 1),$$

$$(\text{resp. } \mathrm{Ind}_G(K_{c_j}(n_{j+p})) \ge p + 1).$$

(3) If $c_j = c_{j+p}$ for $M < j \le j + p \le N$, then for every min-maxing sequence $(A_n)_n$ in $\mathcal{F}_{j+p}$,

$$\mathrm{Ind}_G(K_{c_j}^-(n_j) \cap A_\infty) \ge p + 1,$$

$$(\text{resp. } \mathrm{Ind}_G(K_{c_j}^+(n_j) \cap A_\infty) \ge p + 1),$$

$$(\text{resp. } \mathrm{Ind}_G(K_{c_j}(n_j) \cap A_\infty) \ge p + 1).$$

We shall split the proof of the above theorems into several parts which might have an independent interest.

10.2 Restrictions of a homotopy-stable class to a neighborhood

Let $\mathcal{F}$ (resp. $\bar{\mathcal{F}}$) (resp. $\tilde{\mathcal{F}}$) be a homotopic (resp. cohomotopic) (resp. a homological) family of dimension n and with boundary B. Let N be an open set such that $\bar{N} \cap B = \emptyset$. Let $\mathcal{L}$ be a class of subsets of $\mathcal{F}$ (resp. $\bar{\mathcal{F}}$)(resp. $\tilde{\mathcal{F}}$) consisting of sets A such that $A \setminus N \notin \mathcal{F}$ (resp. $\bar{\mathcal{F}}$) (resp. $\tilde{\mathcal{F}}$). We define the *homotopic* (resp. *cohomotopic*)(resp. *homological*) restriction of $\mathcal{L}$ to N in the following fashion:

(1) Since $A \in \mathcal{F}$ is of the form $f(D)$ with D being an n-dimensional manifold and since $A \cap N \neq \emptyset$, let $D^A = f^{-1}(A \cap \overline{N})$, $D_0^A = f^{-1}(A \cap \partial N)$, $\sigma_A = f_{|D_0^A}$ and consider the class

$$\mathcal{L}^A = \{C; \exists g : D^A \to X \text{ continuous equivariant with}$$
$$g = \sigma_A \text{ on } D_0^A \text{ and } C = g(D^A)\}.$$

Let $\mathcal{L}(N) = \bigcup_{A \in \mathcal{L}} \mathcal{L}^A$. It is clearly a generalized homotopic family of dimension n and with boundary equal to $\bigcup_{A \in \mathcal{L}} A \cap \partial N$.

(2) Let now $A \in \mathcal{L} \subset \bar{\mathcal{F}}$. Since $A \backslash N \notin \bar{\mathcal{F}}$ there is $f : A \backslash N \to S^{n-1}$ continuous, G-invariant such that $f = \bar{\sigma}$ on B. Let $\bar{\sigma}_A = f_{|A \cap \partial N}$ and consider the class

$$\bar{\mathcal{L}}_A = \{C \subset X; C \text{ compact invariant containing} A \cap \partial N \text{ and}$$
$$\gamma_G(C, A \cap \partial N, \bar{\sigma}_A) \geq n\}.$$

Let $\bar{\mathcal{L}}(N) = \bigcup_{A \in \mathcal{L}} \bar{\mathcal{L}}_A$. It is clearly a generalized cohomotopic family of dimension n and with boudary $\bigcup_{A \in \mathcal{L}} A \cap \partial N$.

(3) Let now $A \in \mathcal{L} \subset \tilde{\mathcal{F}}$. Since $A \backslash N \notin \tilde{\mathcal{F}}$, we get from the exact sequence

$$H_n(A \backslash N, B) \to H_n(A, B) \to H_n(A, A \backslash N)$$

that $H_n(A, A \backslash N) \neq \{0\}$ and since by excision, this class is isomorphic to $H_n(A \cap \bar{N}, A \cap \partial N)$, we shall denote by α_A the non-trivial induced class in the latter. Again, we let $\tilde{\mathcal{L}}(N) = \bigcup_{A \in \mathcal{L}} \mathcal{F}(\alpha_A)$ where $\mathcal{F}(\alpha_A)$ is the n-dimensional homological family with boundary $A \cap \partial N$ associated to the class α_A.

The following lemma summarizes the above remarks and compares the homotopic, cohomotopic and homological restrictions to the set-theoretic restriction of $\mathcal{L}$ to N. That is $\mathcal{L}_N = \{A \cap \overline{N}; A \in \mathcal{L}\}$.

The proof is straightforward and is left to the reader.

Lemma 10.9: Let $\mathcal{F}$ (resp. $\bar{\mathcal{F}}$) (resp. $\tilde{\mathcal{F}}$) be a homotopic (resp. cohomotopic) (resp. a homological) family of dimension n with boundary B. Let N be an open set such that $\overline{N} \cap B = \emptyset$. Let $\mathcal{L}$ be a class of subsets of $\mathcal{F}$ (resp. $\bar{\mathcal{F}}$)(resp. $\tilde{\mathcal{F}}$) consisting of sets A such that $A \backslash N \notin \mathcal{F}$ (resp. $\bar{\mathcal{F}}$) (resp. $\tilde{\mathcal{F}}$). Then,

(i) $\mathcal{L}(N)$ (resp. $\bar{\mathcal{L}}(N)$(resp. $\tilde{\mathcal{L}}(N)$) is a generalized homotopic (resp. cohomotopic) (resp. homological) family of dimension n and with boundary $B' = \bigcup_{A \in \mathcal{L}} A \cap \partial N$.

(ii) $\mathcal{L}_N \subset \mathcal{L}(N)$ (resp. $\bar{\mathcal{L}}(N)$(resp. $\tilde{\mathcal{L}}(N)$)

(iii) For each $C \in \mathcal{L}(N)$ (resp. $\bar{\mathcal{L}}(N)$ (resp. $\tilde{\mathcal{L}}(N)$), there is $A \in \mathcal{L}$ such that $(A \backslash N) \cup C \in \mathcal{F}$ (resp. $\bar{\mathcal{F}}$) (resp. $\tilde{\mathcal{F}}$).

10.3 The Marino – Prodi perturbation method

We summarize the properties of this perturbation in the following.

Proposition 10.10: *Let φ be a functional in $C^2(X, \mathbb{R})$ and let K be a compact subset of K_c on which $d^2\varphi$ is a Fredholm operator. For $\delta > 0$ and $\varepsilon > 0$ small enough, there exists a function ψ in $C^2(X, \mathbb{R})$ that can be taken to be comparable to φ and which verifies the following:*

(i) $\|\varphi - \psi\|_{C^2} \leq \varepsilon$.

(ii) $\varphi(x) = \psi(x)$ if $x \notin K^{2\delta}$.

(iii) All the critical points of ψ in K^δ (if any) are non-degenerate and finite in number.

(iv) $d\psi$ is a proper map on $K^{4\delta}$.

Moreover, if K is an isolated critical set in K_c, then δ and ε can be chosen in such a way that

(v) ψ has no critical points in $\overline{K}^{3\delta} \backslash K^\delta$ at any level in $[c - \varepsilon, c + \varepsilon]$. More precisely, for $\delta > 0$ small enough, there exists $\rho = \rho(\delta) > 0$ such that for ε small enough, ψ can be chosen in such a way that for every $x \in X$

$$\max\{\|d\psi(x)\|, |\psi(x) - c|, \operatorname{dist}(x, K^{3\delta} \backslash K^\delta); x \in X\} \geq \rho.$$

For the proof we shall need the following two lemmas. For simplicity, we shall assume that X is a Hilbert space.

Lemma 10.11: *If K is a compact subset of X, then for every $\mu > 0$ there exists a C^∞ function $\ell : X \to [0, 1]$ with all its derivatives bounded such that*

$$\ell(x) = 1 \quad \text{for all } x \in K^\mu \tag{1}$$

$$\ell(x) = 0 \quad \text{for all } x \in X \backslash K^{2\mu}. \tag{2}$$

Proof: Fix a finite number of points $x_1, x_2, \ldots x_m \in K$ such that

$$K \subset \cup_{i=1}^m B(x_i, \frac{\mu}{2}).$$

We get C^∞-functions ℓ_i on X with bounded derivatives such that

$$\ell_i(x) = 1 \quad \text{if} \quad x \in B(x_i, \frac{3}{2}\mu)$$

$$\ell_i(x) = 0 \quad \text{if} \quad x \notin B(x_i, 2\mu)$$

by simply taking a real function $\alpha : R \to [0, 1]$ such that

$$\alpha(s) = 1 \quad \text{if} \quad s \leq \frac{9}{4}$$

$$\alpha(s) = 0 \quad \text{if} \quad s \geq 4$$

and letting

$$\ell_i(x) = \alpha\left(\frac{\|x - x_i\|^2}{\mu^2}\right).$$

Then we take $\beta : \mathbb{R} \to [0,1]$ such that

$$\beta(s) = 0 \quad \text{if} \quad s \leq 0,$$
$$\beta(s) = 1 \quad \text{if} \quad s \geq 1.$$

The function $\ell : X \to \mathbb{R}$ defined by

$$\ell(x) = \beta(\sum_{i=1}^{m} \ell_i(x))$$

is the required function.

Lemma 10.12: *Assume that for every x in the compact set K, $d^2\varphi(x)$ is a Fredholm operator. Then there exist a neighbourhood N of K and $\varepsilon > 0$, such that every functional $\psi \in C^2$, with $\|\varphi - \psi\|_{C^2} \leq \varepsilon$ has a gradient $d\psi$ that is a proper map on N.*

Proof: It is clear that we can restrict ourselves to the case where K is a singleton $\{\bar{x}\}$. For a given ε, take a closed bounded neighborhood N of $\bar{x}$ so that $d\varphi - d^2\varphi(\bar{x})$ is a Lipschitz map on N, with a Lipschitz constant less or equal than ε. If $\psi \in C^2$ is such that $\|\varphi - \psi\|_{C^2} \leq \varepsilon$, we consider the map $r(x) = d\psi(x) - d^2\varphi(\bar{x})(x)$ which is also a Lipschitz map on N with Lipschitz constant less or equal to 2ε. Since $d^2\varphi(\bar{x})$ is a Fredholm map, we can therefore split the space E into the sum of two closed subspaces V and W such that V is finite dimensional and such that for some positive constant α

$$\|d^2\varphi(\bar{x})(w)\| \geq \alpha\|w\| \quad \text{for all } w \in W. \tag{3}$$

Suppose now $(x_n)_n$ is a sequence in N such that $d\psi(x_n)$ is convergent. For every n, write $x_n = v_n + w_n$ with $v_n \in V$ and $w_n \in W$. Since V_n is finite dimensional we can assume that the sequence $(v_n)_n$ is convergent. For integers n and m, we have :

$$\begin{aligned}
\|w_n - w_m\| &\leq \alpha^{-1}\|(d^2\varphi(\bar{x}))(w_n - w_m)\| \\
&\leq \alpha^{-1}\|r(x_n) - r(x_m)\| + \alpha^{-1}\|(d^2\varphi(\bar{x}))(v_n - v_m)\| \\
&\quad + \alpha^{-1}\|d\psi(x_n) - d\psi(x_m)\| \\
&\leq 2\varepsilon\alpha^{-1}(\|v_n - v_m\| \\
&\quad + \|w_n - w_m\|) + \alpha^{-1}\|d^2\varphi(\bar{x})\|\,\|v_n - v_m\| \\
&\quad + \alpha^{-1}\|d\psi(x_n) - d\psi(x_m)\|.
\end{aligned}$$

So, if we take $\varepsilon < \frac{1}{2}\alpha$, we have from (3) that for a suitable constant c,

$$\|w_n - w_m\| \leq c(\|v_n - v_m\| + \|d\psi(x_n) - d\psi(x_m)\|). \tag{4}$$

Hence, since $(v_n)_n$ and $(d\psi(x_n))_n$ are two Cauchy sequences, we see from (4) that $(x_n)_n$ is a Cauchy sequence. Therefore the map $d\psi$ is proper on N.

Proof of Proposition 10.10: Assume that $d^2\varphi(x)$ is a Fredholm map for every $x \in K$ where K is a given compact subset of K_c. Let $N = K^{4\mu}$ ($\mu > 0$) be the neighborhood of K given by Lemma 10.12, and let ℓ be the function associated with φ, K and μ by Lemma 10.11. Let $M = \sup\{\|x\|; x \in K^{2\mu}\}$. Fix $\varepsilon > 0$. Since $d\varphi$ is Fredholm, we may use the Sard-Smale theorem [Sm] to find $y \in X^*$, with $\|y\| \leq \varepsilon/2M$ such that $-y$ is a regular value for $d\varphi$.

For any $x_0 \in K^{2\mu}$, the function defined by

$$\psi(x) = \varphi(x) + \ell(x)\langle y, x - x_0\rangle$$

clearly verifies (i) and (ii) of Proposition 10.10. Moreover, since on K^μ we have that $d\psi = d\varphi + y$, it follows that all critical points of ψ in K^μ are nondegenerate.

To obtain $\varphi \leq \psi$ or $\psi \leq \varphi$, it is enough to choose the x_0 that minimizes or maximizes the function $x \to \langle y, x\rangle$ on $K^{2\mu}$

Suppose now that K is isolated in K_c. There is a $\mu > 0$ such that φ has no critical points at level c in $K^{3\mu}\backslash K$. By Lemma 10.12, we can assume $d\varphi$ is a proper mapping on $K^{4\mu}$. It follows that the function

$$x \to \|d\varphi(x)\| + |\varphi(x) - c|$$

is bounded below by a positive constant b on the set $K^{3\mu}\backslash K^\mu$. It is now enough to apply the first part of the proposition with $\varepsilon < \frac{b}{2}$ to get assertion (v).

If now $(A_n)_n$ is a min-maxing sequence in $\mathcal{F}$ and $(F_n)_n$ is a suitable max-mining sequence of dual sets to $\mathcal{F}$, we shall need to isolate the set $F_\infty \cap K_c \cap A_\infty$ in K_c. The next two propositions shows that we can sometimes do that provided we perturb the function appropriately.

Proposition 10.13: *Let φ be a C^2-functional on X and consider a homotopy-stable family $\mathcal{F}$ with closed boundary B. Let $c = c(\varphi, \mathcal{F})$ and assume that φ verifies $(PS)_c$ along a min-maxing sequence $(A_n)_n$ in $\mathcal{F}$ and a suitable max-mining sequence $(F_n)_n$ of dual sets. Let M be a closed subset of K_c such that $M \cap F_\infty$ is an isolated compact subset of K_c on which $d^2\varphi$ is Fredholm. Assume $L := M \cap F_\infty \cap A_\infty$ is non-empty, then for $\delta > 0$ and $\varepsilon > 0$ small enough, there exist $\varphi_1 \in C^2(X)$ and an isolated compact subset K of $K_c(\varphi_1)$ such that:*
(i) $L \subset K_c(\varphi_1) \cap F_\infty \cap A_\infty \subset K \subset L^{2\delta}$.
(ii) $\|\varphi_1 - \varphi\| \leq \varepsilon$, $\varphi \leq \varphi_1$ on X and $\varphi = \varphi_1$ in a neighborhood of L.
(iii) $c(\varphi_1, \mathcal{F}) = c(\varphi, \mathcal{F})$, $(A_n)_n$ is a min-maxing sequence for φ_1 on $\mathcal{F}$ and $(F_n)_n$ is a max-mining sequence for φ_1 at level c.
(iv) φ_1 verifies $(PS)_c$ along $(A_n)_n$ and $(F_n)_n$ and $d^2\varphi_1$ is Fredholm on K.

Proof: Let $\delta_0 > 0$ and $\varepsilon > 0$ be such that any functional ψ with $\|\psi - \varphi\| \leq \varepsilon$ is automatically Fredholm on $M_\infty^{3\delta_0}$ where $M_\infty := M \cap F_\infty$. Since the latter is an isolated compact subset of K_c, there is $\delta_1 > 0$ such that $(M_\infty^{\delta_1} \setminus M_\infty) \cap K_c = \emptyset$. We may assume that for some $0 < \delta < \delta_1/4$ the set $M_\infty \cap (L^{2\delta} \setminus L^\delta) \neq \emptyset$, since otherwise there is nothing to prove. Let $L_1 = \{x \in M_\infty; \delta \leq \operatorname{dist}(x, L) \leq 2\delta\}$ with δ to be chosen small enough according to what is needed later. Note that since $d\varphi$ is proper on a neighborhood of L, we can assume that L_1 is compact. It is far from L, hence for δ_2 small enough we have that for large k, $L_1^{2\delta_2} \cap A_k = \emptyset$. We can also choose δ_2 in such a way that $L_1^{2\delta_2} \subset M_\infty^{\delta_1}$. Apply Proposition 10.10 to L_1, ε, and δ_2 to get a perturbation $\varphi_1 \geq \varphi$ and which satisfies (i), (ii) and (iii) of that proposition. Since $d\varphi$ is proper on $L^{3\delta}$, we can also assume that the critical points of φ_1 at level c in $L^{2\delta} \setminus L^\delta$ are contained in $L_1^{\delta_2}$ and therefore are in finite number. We can now easily find an isolated subset K of $K_c(\varphi_1)$ satisfying (i). Assertion (ii) follows from Proposition 10.10 while we get (iii) by noting that $\varphi = \varphi_1$ on A_k for k large enough and hence

$$c(\varphi_1, \mathcal{F}) \leq \operatorname{liminf}_n \sup \varphi_1(A_n)$$
$$= \lim_n \sup \varphi(A_n)$$
$$= c(\varphi, \mathcal{F}) \leq c(\varphi_1, \mathcal{F}).$$

On the other hand, since $\varphi \leq \varphi_1$, we have :

$$c = c(\varphi_1, \mathcal{F}) \geq \liminf_n \varphi_1(F_n) \geq \liminf_n \varphi(F_n) = c.$$

Assertion (iv) follows from the fact that $d\varphi_1$ is a proper map on the set where φ_1 is different from φ.

The same reasoning as above gives a similar result when $M \cap A_\infty$ is assumed to be an isolated compact subset of K_c. One can then isolate $M \cap A_\infty \cap F_\infty$ in K_c by considering a perturbation φ_1 as above but which is smaller than φ. The details are left to the reader. See also the proof of the next proposition.

Proposition 10.14: *Let φ be a C^2-functional on X and consider a homotopy-stable family $\mathcal{F}$ with closed boundary B. Let $(F_n)_n$ be a sequence of dual sets to $\mathcal{F}$ such that $\lim_n \inf \varphi(F_n) = c := c(\varphi, \mathcal{F})$ and $\lim_n \operatorname{dist}(F_n, B) > 0$. Assume that φ verifies $(PS)_c$ along $(F_n)_n$ and that $d^2\varphi$ is Fredholm on $L = K_c \cap F_\infty$. Then, for $\delta > 0$ and $\varepsilon > 0$ small enough, there exist a C^2-functional φ_1 on X and an isolated compact subset K of $K_c(\varphi_1)$ such that:*

(i) $L \subset K_c(\varphi_1) \cap F_\infty \subset K \subset L^{2\delta}$.

(ii) $\|\varphi_1 - \varphi\| \leq \varepsilon$ and $\varphi = \varphi_1$ in a neighborhood of L.

(iii) $c(\varphi_1, \mathcal{F}) = c(\varphi, \mathcal{F}) = c$ *and* $(F_n)_n$ *is still a suitable max-mining sequence for* φ_1.

(iv) φ_1 *verifies* $(PS)_c$ *along* $(F_n)_n$ *and* $d^2\varphi_1$ *is Fredholm on* K.

Proof: Let $\delta_0 > 0$ and $\varepsilon > 0$ be such that any functional ψ with $\|\psi - \varphi\| \leq \varepsilon$ is automatically Fredholm on $L^{3\delta_0}$. Let

$$T = \{x \in K_c; \delta_1 \leq \text{dist}(x, L) \leq 2\delta_1\}$$

for δ_1 sufficiently small for what will be later needed. Note that since $d\varphi$ is proper on a neighborhood of L, we can assume that T is compact. Moreover, for δ_2 small enough, we have that $T^{2\delta_2} \cap F_k = \emptyset$ for large k. We can also choose δ_2 in such a way that $T^{2\delta_2} \subset L^{\delta_0}$. Apply Proposition 10.10 to T, $\varepsilon/2$, and δ_2 to get a perturbation $\varphi_1 \leq \varphi$ and satisfying (i), (ii) and (iii) of that proposition. Since $d\varphi$ is proper on $L^{3\delta}$, we can also assume that the critical points of φ_1 at level c in $L^{2\delta_1} \backslash L^{\delta_1}$ are in finite number and contained in T^{δ_2}. We can now easily find an isolated subset K of $K_c(\varphi_1)$ satisfying (i). Assertion (ii) follows immediately from Proposition 10.10. To get (iii), we notice that $\varphi = \varphi_1$ on F_k for large k and $\varphi \geq \varphi_1$ from which follows that

$$c(\varphi_1, \mathcal{F}) \geq \liminf_k \varphi_1(F_k) = \liminf_k \varphi(F_k) = c(\varphi, \mathcal{F}) \geq c(\varphi_1, \mathcal{F}).$$

Finally, (iv) follows from the fact that $d\varphi_1$ is a proper map on the set where φ_1 is different from φ.

10.4 The reduction to the non-degenerate case

Proof of Theorem 10.2: Let A_n be a min-maxing sequence in $\mathcal{F}$ (or $\bar{\mathcal{F}}$) and let $(F_n)_n$ be a suitable max-mining sequence in $\mathcal{F}^*$ (or $\bar{\mathcal{F}}^*$). Let L denote the set $K_c \cap F_\infty \cap A_\infty$.

Let $s = \inf\{m(x); x \in L\}$, $t = \sup\{m^*(x); x \in L\} + 1$ and let $\lambda_1(x) \leq \lambda_2(x) \leq \cdots \leq \lambda_k(x) \leq \cdots$ be the eigenvalues of $d^2\varphi(x)$. Note that $\lambda_s(x) < 0 < \lambda_t(x)$ for every $x \in L$. Since the latter is compact and the λ_k's are continuous, we can find $\delta_0 > 0$ so that $\lambda_s(x) < 0 < \lambda_t(x)$ for every x in a δ_0-neighborhood of L. Choose ε_0 small enough so that for φ_ε with $\|\varphi_\varepsilon - \varphi\|_{C^2} \leq \varepsilon \leq \varepsilon_0$, the eigenvalues $(\lambda_k^\varepsilon(x))_k$ of $d^2\varphi_\varepsilon(x)$ will satisfy

$$\lambda_s^\varepsilon(x) < 0 < \lambda_t^\varepsilon(x) \text{ for every } x \text{ in } L^{\delta_0}.$$

In other words, for every critical point x of φ_ε in L^{δ_0} we have :

$$s \leq m(x) \leq t - 1. \tag{1}$$

The remainder of the proof will consist of finding such a φ_ε that will have a critical point of the right index in L^{δ_0}

First we consider the following situation.

Case 1: $L = K_c \cap F_\infty \cap A_\infty \subset K \subset L^{\delta_0}$ where K is a closed isolated subset of K_c on which $d^2\varphi$ is Fredholm.

We can then apply Proposition 10.10 to obtain a function ψ with $\psi \leq \varphi$ and satisfying (i)–(v) of that proposition. We can also assume that δ is such that $K^{4\delta} \subset L^{\delta_0}$. Unfortunately, we cannot apply Theorem 9.4 to ψ and the homotopic family $\mathcal{F}(K^{2\delta})$ because we cannot ensure that the function at the boundary is strictly below the min-max value. To remedy that, we let $N = K^{2\delta}$, $R = K^{3\delta}\backslash K^{\delta}$ and we consider the class

$$\mathcal{C} = \{A \in \mathcal{F}; \varliminf_n \sup_{A \cap F_n} \psi < c \text{ and } \varliminf_n \sup_{(A\backslash F_n)\cap R} \psi < c\}.$$

We distinguish two cases:

If $\mathcal{C}$ is empty then, since $\psi \leq \varphi$ we have that $c(\psi, \mathcal{F}) = c$, the sequence $(A_n)_n$ is min-maxing and $F'_n = F_n \cup R$ is a suitable max-mining sequence for ψ relative to the class $\mathcal{F}$. Moreover

$$K_c(\psi) \cap F'_\infty \cap A_\infty \subset [N \cup (K_c(\varphi) \cap F_\infty \cap A_\infty)] \cup (K_c(\psi) \cap R) \subset N$$

and hence all the critical points of ψ on it are non-degenerate. Since ψ' is proper in $K^{4\delta}$, ψ necessarily verifies condition $(PS)_c$ along $(A_n)_n$ and $(F'_n)_n$. Since the critical points of ψ in N are non-degenerate, Theorem 9.4 then apply to ψ and the same class $\mathcal{F}$ to yield the required result.

Suppose now $\mathcal{C}$ is not empty and let

$$c_1 = \inf_{A \in \mathcal{C}} \sup \psi(A \cap \overline{N}) = c(\psi, \mathcal{C}_N).$$

We have that

$$c - \varepsilon \leq c_1 < c. \tag{2}$$

Indeed, the second inequality is obvious while the first follows from

$$c \leq \varliminf_n \sup \varphi(A \cap F_n)$$
$$\leq \varliminf_n \max\{\sup \psi(A \cap F_n \setminus N), \sup \varphi(A \cap F_n \cap \overline{N})\}$$
$$\leq \max\{\varliminf_n \sup \psi(A \cap F_n), \sup \varphi(A \cap \overline{N})\}$$

so that if $A \in \mathcal{C}$, then $\varliminf_n \sup \psi(A \cap F_n) < c$ and hence $c \leq \sup \varphi(A \cap \overline{N}))$. Since $\|\varphi - \psi\| \leq \varepsilon$, (2) is proved.

Let now $S = K^{5\delta/2} \setminus K^{3\delta/2}$ in such a way that

$$\partial N \subset S \subset S^{\delta/4} \subset R.$$

We shall need the following deformation lemma. Note that in view of Proposition 10.10.(v), we can assume δ, ε and ρ chosen in such a way that $0 < \rho < \delta$, $\varepsilon < \rho^3$ and for every $x \in X$

$$\max\{\|\psi'(x)\|, |\psi(x) - c|, \text{dist}(x, K^{3\delta}\backslash K^{\delta}); x \in X\} \geq \rho.$$

Lemma 10.15: *Let $\tilde{c}$ be such that $\tilde{c} + \rho^3 < c$. There exists then a homeomorphism $\eta : X \to X$ verifying:*
  (i) $\psi(\eta(x)) \leq \psi(x)$ *for all $x \in X$.*
  (ii) $\rho(x, \eta(x)) \leq \rho$.
  (iii) $\eta(x) = x$ *whenever $x \in \psi^{\tilde{c}} \cup (X \setminus R)$.*
  (iv) η *maps $\psi^c \cap S$ into $\psi^{c-\rho^3}$.*

Proof of Lemma 10.15: Let $\ell : \mathbb{R} \to [0,1]$ be a smooth function such that

$$\ell(t) = \begin{cases} 0 & \text{if } t \leq \tilde{c} \\ 1 & \text{if } t \geq c - \rho^3. \end{cases}$$

Let $d : X \to [0,1]$ be the lipschitz function defined by:

$$d(x) = \frac{\text{dist}(x, X \setminus R)}{\text{dist}(x, X \setminus R) + \text{dist}(x, S^{\frac{\tilde{c}}{4}})}.$$

Let now G be the vector field on X defined by

$$G(x) = \begin{cases} \nabla\psi(x) & \text{if } \|\nabla\psi(x)\| \leq 1 \\ \frac{\nabla\psi(x)}{\|\nabla\psi(x)\|} & \text{if } \|\nabla\psi(x)\| \geq 1. \end{cases}$$

For every $x \in X$, define $\eta(x)$ to be $u(\rho, x)$ where $u(t,x)$ is the unique solution of the Cauchy problem

$$\begin{cases} \dot{u} = -\ell(\psi(u)d(u)F(u) \\ u(0) = x \end{cases}$$

That η verifies (i), (ii), (iii) and (iv) is straightforward and is left to the interested reader.

Back to the proof of the theorem. Consider the family

$$\mathcal{L} = \{\eta(A); A \in \mathcal{C}\}$$

and the associated homotopic (resp. cohomotopic) (resp. homological) classes that we shall all denote by $\mathcal{L}(\mathcal{N})$. We shall prove that $c' = c(\psi, \mathcal{L}(\mathcal{N}))$ is a critical level for ψ.

First note that since $\mathcal{L}_N \subset \mathcal{L}(N)$, we have :

$$c' \leq c_2 = c(\psi, \mathcal{L}_N). \tag{3}$$

We now show that

$$\mathcal{L} \subset \mathcal{C} \quad \text{and therefore } c_1 \leq c_2. \tag{4}$$

Indeed, we have for any set A,

$$\underline{\lim}_n \sup \psi(\eta(A) \cap F_n) \leq \underline{\lim}_n \sup \psi(A \cap \eta^{-1}(F_n))$$
$$\leq \max\{\underline{\lim}_n \sup \psi(A \cap F_n), \underline{\lim}_n \sup \psi((A \setminus F_n) \cap \eta^{-1}(F_n))\}.$$

If now $A \in \mathcal{C}$ and since $(\eta^{-1}(F_n) \setminus F_n)) \setminus R = \emptyset$, we get that the above is less than

$$\max\{\underline{\lim}_n \sup \psi(A \cap F_n), \underline{\lim}_n \sup \psi((A\backslash F_n) \cap R)$$
$$, \underline{\lim}_n \sup \psi((A\backslash F_n) \cap \eta^{-1}(F_n) \setminus R)\} < c.$$

On the other hand, since $\eta^{-1}(R) \subset R$, we have :

$$\underline{\lim}_n \sup \psi((\eta(A)\backslash F_n) \cap R) \le \sup \psi(A \cap \eta^{-1}(R)) \le \sup \psi(A \cap R)$$

$$\le \max\{\underline{\lim}_n \sup(\psi(A \cap F_n \cap R), \underline{\lim}_n \sup \psi((A\backslash F_n) \cap R)\} < c.$$

This proves assertion (4). We now prove that

$$c_2 = c_1. \tag{5}$$

Since $\overline{N} \subset \eta(\overline{N} \cup S)$, we have

$$\sup \psi(\eta(A) \cap \overline{N}) \le \sup \psi(\eta(A \cap (\overline{N} \cup S)))$$
$$\le \max\{\sup \psi(A \cap \overline{N}),$$
$$\underline{\lim}_n \sup \psi(\eta(A \cap F_n \cap S),$$
$$\underline{\lim}_n \sup \psi(\eta((A \setminus F_n) \cap R \cap S))\}.$$

If $A \in \mathcal{C}$, we get in view of Lemma 10.14 that

$$\sup \psi(\eta(A) \cap \overline{N}) \le \max\{\sup \psi(A \cap \overline{N}), c - \rho^3, c - \rho^3\}.$$

On the other hand, (2) gives that $\sup \psi(A \cap \overline{N}) \ge c_1 \ge c - \varepsilon > c - \rho^3$ and it follows that

$$\sup \psi(\eta(A) \cap \overline{N}) \le \sup \psi(A \cap \overline{N}).$$

This clearly proves (5). We now prove that

$$c' = c_1. \tag{6}$$

First note that by (3) and (5) we have that $c' \le c_2 = c_1$. For the reverse inequality, let $C \in \mathcal{L}(N)$ and assume without loss of generality that $\sup \psi(C) \le c_1$. By Lemma 10.9.(iii), there exists $A \in \mathcal{L} \subset \mathcal{C}$ such that the set $A' = (A \setminus N) \cup C \in \mathcal{F}$. We claim that $A' \in \mathcal{C}$. Indeed,

$$\underline{\lim}_n \sup \psi(A' \cap F_n) \le \max\{\underline{\lim}_n \sup \psi(A \cap F_n), \sup \psi(C)\}$$
$$\le \max\{\underline{\lim}_n \sup \psi(A \cap F_n), c_1\} < c.$$

Similarly

$$\underline{\lim}_n \sup \psi((A' \setminus F_n) \cap R) \le \max\{\underline{\lim}_n \sup \psi((A \setminus F_n) \cap R), \sup \psi(C)\}$$
$$\le \max\{\underline{\lim}_n \sup \psi((A \setminus F_n) \cap R), c_1\} < c.$$

It follows that $C \cap \overline{N} = A' \cap \overline{N} \in \mathcal{C}_N$, which clearly implies that $c' \ge c_1$ and hence we have equality.

Finally, we claim that we have the boundary condition

$$\sup_{A \in \mathcal{C}} \sup \psi(\eta(A) \cap \partial N) < c'. \tag{7}$$

Indeed, since $\partial N \subset \eta(S)$, we have in view of Lemma 10.15.(iv)

$$\sup \psi(\eta(A) \cap \partial N) \leq \sup \psi(\eta(A \cap S))$$
$$\leq \max\{\underline{\lim}_n \sup \psi(\eta(A \cap F_n \cap S),$$
$$\underline{\lim}_n \sup \psi(A \setminus F_n) \cap S)\}$$
$$\leq c - \rho^3 < c - \varepsilon \leq c_1 = c'.$$

But (7) means that the set $\{\psi \geq c'\}$ verifies conditions (F1) and (F2) with respect to the class $\mathcal{L}(N)$ and the function ψ. Moreover, since $c_2 = c'$ and $\mathcal{L}_N \subset \mathcal{L}(N)$, we can find a min-maxing sequence $(C_n)_n$ consisting of subsets of $\overline{N}$. Since $d\psi$ is proper in N, ψ necessarily verifies $(PS)_{c'}$ along the sequence $(C_n)_n$. It follows that $C_\infty \subset \overline{N}$ and $K_c(\psi) \cap C_\infty \subset K^\delta$ on which the critical points are non-degenerate. Theorem 9.4 and Remark 9.11 now apply and we get a critical point x for ψ in K^{δ_0} with $m(x) \leq n$ (resp. $n \leq m(x)$) (resp. $n = m(x)$). This coupled with (1) above gives the claims in Theorems 10.2.

To finish the proof of these Theorems, it remains to deal with the case where we only assume that:

Case 2: $F_\infty \cap K_c$ is a compact isolated set in K_c on which $d^2\varphi$ is Fredholm.

Since $L = F_\infty \cap K_c \cap A_\infty$ is not necessarily isolated in K_c, we shall use Proposition 10.13 to isolate it in the critical set of an appropriate perturbation of the functional.

Let δ_0 and ε_0 be as in the begining of the proof. Apply Proposition 10.13 with $M = K_c$, $0 < \delta < \delta_0/2$ and $0 < \varepsilon < \varepsilon_0$ to get a perturbation φ_1 that verifies assertions (i)–(iv) of that proposition. Since $K_c(\varphi_1) \cap F_\infty \cap A_\infty = L_1$ is contained in an isolated subset K of $K_c(\varphi_1)$, and since φ_1 verifies the same properties as φ with respect to the class $\mathcal{F}$, we can apply the conclusion of the first case to φ_1 to get $x \in K_c(\varphi_1) \cap F \cap A_\infty = L_1$ with Morse index at most n. Since $\|\varphi_1 - \varphi\| \leq \varepsilon_0$ and $L \subset L_1 \subset L^{\delta_0}$, the claim follows from assertion (1). The proof of the cohomotopic case is identical.

Proof of Theorem 10.3: Suppose now that Ind_G is an index associated to the group G and assume that

$$\text{Ind}_G(L) \leq \inf\{\text{Ind}_G(A \cap F); A \in \mathcal{F}, F \in \mathcal{F}^*\} - 1 \qquad (*)$$

where $L := K_c^-(n) \cap F_\infty \cap A_\infty$. We shall work toward a contradiction.

Assume first that $L = K_c^-(n) \cap F_\infty \cap A_\infty$ is contained in a compact set K which is isolated in K_c, and such that $\text{Ind}_G(K) = \text{Ind}_G(L)$. By property (I3) of the index, there exists an invariant neighborhood

U of K such that $\text{Ind}_G(\overline{U}) = \text{Ind}_G(K)$. Since the latter set is isolated in K_c, we can also suppose that $U \setminus K$ contains no critical points at the level c. Let $F' = X \setminus U$. It is clearly closed and invariant. By property (I4) of the index, we have for any A in $\mathcal{F}$ and any $F \in F^*$ that

$$\text{Ind}_G(A \cap F) \leq \text{Ind}_G((A \cap F) \setminus U) + \text{Ind}_G(\overline{U})$$
$$= \text{Ind}_G(A \cap F \cap F') + \text{Ind}_G(K).$$

This combined with $(*)$ yield that $\text{Ind}_G(A \cap F \cap F') \geq 1$ and in particular $F_n \cap F'$ is a suitable max-mining sequence of dual sets. Moreover, it is easy to see that $K_c \cap F_\infty \cap F'$ is also isolated in K_c. Hence Theorem 10.2 applies to the sequence $F_n \cap F'$ and we get that $K_c^-(n) \cap F_\infty \cap F' \cap A_\infty \neq \emptyset$ which is clearly a contradiction.

To prove the general case assume now that only $F_\infty \cap K_c^-(n)$ is isolated in $F_\infty \cap K_c$. Find an open invariant set V containing L and such that $\text{Ind}_G(L) = \text{Ind}_G(\overline{V})$. Let $\delta > 0$ be such that $L^{2\delta} \subset V$ and apply Proposition 10.13 with $M = K_c^-(n)$ to get φ_1 satisfying the same conditions as φ with respect to $\mathcal{F}$ at the level c and such that $L \subset K \subset L^{2\delta}$ and K is isolated in $K_c(\varphi_1)$. Note that $\text{Ind}_G(K) = \text{Ind}_G(L)$. Our claim follows immediately from the previous case applied to φ_1.

The same proof applies to the cohomotopic case.

Proof of Theorem 10.4: It is identical to the proof of Theorem 10.2 provided one uses Theorem 9.5 for the non-degenerate (homological) case. This is also the reason why it is stated for a single dual set.

Proof of Theorem 10.6: It is identical to the proof of Theorem 10.2 provided one uses Proposition 10.14 to isolate $K_c \cap F_\infty$ in K_c.

Proof of Theorem 10.8: 1) Assume $1 \leq M$ since otherwise there is nothing to prove. In view of Theorem 10.6, it is enough to show that $\text{Ind}_G(A \cap F) \geq M$ for every A in $\mathcal{F}_M$. Suppose it was false for some A in $\mathcal{F}_M$. By property (I3) of the index and since $F \cap B_1 = \emptyset$ we can find U open and invariant such that $A \cap F \subset U$, $\overline{U} \cap B_1 = \emptyset$ and $\text{Ind}_G(\overline{U}) = \text{Ind}_G(A \cap F) \leq M - 1$. The excision property implies that $A \setminus U \in \mathcal{F}_1$ and hence that $(A \setminus U) \cap F \neq \emptyset$ which clearly contradicts the fact that F is dual to $\mathcal{F}_1$.

(2) Suppose that $c_j = c_{j+p}$ for some $M < j < j + p \leq N$ and that $\text{Ind}_G(K_{c_j}^-(n_{j+p})) \leq p$. Since $\sup \varphi(B_j) \leq \inf \varphi(F) < c_j$, we can find, as above, an open invariant neighborhood U of $L = K_{c_j}^-(n_{j+p})$ such that $\overline{U} \cap B_j = \emptyset$ and $\text{Ind}_G(\overline{U}) \leq p$. By property (E) we have $A \setminus U \in \mathcal{F}_j$ for every A in $\mathcal{F}_{j+p}$ which means that the set $F' = \{\varphi \geq c_j\} \setminus U$ is dual to $\mathcal{F}_{j+p}$. It follows from Theorem 10.6, that $F' \cap K_{c_j}^-(n_{j+p}) \neq \emptyset$ which is clearly a contradiction.

(3) Suppose again that $c_j = c_{j+p}$ for $M < j < j + p \leq N$. If $\text{Ind}_G(K^-_{c_j}(n_j) \cap A_\infty)) \leq p$ and since $\sup \varphi(B_j) \leq \inf \varphi(F) < c_j$ we can find, as above, an open invariant neighborhood U of $L = K^-_{c_j}(n_j) \cap A_\infty$ such that $\overline{U} \cap B_j = \emptyset$ and $\text{Ind}_G(\overline{U}) \leq p$. Again, by property (E) we have that $A \setminus U \in \mathcal{F}_j$ for every A in $\mathcal{F}_{j+p}$. If now $(A_k)_k$ is a min-maxing sequence for φ in $\mathcal{F}_{j+p}$, then the sequence $(A'_k)_k = (A_k \setminus U)_k$ is also a min-maxing sequence for φ in $\mathcal{F}_j$. It follows from Corollary 10.5, that $A'_\infty \cap K^-_{c_j}(n_j) \neq \emptyset$ which is a contradiction.

10.5 Application to some standard variational settings

We start with the saddle point theorem and we let again $X = Y \oplus Z$ with $\dim(Y) = n$ and consider the homotopic (resp. cohomotopic, resp. homological) class $\mathcal{F}$ (resp. $\bar{\mathcal{F}}$, resp. $\tilde{\mathcal{F}}$) with boundary S_Y described in Corollary 9.12.

Corollary 10.16: *Let φ be a C^2-functional on the Hilbert space X such that*

$$\alpha := \inf \varphi(Z) \geq 0 \geq \sup \varphi(S_Y).$$

Let $c = c(\varphi, \mathcal{F})$, $\bar{c} = c(\varphi, \bar{\mathcal{F}})$ and $\tilde{c} = c(\varphi, \tilde{\mathcal{F}})$. Assume that φ verifies (PS) and that $d^2\varphi$ is Fredholm on the critical set.

If $0 < \bar{c}$, then

(1) there exists x_1 in K_c with $m(x_1) \leq n$;

(2) there exists x_2 in $K_{\bar{c}}$ with $m^*(x_2) \geq n$;

(3) there exists x_3 in $K_{\tilde{c}}$ with $m(x_3) \leq n \leq m^*(x_3)$.

(4) If $c = \bar{c}$, there exists x_4 in K_c with $m(x_4) \leq m \leq m^*(x_4)$.

On the other hand,

(5) if $\bar{c} = \alpha \geq 0$, there exists x_5 in $K_{\bar{c}} \cap Z$ with $m(x_5) \leq n \leq m^*(x_5)$;

(6) if $\tilde{c} = \alpha \geq 0$, there exists x_6 in $K_{\tilde{c}} \cap Z$ with $m(x_6) \leq n \leq m^*(x_6)$.

Proof: (1), (2), (3) and (4) follow immediately from Corollary 10.5, while (6) is a consequence of Theorem 10.4. Suppose now $\bar{c} = \alpha$, then by Theorem 10.6, there exists x_5 in $K_{\bar{c}} \cap Z$ with $n \leq m^*(x_5)$. It remains to notice that since $\varphi \geq \alpha = \bar{c} = \varphi(x_5)$ on Z, we have that $m(x_5) \leq \text{codim} \, (Z) = n$.

Morse indices for the $\mathbb{Z}_2$-symmetric classes

Let S be the unit sphere of a Hilbert space X. For any compact symmetric subset A of S, we write

$$\gamma(A) = \inf\{k \in \mathbb{N}; \exists h : A \to S^k \text{ odd and continuous}\}$$

and

$$\tau(A) = \sup\{k \in \mathbb{N}; \exists h : S^k \to A \text{ odd and continuous}\}.$$

Now consider

$$\mathcal{F}_k = \{A \subset S; A \text{ compact symmetric and } \tau(A) \geq k\}$$

and

$$\bar{\mathcal{F}}_k = \{A \subset S; A \text{ compact symmetric and } \gamma(A) \geq k\}.$$

By Borsuk-Ulam's theorem, $\mathcal{F}_k \subset \bar{\mathcal{F}}_k$ for each k. If now φ is a functional on S, denote by $c_k = c(\varphi, \mathcal{F}_k)$ and $\bar{c}_k = c(\varphi, \bar{\mathcal{F}}_k)$. It is clear that $\bar{c}_k \leq c_k$ for each k. We assume in the sequel that $\bar{c}_k > -\infty$ for each k.

Corollary 10.17: *Let φ be a C^2-functional on S satisfying (PS). Assume that $d^2\varphi$ is Fredholm on the critical sets K_{c_k} and $K_{\bar{c}_k}$, then for each $k \in \mathbb{N}$, the following hold:*

(1) There exists x_k in K_{c_k} with $m(x_k) \leq k$.

(2) There exists y_k in $K_{\bar{c}_k}$ with $m^*(y_k) \geq k$.

(3) If $c_k = \bar{c}_k$, there exists z_k in K_{c_k} with $m(z_k) \leq k \leq m^*(z_k)$.

(4) If $\bar{c}_{n+k} = \bar{c}_k$, then $\gamma(K_{\bar{c}_k}^+(n+k)) \geq n+1$.

(5) If $\bar{c}_{n+k} = \bar{c}_k = c_k$, then $\gamma(K_{\bar{c}_k}(k)) \geq n+1$.

(6) If $c_{n+k} = \bar{c}_k$, then $\gamma(K_{c_k}(n+k)) \geq n+1$.

Proof: For (1), (2) and (3) it is enough to notice that $\mathcal{F}_k$ (resp. $\bar{\mathcal{F}}_k$) is a homotopic (resp. cohomotopic) class of dimension k and to apply Corollary 10.5. Note that here $B = \emptyset$ and $F = \{\varphi \geq c_k\}$.

(4) follows from Theorem 10.8, since the sequence $(\bar{\mathcal{F}}_k)_k$ clearly verifies the excision property.

(5) Assume $\bar{c}_{n+k} = \bar{c}_k = c_k$ and let $L = \{x \in K_{c_k}; m(x) \leq k \leq m^*(x)\}$. If $\gamma(L) \leq n$, find U open symmetric containing L and such tht $\gamma(\bar{U}) = \gamma(L) \leq n$. It follows that for any A in $\bar{\mathcal{F}}_{n+k}$, $A \setminus U \in \bar{\mathcal{F}}_k$ and if $(A_i)_i$ is a min-maxing sequence in $\bar{\mathcal{F}}_{n+k}$, the sequence $(A_i')_i = (A_i \setminus U)_i$ will then be min-maxing in $\bar{\mathcal{F}}_k$. Since $\mathcal{F}_k \subset \bar{\mathcal{F}}_k$ and $c_k = \bar{c}_k$, it follows from Corollary 10.5 that $A_\infty' \cap K_{\bar{c}_k}$ contains x with $m(x) \leq k \leq m^*(x)$. This is a contradiction since that set is disjoint from L.

(6) We always have $\bar{c}_k \leq \bar{c}_{n+k} \leq c_{n+k}$ and $\bar{c}_k \leq c_k \leq c_{n+k}$. So, if we assume that $c_{n+k} = \bar{c}_k$, we necessarily have $c = c_{n+k} = c_k = \bar{c}_k = \bar{c}_{n+k}$. Let $L = \{x \in K_c; m(x) \leq n+k \leq m^*(x)\}$. If $\gamma(L) \leq n$, proceed as in (5) to find U open symmetric containing L and such that $\gamma(\bar{U}) = \gamma(L) \leq n$. It follows that for any A in $\bar{\mathcal{F}}_{n+k}$, the set $A \setminus U \in \bar{\mathcal{F}}_k$ and consequently $(A \setminus U) \cap \{\varphi \geq c\} \neq \emptyset$. Let $F = \{\varphi \geq c\} \setminus U$. It verifies (F1) and (F2) with respect to $\bar{\mathcal{F}}_{n+k}$ and $\mathcal{F}_{n+k} \subset \bar{\mathcal{F}}_{n+k}$ while $c_{n+k} = \bar{c}_{n+k}$. It follows from Theorem 10.6 that there exists $x \in F \cap K_c$ with $m(x) \leq n+k \leq m^*(x)$. This is a contradiction, since $F \cap L = \emptyset$.

Morse indices in the symmetric mountain pass theorem

Let X be a Banach space and assume $X = Y \oplus Z$ where Y is a finite dimensional subspace of dimension k. Let E_n be a vector subspace of X containing Y with $\dim(E_n) = n > k$. For $R > 0$, consider the family

$$\mathcal{L}_n = \{h \in C(\bar{B}_R(E_n), X); h \text{ odd and equal the identity on } S_R(E_n)\}.$$

We shall write D_n for $B_R(E_n)$ and S_n for $S_R(E_n)$.

For $k < j \leq n$, define

$$\mathcal{F}_j = \mathcal{F}_j(E_n, R) = \{h(\overline{D_n \setminus K}); h \in \mathcal{L}_n, \gamma_{Z_2}(K) \leq n - j\}.$$

Lemma 10.18: *The classes $\mathcal{F}_j$ ($k < j \leq n$) are non-empty and they satisfy the following properties:*

(a) *(Monotonicity)* $\mathcal{F}_{j+1} \subset \mathcal{F}_j$.

(b) *(Stability)* $\mathcal{F}_j$ *is a $\mathbb{Z}_2$-homotopy stable class with boundary S_n.*

(c) *(Excision)* *For any A in $\mathcal{F}_{j+p}$, any open symmetric set U such that $\bar{U} \cap S_n = \emptyset$ and $\gamma_{\mathbf{z}_2}(\bar{U}) \leq p$ we have $A \setminus U \in \mathcal{F}_j$.*

(d) *(Linking)* *For all A in $\mathcal{F}_j$, $\gamma_{Z_2}(A \cap S_\rho(Z)) \geq j - k$ provided $\rho < R$.*

Proof: Since $\mathrm{id} \in \mathcal{L}_n$, it follows that $\mathcal{F}_j \neq \emptyset$ for all j ($k < j \leq n$).

(a) If $A = h(\overline{D_n \setminus K}) \in \mathcal{F}_{j+1}$, then $n \geq j + 1 \geq j$, $h \in \mathcal{L}_n$, $K \in \Sigma$, and $\gamma_{\mathbf{z}_2}(K) \leq n - (j + 1) < n - j$. Therefore $A \in \mathcal{F}_j$.

(b) Suppose $A = h(\overline{D_n \setminus K}) \in \mathcal{F}_j$ and $k \in C(E, E)$ is odd, and $k = \mathrm{id}$ on S_n, then $k \circ h$ is odd, belongs to $C(D_n, E)$, and $k \circ h = \mathrm{id}$ on S_n. Therefore $k \circ h \in \mathcal{L}_n$ and $k \circ h(\overline{D_n \setminus K}) = k(A) \in \mathcal{F}_j$.

(c) Let $A = h(\overline{D_n \setminus K}) \in \mathcal{F}_{j+p}$ and $H \in \Sigma$ with $\gamma_{\mathbf{z}_2}(H) \leq p$. We claim

$$\overline{A \setminus H} = h(\overline{D_n \setminus (K \cup h^{-1}(H))}). \tag{1}$$

Indeed, suppose $b \in h(D_n \setminus (K \cup h^{-1}(H)))$. Then $b \in h(D_n \setminus K) \setminus H \subset A \setminus H \subset \overline{A \setminus H}$. Therefore

$$h(D_n \setminus (K \cup h^{-1}(H))) \subset \overline{A \setminus H}. \tag{2}$$

On the other hand if $b \in A \setminus H$, then $b = h(w)$ where

$$w \in \overline{D_n \setminus K} \setminus h^{-1}(H) \subset \overline{D_n \setminus (K \cup h^{-1}(H))}.$$

Thus

$$A \setminus H \subset h(\overline{D_n \setminus (K \cup h^{-1}(H))}). \tag{3}$$

It is clear that (2)-(3) yield (1) since h is continuous.

Note now that since h is odd and continuous and $H \in \Sigma$, we have that $h^{-1}(H) \in \Sigma$ and by properties (I2)-(I4) of the index

$$\gamma_{\mathbf{z}_2}(K \cup h^{-1}(H)) \leq \gamma_{\mathbf{z}_2}(K) + \gamma_{\mathbf{z}_2}(h^{-1}(H)) \leq \gamma_{\mathbf{z}_2}(K) + \gamma_{\mathbf{z}_2}(H) \leq n - j.$$

Hence $\overline{A \setminus H} \in \mathcal{F}_j$.

d) Again let $A = h(\overline{D_n \setminus K})$ where $n \geq j$ and $\gamma_{\mathbf{z}_2}(K) \leq n - j$. Let

$\tilde{\mathcal{O}} = \{x \in D_n; h(x) \in B_\rho\}$. Since h is odd, $0 \in \tilde{\mathcal{O}}$. Let O denote the component of $\tilde{\mathcal{O}}$ containing 0. Since D_n is bounded, $\mathcal{O}$ is a symmetric bounded neighborhood of 0 in E_n. Therefore by Lemma 7.20.(b), $\gamma(\partial\mathcal{O}) = m$. We claim

$$h(\partial\mathcal{O}) \subset \partial B_\rho. \tag{4}$$

Indeed, suppose $x \in \partial\mathcal{O}$ and $h(x) \in B_\rho$. If $x \in D_n$, there exists a neighborhood V of x such that $h(V) \subset B_\rho$, but then $x \notin \partial\mathcal{O}$. It follows that x is necessarily in ∂D_n. But if $x \in \partial D_n$ and $h(x) \in B_\rho$, then $h(x) = x$ and $\|h(x)\| = \|x\| = R < \rho$, thus (4) holds.

Set now $W \equiv \{x \in D_n; h(x) \in \partial B_\rho\}$. (4) implies that $W \supset \partial\mathcal{O}$. Hence by Lemma 7.20.b, $\gamma_{\mathbf{Z}_2}(W) = m$ and by property (I4) of the index, $\gamma(\overline{W\backslash K}) \geq m-(m-j) = j$. Thus by (I2), $\gamma_{\mathbf{Z}_2}(h(\overline{W\backslash K})) \geq j > k$. Since codim $Z = k$, we get from Lemma 7.20.(c) that $\gamma_{\mathbf{Z}_2}(h(\overline{W\backslash K})\cap Z \geq j-k$. But $h(\overline{W\backslash K}) \subset (A\cap\partial B_\rho)$. This finishes the proof of the lemma.

Now we can prove the following.

Corollary 10.19: Let φ be an even C^2-functional satisfying (PS) on a Banach space $X = Y \oplus Z$ and such that $d^2\varphi$ is Fredholm on the critical set. Assume $\varphi(0) = 0$ as well as the following conditions:
(a) There is $\rho > 0$ and $\alpha \geq 0$ such that $\inf \varphi(S_\rho(Z)) \geq \alpha$.
(b) There exists $R > \rho$ such that $\sup \varphi(S_R(E)) \leq 0$.
  Setting $c_j = c(\varphi, \mathcal{F}_j)$, we have :
(i) If $0 < c_i$ for some i $(k < i \leq n)$, there exists $x \in K_{c_i}$ with $m(x) \leq n$.
(ii) If $0 < c_i = c_{i+l}$ for some $k < i \leq i+l \leq n$, then $\gamma_{\mathbf{Z}_2}(K_{c_i}^-(n)) \geq l+1$.
(iii) If $c_{k+l} = \alpha \geq 0$ for some $1 \leq l \leq n - k$, then

$$\gamma_{\mathbf{Z}_2}(K_\alpha^-(n) \cap S_\rho(Z)) \geq l.$$

In particular φ has at least $n - k$ distinct pairs of non-trivial critical points of Morse index less than or equal to n.

Proof: Follows immediately from Lemma 10.18, Corollary 10.5 and Theorem 10.8.

Morse indices for the S^1-cohomology classes

Let $E = S^\infty$ be the *classifying space for the group* S^1. The cohomology ring (with rational coefficients) of $\mathbb{C}P^\infty = E/S^1$ is then generated by an element ω_2 of degree 2, in such a way that $H^{2k}(\mathbb{C}P^\infty) = \mathbb{Q}\omega_2^k$ and $H^{2k+1}(\mathbb{C}P^\infty) = \{0\}$. Suppose now X is a C^2-manifold with a free S^1-action. For any S^1-invariant compact subset A, define the index

$$\mathrm{Ind}_{S^1}(A) = \sup\{k; \tilde{f}^*(\omega_2)^{k-1} \neq 0\}$$

where $f : A \to E$ is a classifying map, $\tilde{f} : A/S^1 \to E/S^1$ is the map it induces on the orbit-space, and $\tilde{f}^* : H^*(\mathbb{C}P^\infty) \to H^*(A/S^1)$ is the corresponding homomorphism between the cohomology algebras. It is shown in ([F-R]), that Ind_{S^1} enjoys the properties of an index listed in Chapter 7. For each n, we let

$$\mathcal{F}_n = \{A; A \text{ compact invariant with } \text{Ind}_{S^1}(A) \geq n\}.$$

It is clearly an S^1-homotopy stable family. We claim that it is an S^1-cohomological family of dimension $2(n-1)$. Indeed, it suffices to realize that $\mathcal{F}_n = \mathcal{F}(\alpha_2^{n-1})$ where $\alpha_2 = \tilde{f}^*(\omega_2)$ and $f : X \to E$ is a classifying map. Note that α_2^{n-1} is a cohomology class in $H^{2(n-1)}(X/S^1)$.

Corollary 10.20: *Let φ be a C^2-functional on X and suppose that for each k $(1 \leq k \leq N)$ $c_k = c(\varphi, \mathcal{F}_k)$ is finite, that φ verifies $(PS)_{c_k}$ and that $d^2\varphi$ is Fredholm on the critical sets K_{c_k}. Then for each k $(1 \leq k \leq N)$, the following hold:*
(1) There exists z_k in K_{c_k} with $m(z_k) \leq 2(k-1) \leq m^*(z_k)$.
(2) If $c_{n+k} = c_k$ for some $1 \leq k \leq n+k \leq N$, then

$$\text{Ind}_{S^1}(K_{c_k}(2(n+k-1))) \geq n+1$$

and for any min-maxing sequence $(A_i)_i$ in $\mathcal{F}_{n+k}$, we have :

$$\text{Ind}_{S^1}(K_{c_k}(2(k-1)) \cap A_\infty) \geq n+1.$$

Notes and Comments: It is standard to reduce the degenerate (but Fredholm) case to the non-degenerate case, via the perturbation method of Marino-Prodi [M-P]. However, if we only assume the Palais-Smale condition on a given min-maxing sequence of sets, the situation is more delicate and Solimini [S 1] was the first to deal with such a difficulty. Yet another difficulty in the reduction arises when we require the extra information concerning the position of the critical point on a given dual set. This was done in [G 1] by a careful combination of the strong form of the min-max principle with the ideas of Lazer-Solimini [L-S] and Solimini [S 1]. Once such a result is obtained, a whole bunch of new multiplicity results about critical points with a given Morse index follow. Corollary 10.16 is an improvement of a result of Lazer-Solimini [L-S], while Corollary 10.17 extends various results of Bahri [Ba 1] and Bahri-Lions [Ba-L]. Corollary 10.20 is a refinement of results of Ekeland-Hofer [E-H 2].

11

MORSE-TYPE INFORMATION ON

PALAIS-SMALE SEQUENCES

Let φ be a C^2-functional on a Hilbertian manifold H, and suppose c is an *inf-max level* of φ over a homotopic family $\mathcal{F}$ of dimension n. We have seen in the last two chapters that, if φ verifies a (PS)-type condition and if $d^2\varphi$ is Fredholm, then one can find critical points at the level c whose Morse index is at most n. In this chapter, we study what happens when the compactness and non-degeneracy conditions on φ are not satisfied. More specifically, we want to construct a *Palais-Smale sequence* $(x_k)_k$ that carries the topological information given by $\mathcal{F}$, in addition to the properties concerning the location of $(x_k)_k$ obtained in Theorem 4.1. Actually, we are looking for an analytical (second order) property concerning the Hessian $d^2\varphi(x_k)$, which can be viewed as the *asymptotic version* of the information on the Morse index of the limit of $(x_k)_k$ whenever such a limit exists.

As was shown in Chapter 1, this can be done in minimization problems: one then gets an almost critical sequence $(x_k)_k$ that is minimizing and which satisfies

$$\langle d^2\varphi(x_k)w, w\rangle \geq -\frac{1}{k}\|w\|^2 \quad \text{for any } w \in X.$$

To obtain such sequences, we established a *smooth perturbed minimization principle* which showed that an appropriate quadratic perturbation of φ actually attains its minimum.

Analogously, one may try to establish *perturbed variational principles* for problems not involving minimization. The idea is to construct C^2-small perturbations so that the new functional has a true critical point with the Morse index that is expected for the original one. We can then transfer this information to the Hessian of the original function φ at that point, even though it is only pseudo-critical for that functional. Unfortunately, and unlike the minimization case, the problems are quite involved and we run into serious difficulties while trying to execute this program. The main difficulty being in the improvement of the perturbations from being in the space $C^{1,1}$ to C^2 which is relevant for applying the results of Chapter 10.

However, we shall now present a method for constructing directly – without establishing the perturbation result– an almost critical sequence with the appropriate second order information, provided we have an additional, but acceptable, assumption of Hölder continuity on the first and second derivatives. In return, and in the case where a (weaker) (PS) condition holds, we obtain all the results of Chapter 10 without the Fredholm-type assumption on $d^2\varphi$.

11.1 Palais-Smale sequences with second order properties– the homotopic case

Here is the main result of this section.

Theorem 11.1 (Fang-Ghoussoub):   *Let φ be a C^2-functional on a complete C^2-Riemannian manifold X with $d\varphi$ and $d^2\varphi$ Hölder continuous on X. Let $\mathcal{F}$ be a homotopic family of dimension n with boundary B and let $\mathcal{F}^*$ be a family dual to $\mathcal{F}$ such that:*

$$\sup_{F\in\mathcal{F}^*} \inf_{x\in F} \varphi(x) = \inf_{A\in\mathcal{F}} \max_{x\in A} \varphi(x) =: c.$$

Then, for every min-maxing sequence $(A_k)_k$ in $\mathcal{F}$ and every suitable max-mining sequence $(F_k)_k$ in $\mathcal{F}^*$, there exist sequences $(x_k)_k$ in X and $(\rho_k)_k$ in $\mathbb{R}^+$ with $\lim_k \rho_k = 0$ such that:
   (i) $\lim_k \varphi(x_k) = c$,
   (ii) $\lim_k d\varphi(x_k) = 0$,
   (iii) $x_k \in A_k$ for each k,
   (iv) $\lim_k \operatorname{dist}(x_k, F_k) = 0$,
   (v) for each k, $d^2\varphi(x_k)$ has at most n eigenvalues below $-\rho_k$.

By applying Theorem 11.1 in the case where the dual family consists of the singleton $F = \{\varphi \geq c\}$, we obtain the following conclusion, in the setting of the saddle point theorem.

Corollary 11.2:   *Let φ be a real-valued C^2-functional on a Hilbert space H such that φ' and φ'' are Hölder continuous. Assume that $H =$*

$H_1 \oplus H_2$ where $\dim H_1 = n$ and that

$$\max \varphi(S_{H_1}) < \inf \varphi(H_2).$$

Then, for some $c \geq \inf \varphi(H_2)$, there exist sequences $(x_k)_k$ in H and $(\rho_k)_k$ in $\mathbb{R}^+$ with $\lim_k \rho_k = 0$ such that:

(i) $\lim_k \varphi(x_k) = c$,

(ii) $\lim_k \varphi'(x_k) = 0$,

(iii) for each k, $\varphi''(x_k)$ has at most n eigenvalues below $-\rho_k$.

Moreover, if $c = \inf \varphi(H_2)$, then $(x_k)_k$ can be chosen so that it also satisfies

(iv) $\lim_k \operatorname{dist}(x_k, H_2) = 0$.

Proof: The subspace H_2 is dual to the class

$$\mathcal{F} = \{A; \exists h \in C(B_{H_1}; H), \ h(x) = x \text{ on } S_{H_1} \text{ and } A = h(B_{H_1})\}$$

and the latter is a homotopic class of dimension n with boundary S_{H_1}. Therefore, $\max \varphi(S_{H_1}) < \inf \varphi(H_2) \leq c(\varphi, \mathcal{F}) := c$ and Theorem 11.1 applies to yield our claim.

The above theorem also yields, among other things, the following result concerning the existence of some *good paths*. First, recall the notation

$$G_d = \{x \in X; \ \varphi(x) < d\}, \quad \text{and} \quad L_d = \{x \in X; \varphi(x) \geq d\}.$$

For $x \in X$ and $\rho > 0$, we define the following *approximate Morse index*:

$$m_\rho^-(x) = \sup\{\dim(E); \ E \text{ subspace of } T_x(X) \text{ with}$$
$$\langle d^2 \varphi(x) w, w \rangle < -\rho \|w\|^2 \text{ for all } w \in E\}$$

Also let

$$K_c^-(n, \varepsilon) = \{x; c - \varepsilon \leq \varphi(x) \leq c + \varepsilon, \ \|d\varphi(x)\| < \varepsilon, \ m_\varepsilon^-(x) \leq n\}$$

and

$$K_c^-(n) = \{x; \varphi(x) = c, \ d\varphi(x) = 0, \ m^-(x) \leq n\}.$$

Corollary 11.3: Let φ be a C^2-function on a complete C^2-Riemannian manifold X such that $d\varphi$ and $d^2\varphi$ are Hölder continuous and let $\mathcal{F}$ be a homotopic family of dimension n with boundary B such that $\sup \varphi(B) < c(\varphi, \mathcal{F}) =: c$. Then, for every $\varepsilon > 0$, there exist δ $(0 < \delta < \varepsilon)$ and $A \in \mathcal{F}$ such that

$$A \subseteq G_{c-\delta} \cup K_c^-(n, \varepsilon).$$

Proof: If not, then there exists $\varepsilon > 0$ such that for every $\delta > 0$, the set

$$F_\delta = L_{c-\delta} \cap (\{x; \|d\varphi(x)\| \geq \varepsilon\} \cup \{x; m_\varepsilon^-(x) > n\})$$

is dual to $\mathcal{F}$. Since $\lim_{\delta \to 0} \inf \varphi(F_\delta) = c$, Theorem 11.1 yields a sequence $(x_\delta)_\delta$ and a positive function $\rho(\delta)$ with $\lim_{\delta \to 0} \rho(\delta) = 0$ such that:

$$\lim_{\delta \to 0} \varphi(x_\delta) = c, \lim_{\delta \to 0} d\varphi(x_\delta) = 0 \text{ and } m^-_{\rho(\delta)}(x_\delta) \leq n,$$

while in the same time approaching the set

$$\{x; \|d\varphi(x)\| \geq \varepsilon\} \cup \{x; m^-_\varepsilon(x) > n\}.$$

This contradicts the uniform continuity of $d\varphi$ and $d^2\varphi$.

In view of the above result, the following definition is in order

Definition 11.4: A C^2-function on a C^2-Riemannian manifold X is said to have the *Palais-Smale condition at level c, around the set F and of order less than n* (in short $(PS)_{F,c,n^-}$), if a sequence $(x_k)_k$ in H is relatively compact whenever it verifies the following conditions: $\lim_k \varphi(x_k) = c$, $\lim_k d\varphi(x_k) = 0$, $\lim_k \text{dist}(x_k, F) = 0$ and there exists a sequence of positive reals $(\rho_k)_k$ with $\lim_k \rho_k = 0$ such that for every k, $d^2\varphi(x_k)$ has at most n eigenvalues below $-\rho_k$.

After the proof of the main Theorem, we shall give examples of functionals that satisfy $(PS)_{c,n^-}$ but not the classical Palais-Smale condition. Now we can state the following

Corollary 11.5: *Under the hypothesis of Theorem 11.1, assume that $\mathcal{F}^*$ consists of a single set F and that φ satisfies condition $(PS)_{F,c,n^-}$, then $K_c^-(n) \cap F$ is non-empty.*

Moreover, for any $\varepsilon > 0$ and any neighborhood U of $K_c^-(n) \cap F$, there exists $A \in \mathcal{F}$ such that $\sup \varphi(A) < c + \varepsilon$ and $A \subseteq (X \backslash F) \cup U$.

In particular, if $\sup \varphi(B) < c$, then for any $\varepsilon > 0$ and any neighborhood U of $K_c^-(n)$, there exists $A \in \mathcal{F}$ such that $\sup \varphi(A) < c + \varepsilon$ and $A \subseteq G_c \cup U$.

Proof: The first part follows from Theorem 11.1 and the fact that φ satisfies $(PS)_{F,c,n^-}$. Let now U be any open neighborhood of $K_c^-(n) \cap F$ and assume the second assertion not true for some $\varepsilon > 0$. This means that the set $F \backslash U$ is dual to $\mathcal{F}$ and since $\inf \varphi(F) \geq c$, it follows from the first part that $K_c^-(n) \cap (F \backslash U)$ is non-empty, which is a contradiction.

We shall split the proof of Theorem 11.1 into several lemmas. To simplify the exposition, we shall assume that X is a Hilbert space H. Since all the arguments in the proof are of local nature, they extend easily to the case of a C^2-Riemannian manifold modelled on a Hilbert space. Theorem 11.1 will follow immediately from the following *more quantitative* result.

For $u \in H$ and $\varepsilon > 0$, we shall write as $B(u, \varepsilon)$ for the open ball centered at u, with radius ε. For any subset $A \subseteq X$, $\delta > 0$, we shall write $N_\delta(A) = \{u \in X; \text{dist}(u, A) < \delta\}$ for its δ-neighborhood.

Theorem 11.6: *Let φ be a C^2-functional on a Hilbert space H. Suppose that for some $0 < \alpha \leq 1$ and $M \geq 1$, we have for all $x_1, x_2 \in X$,*

$$\|\varphi'(x_1) - \varphi'(x_2)\| \leq M\|x_1 - x_2\|^\alpha \text{ and } \|\varphi''(x_1) - \varphi''(x_2)\| \leq M\|x_1 - x_2\|^\alpha.$$

Let $\mathcal{F}$ be a homotopic family of dimension n with boundary B such that $c := \inf_{A \in \mathcal{F}} \max_{x \in A} \varphi(x)$ is finite, and let F be a dual set such that $\inf_{x \in F} \varphi(x) \geq c - \varepsilon$ for some ε verifying:

$$0 < \varepsilon \leq \min\left\{4^{-(4n+7)} M^{\frac{-4}{\alpha}} (n+1)^{-2n}, \quad M^{\frac{1}{\alpha_1}} (\frac{1}{2}\text{dist}(B; F))^{\frac{\alpha}{\alpha_1}}\right\}$$

where $0 < \alpha_1 \leq \frac{\alpha}{2(\alpha+2)} < 1$. Then for any $A \in \mathcal{F}$ with $\max_{x \in A} \varphi(x) \leq c + \varepsilon$, there exist $x_\varepsilon \in H$ such that:

(i) $c - \varepsilon \leq \varphi(x_\varepsilon) \leq c + \varepsilon$.

(ii) $\|\varphi'(x_\varepsilon)\| \leq 3\varepsilon^{\alpha_1}$.

(iii) $x_\varepsilon \in A$.

(iv) $\text{dist}(x_\varepsilon, F) \leq \varepsilon^{\alpha_1/\alpha}$.

(v) *If $\langle \varphi''(x_\varepsilon)w, w \rangle < -2\varepsilon^{\alpha_1}\|w\|^2$ for all w in a subspace E of H, then $\dim E \leq n$.*

Remark: Note that assertion (v) of Theorem 11.6 necessarily implies that $\varphi''(x_\varepsilon)$ has at most n eigenvalues below $-2\varepsilon^{\alpha_1}$. This is a direct consequence of Lemma 1.22.

Lemma 11.7: *Let φ be a C^2-functional on H, u_0 a vector in H and $\beta > 0$. Suppose there is a non-zero subspace E of H and $\delta > 0$ such that:*

$$\langle \varphi''(u)w, w \rangle < -\beta\|w\|^2 \text{ for all } u \in B(u_0, \delta) \text{ and } w \in E. \quad (1)$$

Let P be the projection from H onto E and set $w_1(u) = -\frac{P\varphi'(u)}{\|P\varphi'(u)\|}$ if $P\varphi'(u) \neq 0$ and $w_1(u) = 0$ otherwise. Then for any $u \in B(u_0, \frac{\delta}{2})$, the following holds:

(1) *If $w_1(u) \neq 0$, then for all $0 < t < \frac{\delta}{2}$, we have*

$$\varphi(u + tw_1(u)) < \varphi(u) - \frac{\beta}{2}t^2.$$

(2) *If $w_1(u) = 0$, then for all $0 < t < \frac{\delta}{2}$, we have*

$$\varphi(u + tw) < \varphi(u) - \frac{\beta}{2}t^2 \text{ for all } w \in E \text{ with } \|w\| = 1.$$

Moreover, for any $0 < t_0 < \delta/2$, there is δ_u with $0 < \delta_u < \frac{\delta}{2} - t_0$ such *that for every $v \in \overline{B}(u, \delta_u)$ and any $t_0 \le t < \frac{\delta}{2}$ we have*

$$\varphi(v + tw) < \varphi(v) - \frac{\beta}{2}t^2 \text{ for all } w \in E \text{ with } \|w\| = 1.$$

Proof: (1) If $w_1(u) \ne 0$ then there is τ $(0 < \tau < 1)$ such that for all $0 < t < \frac{\delta}{2}$ we have

$$\varphi(u + tw_1(u)) = \varphi(u) + t\langle\varphi'(u), w_1(u)\rangle$$
$$+ \frac{t^2}{2}\langle\varphi''(u + t\tau w_1(u))w_1(u), w_1(u)\rangle$$
$$< \varphi(u) + t\langle P\varphi'(u), -\frac{P\varphi'(u)}{\|P\varphi'(u)\|}\rangle - \frac{t^2}{2}\beta$$
$$< \varphi(u) - \frac{t^2}{2}\beta.$$

(2) If $w_1(u) = 0$, then for $0 < t < \frac{\delta}{2}$ and $w \in E$ with $\|w\| = 1$, there is τ $(0 < \tau < 1)$ such that:

$$\varphi(u + tw) = \varphi(u) + t\langle\varphi'(u), w\rangle + \frac{t^2}{2}\langle\varphi''(u + t\tau w)w, w\rangle < \varphi(u) - \frac{t^2}{2}\beta.$$

Moreover, there is δ_u $(0 < \delta_u < \frac{\delta}{2} - t_0)$ such that for all $v \in \overline{B}(u, \delta_u)$ we have that $\|\varphi'(v) - \varphi'(u)\| \le \frac{\beta}{4}t_0^2$. Now again, there is τ $(0 < \tau < 1)$ such that for all $v \in \overline{B}(u, \delta_u)$, $t_0 \le t < \frac{\delta}{2}$ and all $w \in E$ with $\|w\| = 1$, we have

$$\varphi(v + tw) = \varphi(v) + t\langle\varphi'(v), w\rangle + \frac{1}{2}t^2\langle\varphi''(v + t\tau w)w, w\rangle$$
$$< \varphi(v) + \|\varphi'(v) - \varphi'(u)\| - \frac{1}{2}t^2\beta$$
$$\le \varphi(v) - \frac{t^2}{4}\beta.$$

Lemma 11.8: In Lemma 11.7, we further assume that $\dim E \ge n + 1$. Let f be a continuous map from a closed subset D of $\mathbb{R}^n$ into H and let K be a compact subset of D such that $f(K) \subset B(u^0, \frac{\delta}{2})$, then for any $\nu > 0$ and $0 < t_0 < \frac{\delta}{2}$, there is a continuous map $\sigma : \mathbb{R} \times D \to H$ such that:

(i) $\sigma(t, x) = f(x)$ if $(t, x) \in (\{0\} \times D) \cup (\mathbb{R} \times D \backslash N_\nu(K))$.
(ii) $\varphi(\sigma(t, x)) \le \varphi(f(x))$ if $(t, x) \in [0, \delta/2) \times D$.
(iii) $\varphi(\sigma(t, x)) < \varphi(f(x)) - \frac{\beta}{4}t^2$ if $(t, x) \in [t_0, \delta/2) \times K$.
(iv) $\|\sigma(t, x) - f(x)\| \le t$ for all $(t, x) \in \mathbb{R} \times D$.

Proof: We can clearly assume that $f(N_\nu(K) \cap D) \subset B(u^0, \delta/2)$. Let $T = \{x \in N_\nu(K) \cap D; w_1(f(x)) = 0\}$. According to Lemma 11.7, we can find for each $y \in T$, an open ball $B(y, \nu^y)$ in $\mathbb{R}^n$ such that for all $w \in E$

with $\|w\| = 1$ and $z \in B(y, \nu^n) \cap D$ we have that

$$\varphi(f(z) + tw) < \varphi(f(z)) - \frac{\beta}{4}t^2 \qquad t_0 \le t < \frac{\delta}{2}.$$

Put $O = \cup_{y \in T} B(y, \nu^y/2)$ and let $g : \mathbb{R}^n \to [0, 1]$ be a continuous function such that:

$$g(x) = \begin{cases} 1 & x \in K \\ 0 & x \in \mathbb{R}^n \backslash N_\nu(K). \end{cases}$$

Next consider the continuous map $f_1 : D\backslash O \to E$ defined by $f_1(x) = w_1(f(x))$. Clearly $\|f_1(x)\| = 1$ for all $x \in (N_\nu(K) \cap D)\backslash O$. If the latter set is empty, we just let $f_1 \equiv e$ for some fixed $e \in E, \|e\| = 1$. Since $(N_\nu(K) \cap D)\backslash O \subseteq \mathbb{R}^n$ and $\dim E \ge n+1$, there exists a continuous map $f_2 : \mathbb{R}^n \to H$ such that $\|f_2(x)\| = 1$ for all $x \in \mathbb{R}^n$ and $f_2(x) = f_1(x)$ on $(N_\nu(K) \cap D)\backslash O$ (Appendix D.9). Now let $\sigma(t, x) = f(x) + t f_2(x)g(x)$ on $\mathbb{R} \times D$. Clearly $\sigma(t, x)$ verifies the claims of the Lemma.

We shall also need the following combinatorial result.

Lemma 11.9: *For $n \in \mathbb{N}$, there is an integer $N(n) \le (2\sqrt{n+1} + 2)^n$ such that for any compact subset $D \subseteq \mathbb{R}^n$ and any $\varepsilon > 0$, there exist a finite number of distinct points $\{x_i; 1 \le i \le k\}$ with the following properties:*

(i) $D \subseteq \cup_{i=1}^k B(x_i; \varepsilon/4) \subseteq N_{\varepsilon/2}(D)$.

(ii) The intersection of any $N(n)$ elements of the cover $(\overline{B}(x_i, \varepsilon/2))_{i=1}^k$ is empty.

Proof: Let $Q = \{l = (l_1, l_2, \cdots, l_n) \in \mathbb{Z}^n \subset \mathbb{R}^n\}$. Clearly, the family of balls of the form $\{B(\frac{\varepsilon}{2\sqrt{n+1}}l, \frac{\varepsilon}{4}); l \in Q\}$ will cover $\mathbb{R}^n$. By the compactness of D, there are $x_1, ..., x_k \in \mathbb{R}^n$ such that $D \subseteq \cup_{i=1}^k B(x_i; \varepsilon/4) \subseteq N_{\varepsilon/2}(D)$. Note now that for each $l \in Q$, $B(\frac{\varepsilon}{2\sqrt{n+1}}l, \frac{\varepsilon}{2})$ intersects at most $(2\sqrt{n+1} + 2)^n$ distinct balls of that type.

Lemma 11.10: *Let φ be a real-valued C^2-functional on H and let f be a continuous map from a closed subset D of $\mathbb{R}^n$ into H. Suppose that K is a compact subset of D with the following property.*

There exist two constants $\hat{\delta} > 0$, $\beta > 0$ such that for all $y \in K$ there is a subspace E_y of H with $\dim E_y \ge n+1$ so that for all $x \in B(f(y), \hat{\delta})$, we have

$$\langle \varphi''(x)w, w \rangle < -\beta \|w\|^2 \text{ for all } w \in E_y. \tag{2}$$

Then, for any δ $(0 < \delta \le \hat{\delta})$ and $\nu > 0$ there is a continuous map $\hat{f} : D \to H$ such that, if $N := N(n)$ is the number given by Lemma 11.9, we have:

(i) $\hat{f}(x) = f(x)$ for $x \in D \backslash N_\nu(K)$.

(ii) $\varphi(\hat{f}(x)) \leq \varphi(f(x))$ for all $x \in D$.

(iii) If $x \in K$, then $\varphi(\hat{f}(x)) < \varphi(f(x)) - \frac{\beta\delta^2}{16N^2}$.

(iv) $\|\hat{f}(x) - f(x)\| \leq \delta/2$ for all $x \in D$.

Proof: Let $0 < \delta \leq \hat{\delta}$ be fixed. For any $y \in K$ we can choose a ball B_y in $\mathbb{R}^n$ such that $\overline{B}_y \subseteq N_\nu(K)$ and $f(x) \in B(f(y), \frac{\delta}{8N})$ for all $x \in \overline{B}_y \cap D$. Since K is compact, there is a finite subcovering $B_{y_1}, \cdots, B_{y_m}$ of K. Choose $0 < \tau < \nu$ small enough such that $\cup_{i=1}^m N_\tau(B_{y_i}) \subseteq N_\nu(K)$ and for $1 \leq i \leq m$,

$$f(x) \in B(f(y_i), \frac{\delta}{4N}) \text{ if } x \in \overline{N}_\tau(B_{y_i}) \cap D.$$

By using Lemma 11.9, we may assume that any N possible distinct $\overline{N}_\tau(B_{y_i})$'s have an empty intersection.

We shall now define by induction, continuous functions $f_0, f_1, \cdots, f_m : D \to H$ such that for all $1 \leq i \leq m$ we have that

$$\varphi(f_i(x)) < \varphi(f_{i-1}(x)) - \frac{\delta^2}{16N^2} \text{ if } x \in \overline{B}_{y_i} \cap D, \tag{3}$$

$$\varphi(f_i(x)) \leq \varphi(f_{i-1}(x)) \text{ if } x \in D, \tag{4}$$

$$\|f_i(x) - f_{i-1}(x)\| \leq \begin{cases} 0 & x \in D \backslash N_\tau(\overline{B}_{y_i}) \\ \frac{\delta}{2N} & x \in N_\tau(\overline{B}_{y_i}) \cap D \end{cases} \tag{5}$$

Let $f_0 = f$ and suppose that $f_0, f_1, \cdots, f_k$ are well defined and satisfy (3), (4) and (5) for $k < m$. Clearly

$$\|f_i(x) - f(x)\| \leq \frac{i\delta}{2N} \quad \text{if } x \in \cap_{j=1}^i N_\tau(\overline{B}_{y_i}) \cap D.$$

Since any intersection of N distinct sets $N_\tau(B_{y_i})$ is empty, we have that

$$\|f_k(x) - f(x)\| \leq \frac{\delta(N-1)}{2N} \text{ if } x \in D.$$

Since $f : \overline{B}_{y_{k+1}} \cap D \to B(f(y_{k+1}), \frac{\delta}{4N})$, we see that f_k maps $\overline{B}_{y_{k+1}} \cap D$ into $B(f(y_{k+1}), \frac{\delta}{2}(1 - \frac{1}{N}) + \frac{\delta}{4N}) \subset B(f(y_{k+1}), \delta/2)$. By assumption (2), there is some subspace $E_{y_{k+1}}$ of H with dim $E_{y_{k+1}} \geq n + 1$ such that for any $u \in B(f(y_{k+1}), \delta)$ and $w \in H_{y_{k+1}}$ with $\|w\| = 1$, we have that $\langle \varphi''(u)w, w \rangle < -\beta$. Hence, we may apply Lemma 11.8 with f_k and any $0 < t_0 < \frac{\delta}{2N}$, to obtain a continuous deformation $\sigma(t, x)$ satisfying the conclusion of that lemma. Define now $f_{k+1}(x) = \sigma(\frac{\delta}{2N}, x)$ to get a continuous function $f_{k+1} : D \to H$ satisfying

$$\varphi(f_{k+1}(x)) < \varphi(f_k(x)) - \frac{\delta^2}{16N^2} \text{ for } x \in \overline{B}_{y_{k+1}} \cap D,$$

$$\varphi(f_{k+1}(x)) \leq \varphi(f(x)) \text{ for } x \in D,$$

$$\|f_{k+1}(x) - f_k(x)\| \le \begin{cases} 0 & x \in D \backslash N_\tau(\overline{B}_{y_{k+1}}) \\ \frac{\delta}{2N} & x \in N_\tau(\overline{B}_{y_{k+1}}) \cap D. \end{cases}$$

By induction we see that $f_0, \cdots, f_m$ are well defined. Clearly $\hat{f} = f_m$ verifies the claims of the lemma.

Lemma 11.11: *Let φ be a real valued C^1-functional on H and let f be a continuous map from a complete metric space D into H. Suppose K is a compact subset of D such that for two constants $\hat{\delta} > 0$ and $\beta > 0$, the following holds:*

$$\|\varphi'(u)\| > \beta \quad \text{for all } u \in N_{\hat{\delta}}(f(K)). \tag{6}$$

Then, for any δ $(0 < \delta \le \hat{\delta})$ and any $\nu > 0$, there is a continuous map $\hat{f} : D \to H$ such that:
(i) $\hat{f}(x) = f(x)$ for $x \in D \setminus N_\nu(K)$.
(ii) $\varphi(\hat{f}(x)) \le \varphi(f(x))$ for all $x \in D$.
(iii) $\varphi(\hat{f}(x)) \le \varphi(f(x)) - \beta\delta$ for $x \in K$.
(iv) $\|\hat{f}(x) - f(x)\| \le \delta$ for $x \in D$.

Proof: Step 1: We first claim that there exists σ with $0 < \sigma \le \delta$ such that (i), (ii) and (iii) and (iv) hold with σ replacing δ.

Indeed, For any $y \in K$, let B_y be an open ball in D containing y such that for all $x \in B_y$, $f(x) \in B(f(y), \frac{\delta}{4})$. Hypothesis (6) coupled with the continuity of φ' yields $u_y \in H$ with $\|u_y\| = 1$ and a $\sigma_y > 0$ such that for all $v \in H$ with $\|v\| \le \sigma_y$, we have

$$\langle \varphi'(f(x) + v), u_y \rangle < -\beta \quad \text{for all } x \in B_y. \tag{7}$$

Since K is compact, there is a finite subcovering $B_{y_1}, \ldots, B_{y_k}$ of K. We may assume $\cup_{i=1}^k B_{y_i} \subseteq N_\nu(K)$ and let us define $\psi_j : D \to [0,1]$ $(j = 1, 2, \ldots, k)$ by

$$\psi_j(x) = \frac{\text{dist}(x; D \setminus B_{y_j})}{\sum_{i=1}^n \text{dist}(x; D \setminus B_{y_i})}. \tag{8}$$

Clearly ψ_j is continuous on $\cup_{i=1}^k B_{y_i}$ and $\sum_{j=1}^k \psi_j(x) = 1$ for x in $\cup_{i=1}^k B_{y_i}$. Next let $\psi : D \to [0,1]$ be continuous such that:

$$\psi(x) = \begin{cases} 1 & x \in K \\ 0 & x \in D \setminus \cup_{i=1}^k B_{y_i}. \end{cases} \tag{9}$$

Take $\sigma = \min\{\sigma_{y_1}, \ldots, \sigma_{y_k}, \delta\}$ and define

$$\hat{f}(x) = f(x) + \sigma\psi(x) \sum_{j=1}^k \psi_j(x) u_{y_j}.$$

Clearly $\|\hat{f}(x) - f(x)\| \le \sigma$. On the other hand, by the mean value theorem, for any $x \in D$, there is τ $(0 < \tau < 1)$ such that:

$$\varphi(\hat{f}(x)) - \varphi(f(x))$$

$$= \langle \varphi'(f(x) + \tau\sigma\psi(x)\sum_{j=1}^{k}\psi_j(x)u_{y_j}), \sigma\psi(x)\sum_{i=1}^{k}\psi_i(x)u_{y_i}\rangle$$

$$= \sigma\psi(x)\sum_{i=1}^{k}\psi_i(x)\langle\varphi(f(x) + \tau\sigma\varphi(x)\sum_{j=1}^{k}\psi_j(x)u_{y_j}), u_{y_j}\rangle$$

$$< -\beta\sigma\psi(x).$$

In view of (9), we see that $\hat{f}$ verifies (i)-(iv) of the lemma with σ instead of δ.

Step 2: We shall now iterate the above process and construct by transfinite induction a transfinite family of continuous functions $(f_\alpha)_\alpha$ from D into H, and an increasing family of reals $(\delta_\alpha)_\alpha$ with $0 \le \delta_\alpha \le \hat{\delta}$, such that for every α:

(a) $f_\alpha(x) = f(x)$ for $x \in D \setminus N_\nu(K)$ and $\varphi(f_\alpha(x)) \le \varphi(f(x))$ for all $x \in D$.

(b) $\varphi(f_\alpha(x)) < \varphi(f(x)) - \beta\delta_\alpha$ for $x \in K$.

(c) $\|f_{\alpha+1}(x) - f_\alpha(x)\| \le \delta_{\alpha+1} - \delta_\alpha$ for $x \in D$.

(d) $\|\varphi'(u)\| > \beta$ for all $u \in N_{\hat{\delta}-\delta_\alpha}(f_\alpha(K))$.

To do that, we start with $f_0 = f$.

If $\alpha = \gamma + 1$ and if $\delta_\gamma < \hat{\delta}$, apply step (1) to f_γ and $\hat{\delta} - \delta_\gamma$ to find σ_γ with $0 < \sigma_\gamma \le \hat{\delta} - \delta_\gamma$ and the function $\hat{f}_\gamma$ satisfying the above properties (i), (ii), (iii) and (iv). Set now $\delta_\alpha = \delta_\gamma + \sigma_\gamma$ and $f_\alpha = \hat{f}_\gamma$.

If now α is a limit ordinal: i.e $\alpha = \lim_n \gamma_n$ with $\gamma_n < \alpha$, then (iii) implies that $(f_{\gamma_n})_n$ is a Cauchy sequence and we let f_α be its limit. Also set δ_α to be the limit of the increasing and bounded sequence δ_{γ_n}. This completes the construction.

Now note that the increasing family of reals $(\delta_\alpha)_\alpha$ is bounded above by $\hat{\delta}$, hence there is γ_0 before the first uncountable ordinal Ω such that $\delta_\alpha = \delta_{\gamma_0}$ for all $\alpha \ge \gamma_0$. It follows that $\delta_{\gamma_0} = \hat{\delta}$ and therefore $\hat{f} = f_{\gamma_0}$ verifies the claim of Lemma 11.11.

Lemma 11.12: *Let φ be a real-valued C^2-functional on H and let f be a continuous map from a closed subset D of $\mathbb{R}^n$ into H. Let $\beta_1 > 0$, $\beta_2 > 0$ and $\hat{\delta} > 0$ be fixed constants and suppose K is a compact subset of D with the following property.*

For all $x \in K$, either $\|\varphi'(u)\| > \beta_1$ for all $u \in B(f(x), \hat{\delta})$ or there is a subspace E_x of H with $\dim E_x \ge n + 1$ such that for all $u \in B(f(x), \hat{\delta})$ and $w \in E_x$, we have that $\langle \varphi''(u)w, w\rangle < -\beta_2\|w\|^2$.

Then, for any δ $(0 < \delta \le \frac{\hat{\delta}}{2})$ and $\nu > 0$, there is a continuous map $\hat{f}$

from D into H such that, if $N := N(n)$ is the number given in Lemma 11.9, we have

(i) $\hat{f}(x) = f(x)$ for $x \in D\backslash N_\nu(K)$.

(ii) $\varphi(\hat{f}(x)) \leq \varphi(f(x))$ for $x \in D$.

(iii) $\varphi(\hat{f}(x)) < \varphi(f(x)) - \mu\delta$ for $x \in K$ where $\mu = \min\{\frac{7}{16}\beta_1, \frac{\beta_2\delta}{16N^2}\}$.

(iv) $\|\hat{f}(x) - f(x)\| < \delta$ for $x \in D \cap N_\nu(K)$.

Proof: Let $T_1 = \{x \in K; \|\varphi'(u)\| > \beta_1$ for all $u \in B(f(x), \hat{\delta})\}$ and $T_2 = K\backslash T_1$. Clearly, for any $x \in \overline{T}_1$ we have that $\|\varphi'(u)\| > \beta_1$ for all $u \in B(f(x), \frac{15}{16}\hat{\delta})$. Also for any $x \in \overline{T}_2$, there is a subspace E_x of H with $\dim E_x \geq n + 1$ such that for all $u \in B(f(x), \frac{15\hat{\delta}}{16})$ and $w \in E_x$ with $\|w\| = 1$ we have that $\langle\varphi''(u)w, w\rangle < -\beta_2$. Note that $\overline{T}_1, \overline{T}_2$ are compact and $K = \overline{T}_1 \cup \overline{T}_2$. Apply now Lemma 11.11 with $\frac{7}{16}\delta$ and $\frac{\nu}{2} > 0$ to obtain a continuous map $g : D \to H$ such that:

$$g(x) = f(x) \text{ for } x \in D\backslash N_{\frac{\nu}{2}}(\overline{T}_1) \text{ and } \varphi(g(x)) \leq \varphi(f(x)) \text{ for } x \in D, \quad (10)$$

$$\varphi(g(x)) < \varphi(f(x)) - \frac{7}{16}\delta\beta_1 \text{ for } x \in \overline{T}_1, \quad (11)$$

$$\|g(x) - f(x)\| \leq \frac{7}{16}\delta \text{ for } x \in N_{\frac{\nu}{2}}(\overline{T}_1) \cap D. \quad (12)$$

But (12) yields that for $x \in \overline{T}_2$, $B(g(x), \frac{\hat{\delta}}{2}) \subseteq B(f(x), \frac{15}{16}\hat{\delta})$. Apply now Lemma 11.10 with $\delta \leq \frac{\hat{\delta}}{2}$ and $\frac{\nu}{2}$ to obtain a continuous map $\hat{f} : D \to H$ such that

$$\hat{f}(x) = g(x) \text{ for } x \in D\backslash N_{\frac{\nu}{2}}(\overline{T}_2) \text{ and } \varphi(\hat{f}(x)) \leq \varphi(g(x)) \text{ for } x \in D, \quad (13)$$

$$\varphi(\hat{f}(x)) < \varphi(g(x) - \frac{\beta_2\delta^2}{16N^2} \text{ for } x \in \overline{T}_2, \quad (14)$$

$$\|\hat{f}(x) - g(x)\| \leq \frac{\delta}{2} \text{ for } x \in N_{\frac{\nu}{2}}(\overline{T}_2) \cap K. \quad (15)$$

Clearly $\hat{f}$ verifies the claims of the lemma.

Proof of Theorem 11.6: Suppose $\max_{u \in A} \varphi(u) \leq c + \varepsilon$ where A is a set in $\mathcal{F}$. There exists then a continuous function f from $D \subset \mathbb{R}^n$ into H, which is equal to σ on D_0 and such that $A = f(D)$.

Let $\delta := \delta(\varepsilon) = \frac{1}{2}\varepsilon^{\frac{\alpha_1}{\alpha}} M^{-\frac{1}{\alpha}}$ and consider the closed set

$$K = \{x \in D; \varphi(f(x)) \geq c - \varepsilon \text{ and } f(x) \in \overline{N}_{2\delta}(F) \cap A\}. \quad (16)$$

Since $\varepsilon \leq M^{\frac{1}{\alpha_1}}(\frac{1}{2}\text{dist}(F; B))^{\frac{\alpha}{\alpha_1}}$, we have that $2\delta \leq \frac{1}{2}\text{dist}(F; B)$ and hence K is a compact subset of $D\backslash D_0$.

Suppose now that the conclusion of Theorem 11.6 does not hold, then for all $x \in K$, we have that either $\|\varphi'(f(x))\| > 3M\delta^\alpha$ or there is a subspace E_x of H with $\dim E_x \geq n + 1$ such that for all $w \in E_x$, we have that $\langle \varphi''(f(x))w, w \rangle < -2\varepsilon^{\alpha_1} \|w\|^2$.

Let $\nu = \frac{1}{2} \operatorname{dist}(D_0, K)$ in such a way that $N_\nu(K) \subset \mathbb{R}^n \backslash D_0$. In view of the Hölder continuity assumption, the hypotheses of Lemma 11.12 are satisfied with $\hat{\delta} = 2\delta(\varepsilon)$, $\nu = \frac{1}{2} \operatorname{dist}(D_0, K)$, $\beta_1 = M\delta^\alpha$ and $\beta_2 = \varepsilon^{\alpha_1}$. Hence, we can find a continuous map $\hat{f} : D \to H$ such that:

$$\hat{f}(x) = f(x) \text{ for } x \in D \backslash N_\nu(K) \text{ and } \varphi(\hat{f}(x)) \leq \varphi(f(x) \text{ for } x \in D, \quad (17)$$

$$\varphi(\hat{f}(x)) < \varphi(f(x)) - \mu\delta \text{ for } x \in K, \quad (18)$$

where $\mu = \min\{\frac{7}{16} M\delta^\alpha, \frac{\varepsilon^{\alpha_1}}{16N^2} \delta\}$, ($N$ given in Lemma 11.9) and

$$\|\hat{f}(x) - f(x)\| \leq \delta \text{ for all } x \in D. \quad (19)$$

Note that $\hat{f}(D) \in \mathcal{F}$ and $\hat{f}(D) \subseteq N_{2\delta}(A)$. To get a contradiction, we shall estimate $\inf \varphi(\hat{f}(D) \cap F)$. For any $x \in D$ such that $\hat{f}(x) \in F$, we have that

$$f(x) \in N_{2\delta}(F) \cap A \text{ and } \varphi(f(x)) \geq \varphi(\hat{f}(x)) \geq c - \varepsilon.$$

Hence $x \in K$ and by (17) we have that $\varphi(\hat{f}(x)) < \varphi(f(x)) - \mu\delta$.

Note now that

$$\frac{7}{16} M\delta^{1+\alpha} \geq \frac{7}{64} M^{\frac{-1}{\alpha}} \varepsilon^{\alpha_1 + \frac{\alpha_1}{\alpha}}$$

and

$$\frac{\varepsilon^{\alpha_1}}{16N^2} \delta^2 \geq \frac{\varepsilon^{\alpha_1}}{64N^2} M^{-\frac{2}{\alpha}} \varepsilon^{\frac{2\alpha_1}{\alpha}} = \frac{1}{64N^2} M^{\frac{-2}{\alpha}} \varepsilon^{\alpha_1 + \frac{2\alpha_1}{\alpha}}.$$

Since $M \geq 1$, we have that

$$\frac{7}{64} M^{-1/\alpha} \varepsilon^{\alpha_1 + \alpha_1/\alpha} \geq \frac{1}{64N^2} M^{-2/\alpha} \varepsilon^{\alpha_1 + 2\alpha_1/\alpha}$$

and the latter is larger than 2ε, since the hypothesis combined with Lemma II.5 yields that

$$\varepsilon < 4^{-(4n+7)} (n+1)^{-2n} M^{\frac{-4}{\alpha}} < 2^{-14} N^{-4} M^{\frac{-4}{\alpha}}.$$

It follows that $\varphi(\hat{f}(x)) < c + \varepsilon - 2\varepsilon = c - \varepsilon$ which contradicts the assumption that $\inf \varphi(F) \geq c - \varepsilon$. The proof of the theorem is complete.

Remark 11.13: Suppose G is a compact Lie group acting freely and differentiably on a C^2-Riemannian manifold X and suppose that φ is now a G-invariant C^2-functional on X. If $\mathcal{F}$ is now a G-homotopic family of dimension n, we can then get the same result as above in this equivariant case, by just applying Theorem 11.1 to the function $\tilde{\varphi}$ which is induced unambiguously by φ on the orbit manifold X/G.

11.2 Multiple solutions for the Hartree-Fock equations

We recall the Hartree-Fock equations already mentioned in Chapter

1. Consider the purely Coulombic N-body Hamiltonian

$$H = -\sum_{i=1}^{N} \Delta_{x_i} + \sum_{i=1}^{N} V(x_i) + \sum_{i<j} \frac{1}{|x_i - x_j|} \qquad (1)$$

where $V(x) = -\sum_{j=1}^{m} z_j |x - \bar{x}_j|^{-1}$, $m \geq 1, z_j > 0, \bar{x}_j \in \mathbb{R}^3$ are fixed. We write $Z = \sum_{j=1}^{m} z_j$ for the total charge of the nucleii.

We are looking for the critical points of the functional

$$\varphi(u_1, ..., u_N) = \sum_{i=1}^{N} \int_{\mathbb{R}^3} |\nabla u_i|^2 + V|u_i|^2 dx$$

$$+ \frac{1}{2} \iint_{\mathbb{R}^3 \times \mathbb{R}^3} \varrho(x) \frac{1}{|x - y|} \varrho(y) dx dy \qquad (2)$$

$$- \frac{1}{2} \iint_{\mathbb{R}^3 \times \mathbb{R}^3} \frac{1}{|x - y|} |\varrho(x, y)|^2 dx dy$$

over the manifold

$$M = \{(u_1, ..., , u_N) \in H^1(\mathbb{R}^3)^N; \int_{\mathbb{R}^3} u_i u_j^* dx = \delta_{ij}\}, \qquad (3)$$

where z^* denotes the conjugate of the complex number z, while $\varrho(x) = \sum_{i=1}^{N} |u_i|^2(x)$ is the density and $\varrho(x, y) = \sum_{i=1}^{N} u_i(x) u_i^*(y)$ is the density matrix associated to $(u_1, ..., u_n)$.

The Euler-Lagrange equations corresponding to the above problem (after a suitable diagonalization) are the following so-called Hartree-Fock equations: For $1 \leq i \leq N$,

$$-\Delta u_i + V u_i + (\rho * \frac{1}{|x|}) u_i - \int_{\mathbb{R}^3} \rho(x, y) \frac{1}{|x - y|} u_i(y) dy + \varepsilon_i u_i = 0 \quad (4)$$

where for each i, $\lambda_i = -\varepsilon_i$ is the Lagrange multiplier and $(u_1, ..., u_N) \in M$.

Theorem 11.14: *Assume $Z > N$. Then, there exists infinitely many solutions for the Hartree-Fock equation (4).*

We shall first show the following.

Claim 1: The functional φ restricted to M verifies $(PS)_{c,k-}$ for every $c \in \mathbb{R}$ and any $k \in \mathbb{N}$.

Proof: Let $(u^n)_n = (u_1^n, ..., u_N^n) \in M$ be a sequence satisfying $(\varphi(u^n))_n$ bounded, $(\varphi_{|M})'(u^n) \to 0$ and

$$\langle (\varphi_{|M})''(u^n)v, v \rangle \geq -\gamma^n \|v\|^2 \text{ for every } v \text{ in } E_n, \qquad (5)$$

where E_n is a subspace of $T_{u^n}(M)$ of codimension at most k and where $(\gamma_n)_n$ is a sequence of scalars decreasing to 0. The first and second derivative conditions yield the existence of $(\varepsilon_1^n, ..., \varepsilon_N^n) \in \mathbb{R}^N$ such that the following holds in $L^2(\mathbb{R}^3)$:

$$\lim_n \left(-\Delta u_i^n + V u_i^n + \left(\varrho^n * \frac{1}{|x|} \right) u_i^n \right.$$
$$\left. - \int_{\mathbb{R}^3} \varrho^n(x, y) u_i^n(y) \frac{1}{|x-y|} \, dy + \varepsilon_i^n u_i^n \right) = 0 \tag{6}$$

and

$$\sum_{i=1}^N \int_{\mathbb{R}^3} |\nabla w_i|^2 + V|w_i|^2 + \left(\varrho^n * \frac{1}{|x|} \right) |w_i|^2 + (\varepsilon_i^n + \gamma^n)|w_i|^2 \, dx$$
$$- \iint_{\mathbb{R}^3 \times \mathbb{R}^3} \varrho^n(x, y) \frac{1}{|x-y|} w_i(x) w_i(y) \, dx \, dy \tag{7}$$
$$- \frac{1}{2} \iint_{\mathbb{R}^3 \times \mathbb{R}^3} [K^n(x, y) - K^n(x) K^n(y)] \frac{1}{|x-y|} \, dx \, dy \geq 0$$

for all $(w_1, ..., w_N)$ in a subspace of $H^1(\mathbb{R}^3)$ of codimension at most $k + N$, where $K^n(x, y) = \sum_i u_i^n(x) w_i(y) + w_i(x) u_i^n(y)$ and $K^n(x) = K^n(x, x)$.

We shall now use the second order information in (7) to find lower bounds for ε_i^n. Observe first that (7) implies in particular, that we have for each fixed i,

$$\int_{\mathbb{R}^3} (|\nabla w|^2 + V|w|^2 + \left(\varrho^n * \frac{1}{|x|} \right) |w|^2) \, dx + (\varepsilon_i^n + \gamma^n) \int_{\mathbb{R}^3} |w|^2 \, dx \geq 0, \tag{8}$$

for all w in a closed subspace of $H^1(\mathbb{R}^3)$ of codimension at most $k + N$.

By Lemma 1.22, this implies that the Schrödinger operator $H_n = -\Delta + V + \varrho^n * \frac{1}{|x|}$ has at most $k + N$ eigenvalues less than $-(\varepsilon_i^n + \gamma^n)$. On the other hand, since $Z > N$, we may use Lemma 1.23 to find $\delta > 0$ such that H_n admits for all n at least $k + N$ eigenvalues below $(-\delta)$. It follows that $\varepsilon_i^n + \gamma^n \geq \delta$ which means that for n large enough, we have for every i,

$$\varepsilon_i^n \geq \delta/2 > 0. \tag{9}$$

To finish the proof of the claim, one first notices that the Palais -Smale sequence $(u^n)_n$ is bounded in $H^1(\mathbb{R}^3)^N$. Indeed, by the Cauchy-Schwarz inequality we have

$$|\rho(x, y)|^2 \leq \rho(x) \rho(y) \text{ on } \mathbb{R}^3 \times \mathbb{R}^3. \tag{10}$$

On the other hand, the following inequality holds for any $\bar{x} \in \mathbb{R}^3$ and any $u \in H^1(\mathbb{R}^3)$

$$\int_{\mathbb{R}^3} \frac{1}{|x-\bar{x}|} |u(x)|^2 dx \leq C \|u\|_{L^2(\mathbb{R}^3)} \|\nabla u\|_{L^2(\mathbb{R}^3)}. \tag{11}$$

for some C independent of $\bar{x}$ and u. Combining (10) and (11) with the information that $\varphi(u_1^n, ..., u_N^n)$ is bounded above, we can easily deduce the H^1- bound on $(u^n)_n$. We can also deduce that ε_i^n is bounded, and thus we may assume – by extracting subsequences if necessary – that u_i^n converges weakly in $H^1(\mathbb{R}^3)$ (and a.e. in $\mathbb{R}^3$) to some u_i and that ε_i^n converges to ε_i which satisfies $\varepsilon_i \geq \delta/2 > 0$ because of (9). Passing to the limit in (6), we get

$$-\Delta u_i + V u_i + \left(\varrho * \frac{1}{|x|}\right) u_i + \varepsilon_i u_i - \int_{\mathbb{R}^3} \varrho(x, y) \frac{1}{|x - y|} u_i(y) dy = 0. \quad (12)$$

In particular, we find that

$$\limsup_n \sum_i \varepsilon_i^n \int_{\mathbb{R}^3} |u_i^n|^2 dx = -\liminf_n \left\{ \sum_i \int_{\mathbb{R}^3} |\nabla u_i^n|^2 + V|u_i^n|^2 dx \right.$$

$$+ \iint_{\mathbb{R}^3 \times \mathbb{R}^3} \{\varrho^n(x)\varrho^n(y) - |\varrho^n(x, y)|^2\} \frac{1}{|x - y|} dx dy \Bigg\}$$

$$\leq -\left\{ \sum_i \int_{\mathbb{R}^3} |\nabla u|^2 + V|u_i|^2 dx \right.$$

$$+ \iint_{\mathbb{R}^3 \times \mathbb{R}^3} \{\varrho(x)\varrho(y) - |\varrho(x, y)|^2\} \frac{1}{|x - y|} dx dy \Bigg\}$$

$$= \sum_i \varepsilon_i \int_{R^3} |u_i|^2 dx.$$

Hence $\lim_n \|u_i^n\|_2 = \|u_i\|_2$ for every i, and consequently, u_i^n converges strongly in $L^2(\mathbb{R}^3)$ to u_i. The proof of Claim 1 is complete.

To set up the min-max principles, we consider for each $k \in \mathbb{N}$, the following $\mathbb{Z}_2$-homotopy stable family

$$\mathcal{F}_k = \{A; A = f(S^{k-1}) \text{ for some odd and continuous } f : S^{k-1} \to M\}.$$

Note that, according to our terminology, each $\mathcal{F}_k$ is a homotopic class of dimension k. Let $c_k = \inf_{A \in \mathcal{F}_k} \max_{x \in A} \varphi(x)$. We have the following

Claim 2: $-\infty < c_k \leq c_{k+1} < 0$ for each $k \in \mathbb{N}$ and $\lim_k c_k = 0$.

Proof: The monotonicity of $(c_k)_k$ is obvious while $-\infty < c_k$ because φ is bounded below on M (section 1.6). The proof of Lemma 1.23 yields the existence of a k-dimensional subspace V_k of $H^1(\mathbb{R}^3)$ such that for all $u \in V_k$ with $\|u\|_2 = 1$, we have

$$\int_{\mathbb{R}^3} |\nabla u|^2 + V|u|^2 dx \leq -\nu \quad (13)$$

for some $\nu > 0$. The collection of those u yields a sphere homeomorphic to S^{k-1} and which is contained in M. This clearly implies that $c_k < 0$.

For the last assertion, consider a nested sequence E_k of finite dimensional subspaces of $H^1(\mathbb{R}^3)^N$ such that $\dim(E_k) = k$ and $\cup_k E_k$ is dense in $H^1(\mathbb{R}^3)^N$. Since $c_k < 0$, we can find $A_k \in \mathcal{F}_k$ such that

$$c_k \leq \max_{A_k} \varphi < c_k/2. \tag{14}$$

Next, consider the orthogonal complement F_k of E_{k-1} which is dual to the class $\mathcal{F}_k$ and hence we can find $v_k \in A_k \cap F_k$. Note that $\varphi(v_k) \leq c_k/2 < 0 = \varphi(0)$ and $v_k \to 0$ weakly. Since φ is weakly lower semi-continuous, we obtain $0 = \varphi(0) \leq \liminf_k \varphi(v_k) \leq 0$ which, in view of (14) implies that $\lim_k c_k = 0$ and thus completing the proof of Claim 2.

Theorem 11.14 now follows from the combination of Claims 1, 2, Theorem 11.1 and Remark 11.13.

11.3 Palais-Smale sequences with second order properties– the cohomotopic case

We now consider the problem of finding lower estimates for the dimension of the "almost negative" eigenspace. Such an estimate can be obtained by opting for a *sup-inf* procedure. Indeed, straightforward adaptations of the arguments used for Theorem 11.1, will yield the following result. All what is needed is to reverse the direction of the deformations induced by the first and the second derivatives in order to "push up" the critical level. The details will be left for the interested reader.

First define for $x \in H$ and $\rho > 0$,

$$m_\rho^+(x) = \inf\{\operatorname{codim}(E); E \text{ subspace of } H \text{ with}$$
$$\langle \varphi''(x)w, w \rangle > \rho\|w\|^2 \text{ for all } w \in E\}.$$

Theorem 11.15: *Let φ be a C^2-functional on a Hilbert space H such that φ' and φ'' are Hölder continuous. Let $\mathcal{F}$ be a homotopic family of dimension n with boundary B and let $\mathcal{F}^*$ be a family dual to $\mathcal{F}$ such that:*

$$c := \sup_{A \in \mathcal{F}} \inf_{x \in A} \varphi(x) = \inf_{F \in \mathcal{F}^*} \sup_{x \in F} \varphi(x)$$

and is finite. Then, for every max-mining sequence $(A_k)_k$ in $\mathcal{F}$ and every suitable min-maxing sequence $(F_k)_k$ in $\mathcal{F}^*$, there exist sequences $(x_k)_k$ in X and $(\rho_k)_k$ in $\mathbb{R}^+$ with $\lim_k \rho_k = 0$ such that:

(i) $\lim_k \varphi(x_k) = c$,
(ii) $\lim_k \varphi'(x_k) = 0$,
(iii) $x_k \in A_k$,
(iv) $\lim_k \operatorname{dist}(x_k, F_k) = 0$,
(v) for each k, $m_{\rho_k}^+(x_k) \geq \dim(H) - n$.

On the other hand, if one needs to obtain a lower estimate by min-maxing over a cohomotopic class, we have the following cohomotopic counterpart of Theorem 11.1. Unfortunately, we can only prove it in the finite dimensional setting. We only state the result since the proof is quite technical. For more details, we refer the reader to [F-G 2].

Theorem 11.16 (Fang-Ghoussoub):   *Let φ be a C^2-functional on a finite dimensional complete C^2-Riemannian manifold X such that $d\varphi$ and $d^2\varphi$ are Hölder continuous. Let $\mathcal{F}$ be a cohomotopic family of dimension n with boundary B and let $\mathcal{F}^*$ be a family dual to $\mathcal{F}$ such that:*

$$c := \inf_{A \in \mathcal{F}} \max_{x \in A} \varphi(x) = \sup_{F \in \mathcal{F}^*} \inf_{x \in F} \varphi(x)$$

is finite. Then, for every min-maxing sequence $(A_k)_k$ in $\mathcal{F}$ and every suitable max-mining sequence $(F_k)_k$ in $\mathcal{F}^*$, there exist sequences $(x_k)_k$ in X, $(\rho_k)_k$ in $\mathbb{R}^+$ with $\lim_k \rho_k = 0$ such that:
- *(i) $\lim_k \varphi(x_k) = c$,*
- *(ii) $\lim_k \|d\varphi(x_k)\| = 0$,*
- *(iii) $x_k \in A_k$ for each k,*
- *(iv) $\lim_k \mathrm{dist}(x_k, F_k) = 0$,*
- *(v) $m_{\rho_k}^+(x) \geq n$ for each k.*

As above, we set

$$K_c^+(n, \varepsilon) = \{x; c - \varepsilon \leq \varphi(x) \leq c + \varepsilon, \ \|d\varphi(x)\| < \varepsilon, \ m_\varepsilon^+(x) \geq n\}$$

and

$$K_c^+(n) = \{x; \varphi(x) = c, \ d\varphi(x) = 0, \ m^+(x) \geq n\}.$$

A cohomotopic analogue of Corollaries 11.3 and 11.5 can then easily be established in the finite dimensional case. In infinite dimensional situations, we have to settle for results of the following form. Here is the case of the *saddle point theorem*.

Corollary 11.17:   *Under the hypothesis of Corollary 11.2, there exists $\bar{c} \geq \inf \varphi(H_2)$ such that for every increasing sequence $(E_k)_k$ of finite dimensional subspaces containing H_1 and spanning H, there exist sequences $(x_k)_k$ in H and $(\rho_k)_k$ in $\mathbb{R}^+$ with $\lim_k \rho_k = 0$ such that:*
- *(i) $\lim_k \varphi(x_k) = \bar{c}$,*
- *(ii) $x_k \in E_k$ for each k,*
- *ii) $\lim_k (\varphi_{|E_k})'(x_k) = 0$,*
- *(iii) $m_{\rho_k}^+(x_k) \geq n$ for each k.*

Proof: Consider the class

$$\bar{\mathcal{F}} = \{A; A \text{ compact}, A \supset S_{H_1} \text{ and } 0 \in f(A) \text{ whenever } f \in C(A; H_1)$$
$$\text{and } f(x) = x \text{ on } S_{H_1}\}.$$

Also, for each subspace E_k containing H_1, we consider

$$\bar{\mathcal{F}}_k = \{A \subset E_k; A \text{ compact}, A \supset S_{H_1} \text{ and } 0 \in f(A)$$
$$\text{whenever } f \in C(A; H_1) \text{ and } f(x) = x \text{ on } S_{H_1}\}.$$

It is clear that $\bar{\mathcal{F}}$ and each $\bar{\mathcal{F}}_k$ are cohomotopic classes of dimension n with boundary S_{H_1}. Moreover if $\bar{c}$ (resp., $\bar{c}_k$) is the min-max value on $\bar{\mathcal{F}}$ (resp., $\bar{\mathcal{F}}_k$) then $\bar{c}_k \downarrow \bar{c}$. Theorem 11.16 then applies to the restriction of φ to each E_k to yield our claim.

This justifies the following notion.

Definition 11.18: *A C^2-function on a C^2-manifold X is said to have the Palais-Smale condition at level c, around the set F and of order greater than n (in short $(PS)_{F,c,n^+}$), if a sequence $(x_k)_k$ in H is relatively compact whenever it satisfies the following conditions with respect to an increasing sequence $(X_k)_k$ of finite dimensional submanifolds spanning X:*

$$\lim_k \varphi(x_k) = c, \lim_k (\varphi_{|X_k})'(x_k) = 0, \lim_k \text{dist}(x_k, F) = 0 \text{ and there}$$
exists a sequence of positive reals $(\rho_k)_k$ with $\lim_k \rho_k = 0$ such that $m_{\rho_k}^+(x_k) \geq n$ for each k .

Now we can state the following.

Corollary 11.19: *Under the hypothesis of Corollary 11.17, assume that φ satisfies condition $(PS)_{c,n^+}$, then φ has at level c, a critical point of augmented Morse index greater or equal to n.*

In the case where a potential critical level is induced by a homotopic family and simultaneously by a cohomotopic family, we then obtain two-sided second order estimates as the following result asserts. The proof can also be found in [F-G 2].

Theorem 11.20: *Let φ be a C^2-functional on a finite dimensional C^2-Riemannian manifold X such that $d\varphi$ and $d^2\varphi$ are Hölder continuous. Let $\mathcal{F}$ (resp. $\bar{\mathcal{F}}$) be a homotopic (resp. cohomotopic) family of dimension n with boundary B such that $\mathcal{F} \subset \bar{\mathcal{F}}$ and let $\mathcal{F}^*$ be a family dual to $\bar{\mathcal{F}}$. Assume that*

$$c := \inf_{A \in \bar{\mathcal{F}}} \max_{x \in A} \varphi(x) = \inf_{A \in \mathcal{F}} \max_{x \in A} \varphi(x) = \sup_{F \in \mathcal{F}^*} \inf_{x \in F} \varphi(x)$$

and is finite. Then for every min-maxing sequence $(A_k)_k$ in $\mathcal{F}$ and every suitable max-mining sequence $(F_k)_k$ in $\mathcal{F}^*$, there exist sequences $(x_k)_k$ in X, $(\rho_k)_k$ in $\mathbb{R}^+$ with $\lim_k \rho_k = 0$ such that:

(i) $\lim_k \varphi(x_k) = c$,

(ii) $\lim_k \|d\varphi(x_k)\| = 0$,

(iii) $x_k \in A_k$ for each k,

(iv) $\lim_k \text{dist}(x_k, F_k) = 0$,

(v) $m_{\rho_k}^-(x_k) \le n \le m_{\rho_k}^+(x_k)$ for each k.

Other related results can be found in [F-G 2] and in particular the case of functionals invariant under the action of a group acting on the manifold. A topological index associated to that group allows us to give two lower estimates for the size of $K_c^-(n, \varepsilon)$ and $K_c^+(n, \varepsilon)$. One will depend on the size of the intersections with the dual family, while the second is in the spirit of the Ljusternik-Schnirelmann theory, and corresponds to when two families induce the same critical level.

As a possible application to Theorem 11.20, let us reconsider the non-linear ellipic equations involving the limiting Sobolev exponent that were studied in Chapter 8. That is

$$\begin{cases} -\Delta u - \lambda u = u|u|^{2^*-2} & \text{in } \Omega, \\ u = 0 & \text{on } \partial\Omega \end{cases} \tag{1}$$

where Ω is a smoothly bounded domain in $\mathbb{R}^N$, $N > 2$, $2^* = \frac{2N}{N-2}$ and $\lambda < \lambda_1$ the first eigenvalue of $-\Delta$ in $H_0^1(\Omega)$.

We have seen that solving problem (1) is equivalent to finding critical points on $H_0^1(\Omega)$ of the energy functional

$$\varphi_\lambda(u) = \frac{1}{2} \int_\Omega (|\nabla u|^2 - \lambda|u|^2) \, dx - \frac{1}{2^*} \int_\Omega |u|^{2^*} \, dx \tag{2}$$

and that φ_λ satisfies $(PS)_c$ only for $c < \frac{1}{N} S^{N/2}$ where S is the best constant appearing in the Sobolev inequality. By using this fact, we have seen in Chapter 8 that if $N \ge 4$, then problem (1) has a strictly positive solution at a level $c_1 < \frac{1}{N} S^{N/2}$.

In order to get another solution, one can push back the threshold of non compactness to the level $c_1 + \frac{1}{N} S^{N/2}$ provided the Palais-Smale sequences satisfy the appropriate second order conditions. Indeed, as noticed by Robinson [R], one can show that Tarantello's argument in Theorem 8.2 yields that a pseudo-critical sequence $(u_n)_n$ with $\varphi(u_n) \to c < c_1 + \frac{1}{N} S^{N/2}$ is relatively compact in $H_0^1(\Omega)$ provided it satisfies $m_\delta^-(u_n) \ge 2$ for some fixed $\delta > 0$. Note that this is a stronger assumption than what our general Theorem 11.20 gives. However, as shown in Chapter 8, Tarantello managed to find another solution at a level $c_2 < c_1 + \frac{1}{N} S^{N/2}$ by constructing such sequences with the help of the strong form of the min-max principle applied to a clever choice for a dual set. The proximity of pseudo-critical sequences to such a set yielded the required second order conditions.

It is reasonable to expect the level of non-compactness to be pushed back again to $c_2 + \frac{1}{N}S^{N/2}$ provided the Palais-Smale sequences satisfy $m^-_{\delta_n}(u_n) \geq 3$ for some $\delta_n \downarrow 0$. Unfortunately, for technical reasons, Tarantello's method does not extend to this next level and our Theorem 11.20 only yields sequences $(u_n)_n$ with $m^+_{\delta_n}(u_n) \geq 3$ where $\delta_n \to 0$.

On the other hand, one can show that φ verifies $(PS)^+_{c,N+3}$ for any $c < c_2 + \frac{1}{N}S^{N/2}$. Unfortunately, we cannot prove that the associated cohomotopic family of dimension $N+3$ induces a level below $c_2 + \frac{1}{N}S^{N/2}$. For more details, we refer to the papers of Tarantello [T 1,2,4] and the thesis of D. Robinson [Ro].

Notes and Comments: The main results of this chapter are very recent and are due to Fang and Ghoussoub [F-G 2]. The case of the mountain pass theorem was established in [F-G 1]. The first author to emphasize the importance of second order conditions in compactness problems was P. L. Lions [L 1]. Our derivation of the multiple solutions of the Hartree-Fock equations follows his methodology. Actually, P. L. Lions shows that the functional φ restricted to M does not verify the (PS) condition but that, according to our terminology, φ verifies $(PS)_{c,k^-}$ for every $c \in \mathbb{R}$ and any $k \in \mathbb{N}$. In order to get the required Palais-Smale sequences with the appropriate second order information, he approximates φ by the functionals that are associated to the same problem but restricted on suitable approximating bounded domains. Since these functionals satisfy the standard Palais-Smale condition, one can use Morse theory on their true critical points which are essentially the required pseudo-critical points for the original functional φ. Actually, this same ad-hoc argument led us to pursue the general principle (exhibited here) for finding pseudo-critical sequences with these additional properties.

Other examples where second order conditions are relevant for the compactness of Palais-Smale sequences were also given by Coti Zelati, Ekeland and Lions [Ek 2]. The proofs in the cohomotopic case will appear in [F-G 2].

APPENDIX A

RELEVANT FUNCTION SPACES

AND INEQUALITIES

A.1 Sobolev Spaces

Sobolev spaces are, roughly speaking, spaces of p-integrable functions whose derivatives are also p-integrable. There are two basic types of Sobolev spaces we wish to consider. In the first situation, we have a bounded domain $\Omega \subset \mathbb{R}^N$ and we consider functions $u : \Omega \to \mathbb{R}$ for which $u \equiv 0$ on $\partial\Omega$. In the second, we look at functions $u : [0, T] \to \mathbb{R}^N$ satisfying $u(0) = u(T)$. The only problem in straightforwardly defining these spaces is that p-integrable functions need not be differentiable and restricting our attention to those that are differentiable does not provide us with what we want — the resulting spaces are not complete. We must weaken our notion of differentiability.

Definition A.1. *In the following two cases we define an appropriate notion of weak differentiability.*

(a) *Let $\Omega \subset \mathbb{R}^N$ be a domain, i.e., an open and connected set. If $u \in L^1(\Omega; \mathbb{R})$, we say that u has a weak partial derivative w.r.t. x^i provided there is a function $v \in L^1(\Omega; \mathbb{R})$ such that*

$$\int_\Omega \varphi v \, dx = -\int_\Omega u \frac{\partial\varphi}{\partial x^i} \, dx$$

for all $\varphi \in C_0^\infty(\Omega)$. We call v the weak i-partial derivative of u and denote it by $\dfrac{\partial u}{\partial x^i}$. The weak gradient of u is

$$\nabla u = \left(\frac{\partial u}{\partial x^i}, \ldots, \frac{\partial u}{\partial u^N} \right)$$

provided all the partial derivatives exist.

(b)   Let C_T^∞ be the space of infinitely differentiable T-periodic functions from $\mathbb{R}$ to $\mathbb{R}^N$. For $u \in ([0,T]; \mathbb{R}^N)$ we call $v \in L^1([0,T]; \mathbb{R}^N)$ the weak derivative of u if

$$\int_0^T \varphi . v \, dt = -\int_0^T u . \dot\varphi \, dt$$

for all $\varphi \in C_T^\infty$. (Here $\dot\varphi$ denotes $\dfrac{d\varphi}{dt}$.)

Lemma A.2. *Suppose that $u \in L^1([0,T]; \mathbb{R}^N)$ and has a weak derivative, $v \in L^1([0,T]; \mathbb{R}^N)$ then*

$$\int_0^T v(s) \, ds = 0$$

and for some constant $c \in \mathbb{R}^N$

$$u(t) = \int_0^t v(s) \, ds + c \qquad \text{a.e. on } [0,T]$$

Our notion of weak differentiability now in hand, we are able to define Sobolev spaces.

Definition A.3. *We consider each case separately.*

(a)   If Ω is a domain in $\mathbb{R}^N$ and $1 \le p < \infty$, we define the space

$$W^{1,p}(\Omega) = \left\{ u \in L^p(\Omega; \mathbb{R}) \;\Big|\; \frac{\partial u}{\partial x^i} \in L^p(\Omega; \mathbb{R}) \text{ for } i = 1, \dots, N \right\}$$

with norm

$$\|u\|_{W^{1,p}(\Omega)} = \|u\|_p + \left(\sum_{i=1}^N \left\| \frac{\partial u}{\partial x^i} \right\|_p^p \right)^{1/p}$$

$$= \|u\|_p + \|\nabla u\|_p$$

and $H_0^1(\Omega)$ to be the closure (with respect to $\| \cdot \|_{W^{1,2}(\Omega)}$) of $C_0^\infty(\Omega)$ in $W^{1,2}(\Omega)$. We denote the dual of $W^{1,p}(\Omega)$ by $W^{1,-p}(\Omega)$ and that of $H_0^1(\Omega)$ by $H^{-1}(\Omega)$.

(b)   The Sobolev space $W_T^{1,p}$ is defined as

$$W_T^{1,p} = \left\{ u \in L^1([0,T]; \mathbb{R}^N) \;\Big|\; \dot u \in L^1([0,T]; \mathbb{R}^N) \right\}$$

with norm

$$\|u\|_{W_T^{1,p}} = \|u\|_p + \|\dot u\|_p$$

We denote by H_T^1 the space $W_T^{1,2}$.

A.2 Sobolev embedding theorems

The following theorem is of particular importance in the variational approach to solving differential equations as it gives us some control over the nonlinear terms.

Theorem A.4 (Sobolev Embedding). *The following two assertions hold:*

(i)   *Suppose that $\Omega \subset \mathbb{R}^N$ is a bounded domain. Then, the identity is a continuous injection*

$$H_0^1(\Omega) \hookrightarrow L^p(\Omega)$$

   for

(1°) $2 \leq p \leq 2^* := \frac{2N}{N-2}$ *if* $N \geq 3$. *Moreover, the above injection is compact when* $2 \leq p < 2^*$.

(2°) $2 \leq p < \infty$ *if* $N = 1,\ 2$. *In this case, the injection is always compact.*

(ii)   *There is a compact injection*

$$W_T^{1,p} \hookrightarrow C([0,T]; R^N)$$

   whenever, $1 < p < \infty$. *In particular, for each* $u_1 \in W_T^{1,p}$ *there is a continuous function* $u_2 \in W_T^{1,p}$ *so that* $u_1 = u_2$ *a.e. and from Lemma (A.2)* $u_2(0) = u_2(T)$.

The number 2^* is the so-called critical Sobolev exponent. In view of the Sobolev embedding theorem, we have the following.

Proposition A.5. *There exists a constant* $c > 0$ *such that*

(i)   *If* Ω *is a bounded domain in* $\mathbb{R}^N$ *then for every* $u \in H_0^1(\Omega)$

$$\|u\|_p \leq c\|\nabla u\|_2$$

   if $2 \leq p \leq 2^*$ *for* $N \geq 3$ *and* $2 \leq p < \infty$ *for* $N = 1,\ 2$.

(ii)   *If* $u \in W_T^{1,p}$ *then*

$$\|u\|_\infty \leq c\|u\|_{W_T^{1,p}}.$$

   Moreover, if $\int_0^1 u(t)\,dt = 0$, *then*

$$\|u\|_\infty \leq c\|\dot{u}\|_p.$$

For the periodic 1-dimensional Sobolev space, H_T^1, we can obtain the following sharp inequalities.

Proposition A.6. *If* $u \in H_T^1$ *and* $\int_0^1 u(t)\,dt = 0$, *then*

$$\int_0^T |u(t)|^2\,dt \leq (T^2/4\pi^2) \int_0^T |\dot{u}|^2\,dt$$

(Wirtinger's inequality) and

$$\|u\|_\infty^2 \leq (T/12) \int_0^T |\dot{u}(t)|^2\,dt$$

(Sobolev's inequality).

In view of (A.5) there is a constant $c > 0$ such that

$$\|u\|_2 \leq c\|\nabla u\|_2$$

for every $u \in H_0^1(\Omega)$. Hence, we can (and do!) give $H_0^1(\Omega)$ the equivalent norm

$$\|u\|_{H_0^1(\Omega)} = \|\nabla u\|_2. \tag{A.1}$$

If we define an inner product on $H_0^1(\Omega)$ by

$$\langle u, v \rangle = \int_\Omega \nabla u \cdot \nabla v\,dx \tag{A.2}$$

we obtain:

Theorem A.7. *If* $\Omega \subset \mathbf{R}^N$ *is a bounded domain then* $H_0^1(\Omega)$, *with norm and inner product as in (A.1) and (A.2), is a separable, reflexive Hilbert space. Moreover, if we define an inner product on* H_T^1 *by*

$$\langle u, v \rangle = \int_0^T u \cdot v\,dt + \int_0^T \dot{u} \cdot \dot{v}\,dt$$

then H_T^1 *is also a separable, reflexive Hilbert space.*

A.3 Best Sobolev constants

Throughout this section, $\Omega \subseteq \mathbf{R}^N$, $N \geq 3$, will be a domain. From the Sobolev embedding theorem there is a constant $C > 0$ such that

$$C\|u\|_{2^*}^2 \leq \|\nabla u\|_2^2 \qquad \text{for all } u \in H_0^1(\Omega)$$

We let $S(N, \Omega)$ denote the best Sobolev constant i.e., the largest constant C satisfying the above inequality for all $u \in H_0^1(\Omega)$. $S(N, \Omega)$ is given by

$$S(N, \Omega) = \inf_{\substack{u \in H_0^1(\Omega) \\ u \neq 0}} \frac{\|\nabla u\|_2^2}{\|u\|_{2^*}^2} \tag{A.3}$$

We shall typically suppress the N in the above notation, and write $S(\mathbf{R}^N)$ as S.

Lemma A.8. *We have the following facts concerning the best Sobolev constants.*

(1) $S(\Omega)$ *is independent of* Ω *and will henceforth be denoted by* S.

(2) The infimum in (A.3) is never attained when Ω is bounded.

(3) S is attained when $\Omega = \mathbb{R}^N$ by the functions

$$u_\varepsilon^*(x) = \frac{\left[N(N-2)\varepsilon\right]^{(N-2)/4}}{\left[\varepsilon + |x|^2\right]^{(N-2)/2}}$$

so that $\|\nabla u_\varepsilon^*(x)\|_2^2 = S\|u_\varepsilon^*\|_{2^*}^2$. The functions u_ε^* are so-called extremal functions in the Sobolev inequality. These functions also satisfy

$$-\Delta u_\varepsilon^* = |u_\varepsilon^*|^{2^*-2} u_\varepsilon^* \quad \text{on } \mathbb{R}^N$$
$$u_\varepsilon^* \to 0 \qquad \text{as } |x| \to \infty.$$

Hence, $\|\nabla u_\varepsilon^*\|_2^2 = \|u_\varepsilon\|_{2^*}^{2^*}$ and combining this with the above, we have

$$\|\nabla u_\varepsilon^*\|_2^2 = \|u_\varepsilon\|_{2^*}^{2^*} = S^{N/2}.$$

The importance of the best Sobolev constant lies in the fact that when trying to solve various problems involving the critical Sobolev exponent, it provides an energy level above which the Palais-Smale condition, $(PS)_c$, fails for the usual energy function. The extremal functions are useful for making estimates which show that the energy level can be brought down below a point where some kind of Palais-Smale condition can be restored. The next section deals with the very sharp estimates that we have for the various norms of cut-off extremal functions.

A.4 Estimates on the extremal Sobolev functions

Let $u_\varepsilon^*(x) = \dfrac{\left[N(N-2)\varepsilon\right]^{(N-2)/4}}{\left[\varepsilon + |x|^2\right]^{(N-2)/2}}$, $x \in \mathbb{R}^N$ be an extremal function in the Sobolev inequality and fix $\psi \in C^\infty(\mathbb{R}^N; \mathbb{R})$ satisfying

(i) $0 \le \psi \le 1$

(ii) $\psi(x) = \begin{cases} 1 & \text{if } |x| \le R \\ 0 & \text{if } |x| \ge 2R \end{cases}$

where $B_{2R}(0) \subset \Omega$ (assuming, as we may, that $0 \in \Omega$). Setting $u_\varepsilon(x) = \psi(x) \cdot u_\varepsilon^*(x)$ we have:

Lemma A.9. *The following estimates hold:*

(a) $\|\nabla u_\varepsilon\|_2^2 = S^{N/2} + O(\varepsilon^{(N-2)/2})$

(b) $\|u_\varepsilon\|_{2^*}^{2^*} = S^{N/2} + O(\varepsilon^{N/2})$

(c) $\|u_\varepsilon\|_2^2 = \begin{cases} K_1\varepsilon + O(\varepsilon^{(N-2)/2}) & \text{if } N \geq 5 \\ K_1|\log\varepsilon| + O(\varepsilon) & \text{if } N = 4 \end{cases}$

where $K_1 > 0$ is a constant.

(d) $\|u_\varepsilon\|_1 \leq K_2\varepsilon^{(N-2)/4}$, $K_2 > 0$ a constant.

(e) $\|u_\varepsilon\|_{2^*-1}^{2^*-1} \leq K_3\varepsilon^{(N-2)/4}$, $K_3 > 0$ a constant.

Proof: (a) We have

$$\nabla u_\varepsilon(x) = \frac{[N(N-2)\varepsilon]^{(N-2)/4}}{[\varepsilon + |x|^2]^{(N-2)/2}} \cdot \nabla\psi(x)$$

$$- \frac{(n-2)[N(N-2)\varepsilon]^{(N-2)/4}\psi(x)x}{[\varepsilon + |x|^2]^{N/2}}$$

$$= \begin{cases} -\dfrac{(N-2)[N(N-2)\varepsilon]^{(N-2)/4}x}{[\varepsilon + |x|^2]^{N/2}} & \text{if } |x| \leq R \\ 0 & \text{if } |x| > 2R \end{cases}$$

Therefore,

$$\|\nabla u_\varepsilon\|_2^2 = \int_\Omega |\nabla u_\varepsilon(x)|^2 \, dx$$

$$= (N-2)^2 [N(N-2)\varepsilon]^{(N-2)/2} \int_{|x|\leq R} \frac{|x|^2}{[\varepsilon + |x|^2]^N} \, dx$$

$$+ O(\varepsilon^{(N-2)/2})$$

$$= (N-2)^2 [N(N-2)\varepsilon]^{(N-2)/2} \int_{\mathbb{R}^N} \frac{|x|^2}{[\varepsilon + |x|^2]^N} \, dx$$

$$+ O(\varepsilon^{(N-2)/2})$$

Letting $x = \sqrt{\varepsilon}\,w$

$$= (N-2)^2 [N(N-2)]^{(N-2)/2} \int_{\mathbb{R}^N} \frac{|w|^2}{[1 + |w|^2]^N} \, dx$$

$$+ O(\varepsilon^{(N-2)/2})$$

$$= \|\nabla u_1^*\|_2^2 + O(\varepsilon^{(N-2)/2})$$

$$= S^{N/2} + O(\varepsilon^{(N-2)/2})$$

(b)

$$\|u_\varepsilon\|_{2^*}^{2^*} = \int_\Omega \frac{\psi^{2^*}(x)\left[N(N-2)\varepsilon\right]^{N/2}}{\left[\varepsilon+|x|^2\right]^N}\,dx$$

$$= \left[N(N-2)\varepsilon\right]^{N/2}\int_{|x|\leq R}\frac{dx}{\left[\varepsilon+|x|^2\right]^N}+O(\varepsilon^{N/2})$$

$$= \left[N(N-2)\varepsilon\right]^{N/2}\int_{\mathbf{R}^N}\frac{dx}{\left[\varepsilon+|x|^2\right]^N}+O(\varepsilon^{N/2})$$

Let $x = \sqrt{\varepsilon}\,w$

$$= \left[N(N-2)\right]^{N/2}\int_{\mathbf{R}^N}\frac{dw}{\left[1+|w|^2\right]^N}+O(\varepsilon^{N/2})$$

$$= \|u_1^*\|_{2^*}^{2^*}+O(\varepsilon^{N/2})$$

$$= S^{N/2}+O(\varepsilon^{N/2})$$

(c)

$$\|u_\varepsilon\|_2^2 = \int_\Omega \frac{\psi^2(x)\left[N(N-2)\varepsilon\right]^{(N-2)/2}}{\left[\varepsilon+|x|^2\right]^{N-2}}\,dx$$

$$= \left[N(N-2)\varepsilon\right]^{(N-2)/2}\int_{|x|\leq R}\frac{dx}{\left[\varepsilon+|x|^2\right]^{N-2}}+O(\varepsilon^{(N-2)/2})$$

When $N \geq 5$

$$= \left[N(N-2)\varepsilon\right]^{(N-2)/2}\int_{\mathbf{R}^N}\frac{dx}{\left[\varepsilon+|x|^2\right]^{N-2}}+O(\varepsilon^{(N-2)/2})$$

Let $x = \sqrt{\varepsilon}\,w$

$$= \left[N(N-2)\right]^{(N-2)/2}\varepsilon\int_{\mathbf{R}^N}\frac{dw}{\left[1+|w|^2\right]^{N-2}}+O(\varepsilon^{(N-2)/2})$$

Taking $K_1 = \left[N(N-2)\right]^{(N-2)/2}\int_{\mathbf{R}^N}\frac{dw}{\left[1+|w|\right]^{N-2}}$ we obtain the result. When $n = 4$, we have

$$\int_{|x|\leq R}\frac{8\varepsilon}{\left[\varepsilon+|x|^2\right]^2}\,dx \leq \int_\Omega \frac{\psi^2(x)8\varepsilon}{\left[\varepsilon+|x|^2\right]^2}\,dx \leq \int_{|x|\leq 2R}\frac{8\varepsilon}{\left[\varepsilon+|x|^2\right]^2}\,dx$$

and

$$\int_{|x|\le\beta}\frac{8\varepsilon}{\left[\varepsilon+|x|^2\right]^2}\,dx=8\varepsilon\omega\int_0^\beta\frac{r^3}{\left[\varepsilon+r^2\right]^2}\,dr$$

where ω is the area of S^3. Letting $r=\sqrt{\varepsilon}\,s$

$$=8\varepsilon\omega\int_0^{\beta/\sqrt{\varepsilon}}\frac{s^3}{\left[1+s^2\right]^2}\,ds$$

$$=8\varepsilon\omega\int_1^{\beta/\sqrt{\varepsilon}}\frac{s^3}{\left[1+s^2\right]^2}\,ds+O(\varepsilon)$$

$$=8\varepsilon\omega\int_1^{\beta/\sqrt{\varepsilon}}\frac{ds}{s}+O(\varepsilon)$$

$$=8\varepsilon\omega\log\left(\beta/\sqrt{\varepsilon}\right)+O(\varepsilon)$$

$$=4\varepsilon\omega|\log\varepsilon|+O(\varepsilon)$$

Taking $K_1=4\omega$ we obtain the result.

(d)

$$\|u_\varepsilon\|_1=\int_\Omega\frac{\psi(x)\left[N(N-2)\varepsilon\right]^{(N-2)/4}}{\left[\varepsilon+|x|^2\right]^{(N-2)/2}}\,dx$$

$$\le\left[N(N-2)\varepsilon\right]^{(N-2)/4}\omega\int_0^{2R}\frac{r^{N-1}}{\left[\varepsilon+|x|^2\right]^{(N-2)/2}}\,dr$$

Let $r=\sqrt{\varepsilon}\,s$

$$=\left[N(N-2)\right]^{(N-2)/4}\varepsilon^{(N+2)/4}\omega\int_0^{2R/\sqrt{\varepsilon}}\frac{s^{N-1}}{\left[1+s^2\right]^{(N-2)/2}}\,ds$$

$$\le\left[N(N-2)\right]^{(N-2)/4}\varepsilon^{(N+2)/4}\omega\int_0^{2R/\sqrt{\varepsilon}}s\,ds$$

$$=\frac12\left[N(N-2)\right]^{(N-2)/4}(2R)^2\omega\varepsilon^{(N-2)/4}$$

(e)

$$\|u_\varepsilon\|_{2^*-1}^{2^*-1}=\int_\Omega\frac{\psi^{2^*-1}(x)\left[N(N-2)\varepsilon\right]^{(N+2)/4}}{\left[\varepsilon+|x|^2\right]^{(N+2)/2}}\,dx$$

$$\le\left[N(N-2)\varepsilon\right]^{(N+2)/4}\omega\int_0^{2R}\frac{r^{N-1}}{\left[\varepsilon+r^2\right]^{(N+2)/2}}\,dr$$

Let $r=\sqrt{\varepsilon}\,s$

$$= \left[N(N-2)\right]^{(N+2)/4} \varepsilon^{(N-2)/2} \omega \int_0^{2R/\sqrt{\varepsilon}} \frac{s^{N-1}}{\left[1+s^2\right]^{(N+2)/4}} \, ds$$

$$\leq \left[N(N-2)\right]^{(N+2)/4} \varepsilon^{(N-2)/2} \omega \int_0^\infty \frac{s^{N-1}}{\left[1+s^2\right]^{(N+2)/4}} \, ds$$

We take $K_2 = \left[N(N-2)\right]^{(N+2)/4} \omega \int_0^\infty \frac{s^{N-1}}{\left[1+s^2\right]^{(N+2)/2}} \, ds.$

Notes and Comments: A standard reference for the theory of Sobolev spaces is the book of Adams [A]. The recent monograph of Mawhin and Willem [M-W] contains the basics about vector-valued periodic Sobolev spaces. A discussion of the *limiting case* in the Sobolev embedding can be found in the book of Struwe [St]. The estimates on the extremal functions come from the paper [B-N 1] of Brezis and Nirenberg.

APPENDIX B

VARIATIONAL FORMULATION OF

SOME BOUNDARY VALUE PROBLEMS

B.1 Weak Solutions

In the variational approach to solving problems, we associate with the given problem a C^1 functional, φ, on a suitable Banach space (more generally, a manifold) in such a way that the critical points of φ are exactly the solutions to the original problem. Before proceeding with the variational formulation of some differential equations, we need to settle on an appropriate notion of differentiability for functions defined on a Banach space.

Definition B.1. *Suppose that X is a Banach space and a function $\varphi : X \to \mathbb{R}$ is given. We say that φ is Fréchet differentiable at $u \in X$ provided there is a continuous linear functional $\varphi'(u) \in X^*$ satisfying*

$$\varphi'(u)(v) = \lim_{t \to 0^+} \frac{\varphi(u + tv) - \varphi(u)}{t}$$

for all $v \in X$. $\varphi'(u)$ is called the Fréchet derivative of φ. The function φ is called C^1 if it is Fréchet differentiable at all $u \in X$ and the map $u \to \varphi'(u)$ from X to X^* is norm continuous. A critical point of φ is a point $u \in X$ such that $\varphi'(u) = 0$ i.e., $\varphi'(u)(v) = 0$ for all $v \in X$.

We are concerned with two types of problems. The first problem is

$$
\begin{aligned}
-\Delta u &= f(x, u) &&\text{in } \Omega \\
u &= 0 &&\text{on } \partial\Omega
\end{aligned}
\tag{B.1}
$$

where $\Omega \subset \mathbb{R}^N$ is a bounded domain and $f : \Omega \times \mathbb{R} \to \mathbb{R}$. Since we cannot immediately define the Laplacian on $H_0^1(\Omega)$, we need the notion of a weak solution to (B.1). We say that $u \in H_0^1(\Omega)$ is a weak solution to (B.1) provided

$$\int_\Omega \nabla u \cdot \nabla v \, dx = \int_\Omega f(x, u) v \, dx \qquad \text{for all } v \in C_0^\infty(\Omega).$$

This notion of weak solution is motivated by integration by parts. The second problem we want to look at is

$$\ddot{u}(t) = \nabla G(t, u(t)) \qquad \text{a.e. on } [0, T]$$

$$u(0) = u(T) \qquad\qquad\qquad (B.2)$$

$$\dot{u}(0) = \dot{u}(T)$$

where $G : [0, T] \times \mathbb{R}^N \to \mathbb{R}^N$ (here ∇G refers to the gradient of G considering only the last N variables). For this equation, we don't need any notion of a weak solution. A solution to this equation is simply any function $u \in H_T^1$ for which $\dot{u}$ has a weak derivative $\ddot{u}$ satisfying (B.2) for almost every $t \in [0, T]$.

The energy functionals associated with these problems are

$$\varphi(u) = \frac{1}{2} \int_\Omega |\nabla u(x)|^2 \, dx - \int_\Omega F(x, u(x)) \, dx \qquad u \in H_0^1(\Omega) \quad (B.3)$$

where $F(x, s) = \displaystyle\int_0^s f(x, t) \, dt$ and

$$\varphi(u) = \frac{1}{2} \int_0^T |\dot{u}(t)|^2 \, dt + \int_0^T G(t, u(t)) \, dt \qquad u \in H_T^1 \qquad (B.4)$$

respectively. Simply in order that these functionals be properly defined, we must impose some continuity and measurability conditions on f and G.

Definition B.2. *Suppose that $\Omega \subseteq \mathbb{R}^N$ is a domain. A function $f : \Omega \times \mathbb{R} \to \mathbb{R}$ is called a Carathéodory function if:*

(1)   for each fixed $s \in \mathbb{R}$, the function $x \to f(x, s)$ is Lebesgue measurable in Ω.

(2)   for almost every $x \in \Omega$, the function $s \to f(x, s)$ is continuous on $\mathbb{R}$.

Lemma B.3. *If $f : \Omega \times \mathbb{R} \to \mathbb{R}$, is a Carathéodory function, then:*

(1) $x \to f(x, u(x))$ is a measurable function for every measurable function $u : \Omega \to \mathbb{R}$.

(2)   If Ω has finite measure, then the Nemitskii operator $N_f : \mathcal{M} \to \mathcal{M}$ is continuous, where $\mathcal{M}$ is the space of real-valued measurable functions on Ω equipped with the topology of convergence in measure.

(3)   If Ω is a bounded domain and if f satisfies the growth condition

$$|f(x, s)| \le a|s|^{p-1} + b(x),$$

for $p > 1$ and where $b(x) \in L^q(\Omega; \mathbb{R})$ $\left(\frac{1}{p} + \frac{1}{q} = 1\right)$, then N_f

maps continuously L^p into L^q and N_F maps continuously L^p into L^1, where N_F is the Nemitskii map associated with the function

$$F(x,s) = \int_0^s f(x,t)\,dt.$$

Moreover, $\mathcal{F}(u) = \int_\Omega F(x,u(x))\,dx$ defines a continuously Fréchet differentiable functional $\mathcal{F} : L^p \to \mathbb{R}$ such that $\mathcal{F}' = N_f$.

The next two results give sufficient conditions that the above energy functionals are well-defined, C^1 functionals whose critical points are exactly the solutions to their respective problems.

Proposition B.4. *Suppose that $\Omega \subset \mathbb{R}^N$ is a bounded domain and that $f : \Omega \to \mathbb{R}$ is a Carathéodory function satisfying the growth condition*

$$|f(x,s)| \le a|s|^{p-1} + b(x),$$

where $2 \le p \le 2^* = \frac{2N}{N-2}$ if $N \ge 3$, $2 \le p < \infty$ when $N = 1,2$ and where

$$b(x) \in L^q(\Omega;\mathbb{R}) \qquad \left(\frac{1}{p} + \frac{1}{q} = 1\right).$$

Then the functional $\varphi : H_0^1(\Omega) \to \mathbb{R}$ defined in (B.3) is continuously Gâteaux differentiable and

$$\varphi'(u)(v) = \int_\Omega \nabla u \cdot \nabla v\,dx - \int_\Omega f(x,u)v\,dx \qquad \text{for all } u,v \in H_0^1(\Omega).$$

Hence the critical points of φ are precisely the weak solutions to (B.1).

The conditions for the second type of problem are much the same.

Proposition B.5. *Suppose that $G : [0,T] \times \mathbb{R}^N \to R$ satisfies:*

(1) $(t,x) \to G(t,x)$ *is measurable in t for each fixed $x \in \mathbb{R}^N$.*
(2) $x \to G(t,x)$ *is continuously differentiable for almost every $t \in [0,T]$*
(3) *the growth condition*

$$|G(t,x)|, |\nabla G(t,x)| \le a(|x|)b(t)$$

holds for $t \in [0,T]$ a.e., all $x \in \mathbb{R}^N$, where $a \in C(R_+, R_+)$, and $b \in L^1([0,T];\mathbb{R}_+)$.

Then the functional $\varphi : H_T^1 \to \mathbb{R}^N$ of (B.2) is continuously Gâteaux differentiable and

$$\varphi'(u)(v) = \int_0^T \dot{u}(t) \cdot \dot{v}(t)\,dt + \int_0^T \nabla G(t,u(t)) \cdot v(t)\,dt \qquad \text{for all } u,v \in H_T^1$$

Hence, any critical point u of φ satisfies (B.2).

Notes and Comments: The lecture notes of De Figueiredo [De] contain a detailed study of Nemitskii maps. More precision about the variational formulations for boundary value problems can also be found in the appendix of the C.B.M.S. lectures of Rabinowitz [R 1].

APPENDIX C

THE BLOWING UP OF SINGULARITIES

C.1 The failure of the (PS) condition

Let $\Omega \subset \mathbb{R}^N$, $N \geq 3$, be a bounded domain and consider the problem

$$-\Delta u = |u|^{2^*-2}u + \lambda u \quad \text{in } \Omega$$
$$u = 0 \qquad\qquad \text{on } \partial\Omega$$

$(C.1)$

and its corresponding energy

$$\varphi_\lambda(u) = \frac{1}{2}\int_\Omega |\nabla u|^2 - \frac{\lambda}{2}\int_\Omega |u|^2 - \frac{1}{2^*}\int_\Omega |u|^{2^*}$$

for $u \in H_0^1(\Omega)$. By showing that φ_λ satisfies the Palais-Smale condition $(PS)_c$ provided that $c < S^{N/2}/N$, Brezis and Nirenberg managed to show that (C.1) has a positive solution whenever $N \geq 4$ and $0 < \lambda < \lambda_1$ where λ_1 is the first eigenvalue of $-\Delta$ on $H_0^1(\Omega)$. The major impediment to finding further solutions lies in the fact that φ_λ may fail the Palais-Smale condition $(PS)_c$ for $c \geq S^{N/2}/N$. Indeed, if

$$u_\varepsilon^*(x) = \frac{[N(N-2)\varepsilon]^{(N-2)/4}}{[\varepsilon + |x|^2]^{(N-2)/2}} \qquad x \in \mathbb{R}^N$$

denote the extremal Sobolev functions and

$$u_\varepsilon(x) = \psi(x)u_\varepsilon^*(x) \qquad x \in \Omega$$

where $\psi \in C_0^\infty(\Omega)$, then using the estimates from Appendix A, we see that as $\varepsilon \to 0^+$

$$\varphi_\lambda(u_\varepsilon) \to \frac{1}{N}S^{N/2}$$

$$u_\varepsilon \to 0 \qquad\qquad \text{weakly in } H_0^1(\Omega)$$

$$u_\varepsilon \not\to 0 \qquad\qquad \text{in norm}$$

In other words, φ_λ does not satisfy $(PS)_c$ for $c = S^{N/2}/N$. Notice that the functions u_ε^* satisfy the "limiting problem"

$$-\Delta u = |u|^{2^*-2}u \quad \text{on } \mathbb{R}^N$$
$$u(x) \to 0 \quad \text{as } |x| \to \infty. \tag{C.2}$$

The following proposition is useful for providing some sort of limited Palais-Smale condition for (C.1) at levels above $S^{N/2}/N$. In some sense, it tells us that the only obstructions to the global Palais-Smale condition are the solutions to (C.2). If we let

$$\varphi_0(u) = \frac{1}{2}\int_\Omega |\nabla u|^2 - \frac{1}{2^*}\int_\Omega |u|^{2^*}$$

for $u \in H_0^1(\Omega)$, we have the following due to Struwe [St]:

Proposition C.1. *Let $N \geq 3$, $\lambda \in \mathbb{R}$ be given. Suppose $(u_m)_m$ is a sequence in $H_0^1(\Omega)$ that satisfies*

$$\varphi_\lambda(u_m) \leq c, \ \varphi_\lambda'(u_m) \to 0 \text{ strongly in } H^{-1}(\Omega) \text{ as } m \to \infty.$$

Then there is an integer $k \geq 0$, a solution u_0 of (C.1), solutions $u^1, \ldots, u^k$ of (C.2), sequences of points $x_m^1, \ldots, x_m^k \in \mathbb{R}^N$ and radii $r_m^1, \ldots, r_m^k > 0$ such that for some subsequence $m \to \infty$

$$u_m^0 \equiv u_m \rightharpoonup u^0 \qquad \text{weakly in } H_0^1(\Omega)$$
$$u_m^j \equiv (u_m^{j-1} - u^{j-1})_{r_m^j, x_m^j} \rightharpoonup u^j \quad \text{weakly in } H_0^1(\Omega) \text{ for } j = 1, \ldots, k$$

where $u_{r,x_0} = r^{(N-2)/2}u\bigl(r(\cdot - x_0)\bigr)$ and

$$\|u_m\|_1^2 \to \sum_{j=0}^k \|u^j\|_1^2$$

$$\varphi_\lambda(u_m) \to \varphi_\lambda(u^0) + \sum_{j=1}^k \varphi_0(u^j).$$

Notes and Comments: The above proposition can be found in the book of Struwe ([St] p.169). It can be viewed as the *min-max* counterpart of the *concentration compactness* method developed by P. L. Lions for minimization problems. The idea of analyzing the behavior of a (PS) sequence near points of concentration by *blowing up the singularities* is due to Sacks and Uhlenbeck [S-P].

APPENDIX D

ELEMENTS OF DEGREE THEORY

D.1 Existence and properties of a topological degree

The need often arises to solve equations of the type

$$f(x) = y \qquad (D.1)$$

where $f : \Omega \to \mathbb{R}^N$ is continuous, $\Omega \subset \mathbb{R}^N$ is open and $y \in \mathbb{R}^N$ is given. For example, this situation arises when trying to show that a separating set links with the minimax class in some minimax procedures. Once having found a solution to (D.1) we might also want to count the number of solutions and to know how the solutions behave with respect to "small" perturbations of the function f. These questions are answered in part by the notion of the topological degree of f.

As it is somewhat laborious to actually construct the topological degree, we shall only give the axioms for it and proceed to give some interesting applications. Suppose that $\Omega \subset \mathbb{R}^N$ is open and bounded, $f : \bar{\Omega} \to \mathbb{R}^N$ is continuous and that $y \in \mathbb{R}^N \setminus f(\partial\Omega)$ are given. To each such triple we can associate an integer $\deg(f, \Omega, y)$, the topological degree, which satisfies

(d1) (Normalization) If $f = id$ and $y \notin \mathbb{R}^N \setminus \partial\Omega$ then

$$\deg(f, \Omega, y) = 1.$$

This expresses the fact that $f(x) = y$ has the single solution $x = y$.

(d2) (Additivity) If $\Omega_1, \Omega_2 \subset \Omega$ are open and disjoint sets and if $y \notin f(\bar{\Omega} \setminus (\Omega_1 \cup \Omega_2))$ then

$$\deg(f, \Omega, y) = \deg(f, \Omega, y) + \deg(f, \Omega, y)$$

This reflects that if $f(x) = y$ has m solutions is Ω_1, n solutions in

Ω_2 and none in $\bar{\Omega} \setminus (\Omega_1 \cup \Omega_2)$ then $f(x) = y$ has $m + n$ solutions in Ω.

(d3) (Invariance) $\deg\big(g(t,\cdot),\Omega,y(t)\big)$ is independent of $t \in [0,1]$ whenever $g : [0,1] \times \Omega \to \mathbb{R}^n$ and $y : [0,1] \to \mathbb{R}^N$ are continuous. This allows us to calculate $\deg(f,\Omega,y)$ by replacing it with a (hopefully!) easier calculation.

Theorem D.1. *There is a unique function* $\deg(\cdot,\cdot,\cdot)$ *defined on all admissible triples* (f,Ω,y) *which satisfies* (d1)–(d3). *Moreover, as consequences of* (d1)–(d3) $\deg(f,\Omega,y)$ *satisfies:*

(d4) $\deg(f,\Omega,y) \neq 0$ implies that there exists $x \in \Omega$ such that $f(x) = y$.

(d5) $\deg(\cdot,\Omega,y)$ and $\deg(f,\Omega,\cdot)$ are constant on the neighborhoods $\big\{ g \in C(\bar{\Omega}) \mid \|f - g\|_\infty < r \big\}$ and $B_r(y) \subset \mathbb{R}^N$ resp. where $r = \mathrm{dist}\big(y, f(\partial\Omega)\big)$. In addition, $\deg(f,\Omega,\cdot)$ is constant on each connected component of $\mathbb{R}^N \setminus f(\partial\Omega)$.

(d6) $\deg(f,\Omega,y) = \deg(g,\Omega,y)$ whenever $f|_{\partial\Omega} = g|_{\partial\Omega}$.

(d7) $\deg(f,\Omega,y) = \deg(f,\Omega_1,y)$ for every open subset $\Omega_1 \subset \Omega$ such that $y \notin f(\bar{\Omega} \setminus \Omega_1)$.

While (d1)–(d3) effectively define the topological degree, for purposes of calculation it is useful to know the topological degree for a moderately large collection of triples (f,Ω,y).

Definition D.2. *Suppose that* $f : \bar{\Omega} \to \mathbb{R}$ *is in* $C^1(\Omega) \cap C(\bar{\Omega})$ *and that* $y \in \mathbb{R}^N$ *is given. We say that* y *is a regular value of* f *if* $f^{-1}(y) \cap \big\{ x \in \Omega \mid \det f'(x) \neq 0 \big\} \neq \emptyset$. *Otherwise,* y *will be called a singular value of* f. *We let* $S_f = \big\{ x \in \Omega \mid \det f'(x) = 0 \big\}$ *denote the set of singular points of* f.

Using the inverse function theorem one obtains

Proposition D.3. *If* $f \in C^1(\Omega) \cap C(\bar{\Omega})$ *and* $y \notin f(\partial\Omega)$ *is a regular value of* f, *then* $f(x) = y$ *has at most finitely many solutions in* Ω.

If (f,Ω,y) are as above, then $\deg(f,\Omega,y)$ is given by

$$\deg(f,\Omega,y) = \sum_{x \in f^{-1}(y)} \mathrm{sgn}\,\det f'(x)$$

In this case, it should be noted that $\deg(f,\Omega,y)$ does not count the solutions to $f(x) = y$ but does so taking into account their respective "orientations".

To calculate $\deg(f, \Omega, y)$ for an arbitrary triple (f, Ω, y) one needs only to combine (d5) with the following two results.

Proposition D.4. *If* $f \in C(\bar{\Omega})$ *then for every* $\varepsilon > 0$ *there exists a function* $g \in C^1(\Omega) \cap C(\bar{\Omega})$ *with* $\|f - g\|_\infty < \varepsilon$.

Proposition D.5. *If* $\Omega \subset \mathbb{R}^N$ *is open and* $f \in C^1(\Omega) \cap C(\bar{\Omega})$, *then* $f(S_f)$ *has Lebesgue measure zero. In particular, the regular values of* f *are dense in* $\mathbb{R}^N$.

D.2 Brouwer's and Borsuk's theorems

We would now like to give a few useful applications of degree theory.

Theorem D.6 [Brouwer]. *Let* $D \subset \mathbb{R}^N$ *be a nonempty, compact convex set and* $f : D \to D$ *be continuous. Then* f *has a fixed point in* D *i.e., there is an* $x \in D$ *satisfying* $f(x) = x$.

Proof: First, we consider the case $D = \bar{B}_r(0)$. We may, of course, assume that $f(x) \neq x$ for all $x \in \partial D$ for otherwise we are done. Let $h(t, x) = x - tf(x)$ so that $h : [0, 1] \times D \to \mathbb{R}^N$ is continuous and moreover $0 \notin h([0, 1] \times \partial D)$. For the second assertion, note that on $[0, 1) \times \partial D$

$$|h(t, x)| \geq |x| - t|f(x)| \geq (1 - t)r > 0$$

and that $h(1, x) \neq 0$ for all $x \in \partial D$ by assumption that f has no fixed points on ∂D. It follows from (D3) and (D6) that

$$\deg(id - f, D, 0) = \deg(id, D, 0)$$
$$= 1$$

and so by (d4), there exists $x \in D$ with $x - f(x) = 0$ as desired.

Now suppose that D is an arbitrary nonempty, compact, convex subset of $\mathbb{R}^N$. It is easily seen that we can continuously extend $f : D \to D$ to $\tilde{f} : \mathbb{R}^N \to \mathbb{R}^N$ in such a way that $\tilde{f}(\mathbb{R}^N) \subset \overline{\text{conv}} f(D) \subseteq D$. Now, choose $r > 0$ sufficiently large so that $D \subset B_r(0)$. Applying the first case, we find a fixed point, $x \in B_r(0)$ for $\tilde{f}$. But then $x \in D$ and $f(x) = x$.

Another useful result is the following known as Borsuk's theorem.

Theorem D.7. *If* $\Omega \subset \mathbb{R}^N$ *is bounded, open and symmetric (i.e.,* $-\Omega = \Omega$) *with* $0 \in \Omega$ *and* $f : \bar{\Omega} \to \mathbb{R}^N$ *is odd, continuous and* $0 \notin f(\partial \Omega)$ *then* $\deg(f, \Omega, 0)$ *is odd. In particular,* $f(x) = 0$ *for some* $x \in \Omega$.

Proof: First we show that we may assume that $f \in C^1(\Omega) \cap C(\bar{\Omega})$ and that $\det f'(0) \neq 0$. To see this, approximate f by $g \in C^1(\Omega) \cap C(\bar{\Omega})$ and let $g_0 = \frac{1}{2}[g(x) - g(-x)]$ be the odd part of g. Now choose $\delta > 0$ small which is not an eigenvalue of $g_0'(0)$ and let $\tilde{f} = g_0 - \delta id$. The resulting function, $\tilde{f} \in C^1(\Omega) \cap C(\bar{\Omega})$ is odd, $\det \tilde{f}'(0) \neq 0$ and $\|f - \tilde{f}\|_\infty$ can be made arbitrarily small by choosing $\|f - g\|_\infty$ and δ small enough. By d5), $\deg(f, \Omega, 0) = \deg(\tilde{f}, \Omega, 0)$.

Now supposing that $f \in C^1(\Omega) \cap C(\bar{\Omega})$ and $\det f'(0) \neq 0$, we will be done provided we can find an odd function $g \in C^1(\Omega) \cap C(\bar{\Omega})$ for which 0 is a regular value sufficiently close to f, since then $\deg(f, \Omega, 0) = \deg(g, \Omega, 0)$ (by d5)) and

$$\deg(g, \Omega, 0) = \operatorname{sgn} \det g'(0) + \sum_{\substack{x \in g^{-1}(0) \\ x \neq 0}} \operatorname{sgn} \det g'(x)$$

where the sum is even since $g(x) = 0$ if and only if $g(-x) = 0$ and $\det g'(x) = \det g'(-x)$.

We will construct g by induction. Consider

$$\Omega_k = \left\{ x \in \Omega \mid x_i \neq 0 \text{ for some } i \leq k \right\}$$

and let $\varphi \in C^1(\mathbb{R})$ be odd and satisfy $\varphi'(0) = 0$ and $\varphi(t) = 0$ if and only if $t = 0$ e.g., $\varphi(t) = t^3$.

Now, let $\bar{f}(x) = f(x)/\varphi(x_1)$ on the open, bounded set $\Omega_1 = \{ x \in \Omega \mid x_1 \neq 0 \}$. By proposition D.5 there exists $y^1 \notin \bar{f}(S_{\bar{f}}(\Omega_1))$ with $|y^1|$ as small as needed in the sequel. Hence, 0 is a regular value for $g_1(x) = f(x) - \varphi(x_1)y^1$ on Ω_1, since $g_1'(x) = \varphi(x_1)\bar{f}'(x)$ for $x \in \Omega_1$ such that $g_1(x) = 0$. Suppose that we already have an odd function $g_k \in C^1(\Omega) \cap C(\bar{\Omega})$ close to f on Ω such that $0 \in g_k(S_{g_k}(\Omega_k))$ for some $k \leq n$. Then define $g_{k+1}(x) = g_k(x) - \varphi(x_{k+1})y^{k+1}$ with $|y^{k+1}|$ small and such that 0 is a regular value for g_{k+1} on $\{ x \in \Omega \mid x_{k+1} \neq 0 \}$. Clearly, g_{k+1} is odd, in $C^1(\Omega) \cap C(\bar{\Omega})$ and close to f on $\bar{\Omega}$. Moreover, if $x \in \Omega_{k+1}$ and $x_{k+1} = 0$ then $x \in \Omega_k$, $g_{k+1}(x) = g_k(x)$ and $g_{k+1}' = g_k'(x)$. Thus, $\det g_{k+1}'(x) \neq 0$ and so $0 \notin g_{k+1}(S_{g_{k+1}}(\Omega_k))$. Continuing, we see that $g = g_n$ is odd, close to f on $\bar{\Omega}$ and $0 \notin g\left(\left\{ x \in \Omega \setminus \{0\} \mid \det g'(x) = 0 \right\}\right)$ since $\Omega_n = \Omega \setminus \{0\}$. Lastly, we also have $g'(0) = g_1'(0) = f'(0)$ and so 0 is a regular value of g.

Corollary D.8. *Let $\Omega \subset \mathbb{R}^N$ be bounded, open and symmetric with $0 \in \Omega$, $f : \partial\Omega \to \mathbb{R}^M$ be continuous and $M \leq N$. Then $f(x) = f(-x)$ for some $x \in \partial\Omega$.*

Proof: Assume that $f(x) \neq f(-x)$ for all $x \in \partial\Omega$. Let $g(x)$ be any odd extension of $f(x) - f(-x)$ to all of $\bar{\Omega}$ and set

$$\tilde{g}(x) = \big(g(x), \underbrace{0, \ldots, 0}_{N-M}\big) \qquad \text{for all } x \in \Omega$$

Then by (d5) and Borsuk's theorem

$$\deg(\tilde{g}, \Omega, y) = \deg(\tilde{g}, \Omega, 0) \neq 0$$

for all y in some ball $B_r(0)$. But then by d4), we have $B_r(0) \subset \tilde{g}(\Omega)$ which is certainly false.

Remark: The notion of the topological degree can be extended to the case $f : \bar{\Omega} \to X$ where $\Omega \subset X$ is bounded and open, f continuous and X an infinite dimensional Banach space. However, since bounded subsets of infinite dimensional Banach spaces need not be compact, f can no longer be arbitrary; it must be of the form $f = id - K$ where K is a continuous compact map. This extension was done by Leray and Schauder in 1934.

D.3 Extensions of nonlinear mappings

The following basic result about the possibility of extending non-linear mappings between Banach spaces, is used frequently throughout this monograph.

Theorem D.9. *Let V be a finite dimensional space, K a closed subset of $\mathbb{R}^m$ and let θ be a continuous mapping from K into the unit sphere S_V of V. If $m < \dim V$, then θ can be extended to a continuous mapping from $\mathbb{R}^m$ into S_V.*

Notes and Comments: Degree theory can be found in any book on non-linear functional analysis. See for instance the lecture notes of Nirenberg [N 1] or the book of Deimling [Dei]. Theorem D.9 can be found in the monograph of Dugundji [Du].

APPENDIX E

BASIC PROPERTIES OF MARTINGALES

E.1 Martingale inequalities and convergence

We start by recalling the concept of (real-valued) *martingales* as well as some of their elementary properties.

Let (Ω, Σ, P) be a probability space and let $(\Sigma_n)_n$ be an increasing sequence of sub-σ-fields of Σ (i.e., $\Sigma_n \subset \Sigma_{n+1}$).

Definition E.1. *A sequence $(f_n)_n$ of real-valued integrable random variables on Ω is said to be a martingale (resp. a submartingale) with respect to $(\Sigma_n)_n$ if for each $n \in \mathbf{N}$,*

(1) f_n is Σ_n-measurable and
(2) $\int_A f_n\, dP = \int_A f_{n+1}\, dP$ (resp. $\int_A f_n\, dP \leq \int_A f_{n+1}\, dP$) for every $A \in \Sigma_n$.

In other words, $f_n = \mathbf{E}[f_{n+1}; \Sigma_n]$ (resp. $f_n \leq \mathbf{E}[f_{n+1}; \Sigma_n]$) where the latter denotes the *conditional expectation* of f_{n+1} with respect to the σ-field Σ_n.

The following is known as Doob's maximal inequality for martingales. See for instance the book of Neveu [Ne].

Theorem E.2. *Let $(f_n)_n$ be a real-valued submartingale on (Ω, Σ, P):*

(1)   If $\sup_n \int_\Omega |f_n|\, dP < \infty$, then for any $\lambda > 0$, we have

$$P[w \in \Omega; \sup_n |f_n(w)| > \lambda] \leq \frac{1}{\lambda} \sup_n \int_\Omega |f_n|\, dP.$$

(2)   If $\sup_n \int_\Omega |f_n|^p\, dP < \infty$ for some $p > 1$, then $\sup_n |f_n| \in L^p$.

Here is the *martingale convergence theorem.*

Theorem E.2. *Let $(f_n)_n$ be a real-valued submartingale on (Ω, Σ, P) such that $\sup_n \int_\Omega |f_n| \, dP < \infty$. Then, $(f_n)_n$ converges almost surely to a random variable f.*

Moreover, if $\sup_n \int_\Omega |f_n|^p \, dP < \infty$ for some $p > 1$, then $(f_n)_n$ also converges to f in L^p.

E.2 Vector valued martingales

If X is a Banach space, then one can define the *Bochner integral* of an X-valued random variable f as well as its conditional expectation, first by defining them for step functions and then by taking limits in the usual fashion and as in the real valued case. Therefore, one can say that a sequence $(f_n)_n$ of X-valued Bochner integrable random variables is *a vector martingale* if for all $n \in \mathbb{N}$, we have that almost surely $\mathbf{E}[f_{n+1}; \Sigma_n] = f_n$, where both are regarded as X-valued random variables.

The convergence of uniformly bounded Banach space valued martingales does not always hold and it is closely related to the geometry of the space involved. Actually, one can show the following result. The notion of *admissibility* is the one introduced in Chapter 1.

Theorem E.3. *If C is a closed convex bounded subset of a Banach space X, then the following properties are equivalent:*

(1)   Every C-valued martingale converges almost surely.
(2)   The space X^* of continuous linear functionals is an admissible space of perturbations on C.

For a proof, see for instance [G-M 1]).

To extend the above notions to a nonlinear setting, we proceed in the following way.

Suppose that X is a complete metric space and let $\mathcal{A}$ be a convex cone of real valued continuous functions on X.

Definition E.4. *Say that a sequence $(f_n)_n$ of X-valued random variables on Ω is an $\mathcal{A}$-martingale if for any $\varphi \in \mathcal{A}$, the process $(\varphi \circ f_n)_n$ is a real-valued submartingale.*

If X is a Banach space, it is easy to check that the notion of X-valued martingales defined before Theorem E.3 coincides with the notion of $\mathcal{A}$-martingales where $\mathcal{A}$ is the cone of continuous convex (or linear) functionals on X.

On the other hand, if X is a manifold or just a quasi-Banach space (as in the case of L^p ($0 < p < 1$))) the notion of a vector integral is not well defined in general, and the concept of a martingale with respect to a cone of functions is more appropriate.

As stated in Theorem E.3 above (in the case of the space of continuous linear functionals) and as shown in Theorem 2.17 (for the case of Lipschitz plurisubharmonic functions), there is a close relation between the convergence of $\mathcal{A}$-martingales and the admissibility of $\mathcal{A}$ as a cone of perturbations for a minimization principle. These ideas were extensively developed in [G-M 1,2,3] and in [G-M-S].

REFERENCES

[A-Z]   H. Amann, E. Zehnder: *Nontrivial solutions for a class of nonresonance problems and applications to nonlinear differential equations.* Ann. Scuola Normale Superiore di Pisa **7** (1980), p. 539-603.

[A]   R. A. Adams: *Sobolev spaces.* Academic Press, New-York-San Francisco-London (1975).

[A-R]   A. Ambrosetti, P. H. Rabinowitz: *Dual variational methods in critical point theory and applications.* J. Funct. Anal. **14** (1973), p. 349–381.

[A-E]   J. P. Aubin, I. Ekeland: *Applied Non-linear Analysis.* Pure and Applied Math. Wiley Interscience Publications (1984).

[Ba 1]   A. Bahri: *Une méthode perturbative en théorie de Morse.* Thèse d'état. Publications de l'Université Paris VI (1981).

[Ba 2]   A. Bahri: *Critical points at infinity in some variational problems.* Pitman Research Notes, **182** Longman House, Harlow (1989).

[Ba-L]   A. Bahri, P. L. Lions: *Morse index of some min-max critical points I. Application to multiplicity results.* Commun. Pure & App. Math. **41** (1988), p. 1027-1037.

[B 1]   V. Benci: *A geometrical index for the group S^1 and some* *applications to the study of periodic solutions of ordinary differential equations.* Comm. Pure & Appl. Math. **34** (1981), p. 393–432.

[B 2]   V. Benci: *A new approach to the Morse-Conley theory and some applications.* Preprint (1988).

[B-P]   J. Borwein, D. Preiss: *A smooth variational principle with applications to subdifferentiability and to differentiability of convex functions.* Trans. A.M.S. **303** (1987), p. 517-527.

[Bo]   J. Bourgain: *La propriété de Radon-Nikodym.* Publication Mathématiques de l'Université Pierre et Marie Curie - No. **36**, (1979).

[B-N 1]   H. Brezis, L. Nirenberg: *Positive solutions of non linear elliptic equations involving critical Sobolev exponents.* Comm. Pure & Appl. Math. **36** (1983), p. 437–477.

[B-N 2]   H. Brezis, L. Nirenberg: *A minimization problem with critical exponent and non-zero data.* Symmetry in Nature, Scuola Norm. Sup. Pisa (1989), p. 129-140.

[B-N 3]   H. Brezis, L. Nirenberg: *Remarks on finding critical points.* Comm. Pure & Appl. Math. **44**, No. 8–9 (1991) p. 939–963.

[C-S-S]   G. Cerami, S. Solimini, M. Struwe: *Some existence results for superlinear elliptic boundary value problems involving critical exponents,* J. Funct. Anal. 69 (1986), p. 289-306.

[C]   C. V. Coffman: *Ljusternik-Schnirelman theory: Complementary principles and the Morse index.* Nonlinear Analysis, Theory & Applications. **12**, No. 5 (1988), p. 507-529.

[Co]   C. Conley: *Isolated invariant sets and the Morse index.* C.B. M.S. **38**, A.M.S. Providence (1978).

[C-E-L]   M. G. Crandall, L. C. Evans and P. L. Lions: *Some properties of viscosity solutions of Hamilton-Facobi equations.* Trans. A.M.S. **282** (1984), p. 484-502.

[C-L 1]   M. G. Crandall, P. L. Lions: *Viscosity solutions of Hamilton-Jacobi equations.* Trans. A.M.S. **277** (1983), p. 1-42.

[C-L 2]   M. G. Crandall, P. L. Lions: *Hamilton-Jacobi equations in infinite dimensions I.* J. Funct. Anal. **62** (1985), p. 379-398.

[C-L 3]   M. G. Crandall, P. L. Lions: *Hamilton-Jacobi equations in infinite dimensions II*. J. Funct. Anal. **65** (1986), p. 368-405.

[C-L 4]   M. G. Crandall, P. L. Lions: *Hamilton-Jacobi equations in infinite dimensions III*. J. Funct. Anal. **68** (1986), p. 214-247.

[D]   E. N Dancer: *Degenerate critical points, homotopy indices and Morse inequalities*. J. Reine. und Angew. Math. **350** (1984), p. 1-22

[De]   D. G. De Figueiredo: *The Ekeland variational principle with applications and detours*. Tata Institute of fundamental research, Bombay (1989)

[Dei]   K. Deimling: *Nonlinear functional analysis*. Springer Verlag, Berlin-Heidelberg-New York-Tokyo (1985).

[D-G-Z 1] R. Deville, G. Godefroy, V. Zizler: *Un principe variationnel utilisant des fonctions bosses*. C.R. Acad. Sci. Paris, **312**. série I, (1991), p. 281-286.

[D-G-Z 2] R. Deville, G. Godefroy, V. Zizler: *Smoothness and renormings in Banach spaces*. Monographs & Survey Series, Longman. No. 64. (1992).

[Du]   J. Dugundji: *Topology*. Allyn and Bacon, Boston (1973).

[E 1]   G. A. Edgar: *Complex martingale convergence*. Springer-Verlag, Lecture Notes, **1116** (1985), p. 38-59.

[E 2]   G. A. Edgar: *Analytic martingale convergence*. J. Funct. Anal, **69**, No. 2 (1986), p. 268-280.

[Ek 1]   I. Ekeland: *Nonconvex minimization problems*. Bull. Amer. Math. Soc. **1** (1979), p. 443-474.

[Ek 2]   I. Ekeland: *Convexity Methods in Hamiltonian Mechanics*. Springer-Verlag, Berlin, Heidelberg, New-York (1990).

[E-H]   I. Ekeland, H. Hofer: *Convex Hamiltonian energy surfaces and their periodic trajectories*. Comm. Math. Phys. **113** (1987), p. 419–469.

[F-R 1]   E. R. Fadell, P. H. Rabinowitz: *Bifurcation for odd potential operators and an alternative topological index*. J. Funct. Anal. **26** (1977) p. 48–67.

[F-R 2]   E. R. Fadell, P. H. Rabinowitz: *Generalized cohomological index theories for Lie group actions with applications to bifurcation questions for Hamiltonian systems.* Invent. Math. **45** (1978), p. 139–174.

[F-H]   E. Fadell, S. Husseini: *Relative cohomological index theories.* Advances in Math. **64** (1987), p. 1–45.

[F-H-R]   E. Fadell, S. Husseini, P.H. Rabinowitz: *Borsuk-Ulam theorems for arbitrary S^1-actions and applications.* Trans. A.M.S. **275** (1982), p. 345–360.

[F 1]   G. Fang: *The structure of the critical set in the general mountain pass principle,* To appear (1990).

[F 2]   G. Fang: *On the structure of critical points generated by min-max procedures.* Preprint (1993).

[F 3]   G. Fang: *Topics on critical point theory.* PhD dissertation, The University of British Columbia. (1993).

[F-G 1]   G. Fang, N. Ghoussoub: *Second order information on Palais-Smale sequences in the mountain pass theorem.* Manuscripta Math. **75** (1992) p. 81–95.

[F-G 2]   G. Fang, N. Ghoussoub: *Morse-type information on Palais-Smale sequences obtained by min-max principles.* Commun. Pure & App. Math. To appear (1993).

[Fl 1]   A. Floer: *Morse theory for fixed points of symplectic diffeomorphisms.* Bull. Amer. Math. Soc. **16** (1987), p. 279-281.

[Fl 2]   A. Floer: *Witten's complex and infinite dimensional Morse theory.* J. Diff. Geom. **18** (1988), p. 207-221.

[Fl 3]   A. Floer: *Symplectic fixed points and holomorphic spheres.* Comm. Math. Phys. **120** (1989), p. 576-611.

[G-P]   N. Ghoussoub, D. Preiss: *A general mountain pass principle for locating and classifying critical points.* A.I.H.P. Analyse non linéaire, Vol 6, **No. 5** (1989), p. 321-330

[G 1]   N. Ghoussoub: *Location, multiplicity and Morse indices of min-max critical points.* J. Reine. und Angew. Math. **417** (1991), p. 27-76.

[G 2]   N. Ghoussoub: *A Min-Max principle with a relaxed boundary condition.* Proc. A.M.S, **117**, No. 2 (1992) p. 439–447.

[G-L-M]   N. Ghoussoub, J. Lindenstrauss, B. Maurey: *Analytic martingales and plurisubharmonic barriers in complex Banach spaces.* Contemporary Math, **85** (1989), p. 111-130.

[G-M 1]   N. Ghoussoub, B. Maurey: H_δ-*embeddings in Hilbert space and optimization on* G_δ-*sets.* Memoirs of the A.M.S., No. **349**, July (1986).

[G-M 2]   N. Ghoussoub, B. Maurey: *Plurisubharmonic martingales and barriers in complex quasi-Banach spaces.* Ann. Inst. Fourier, **39**, 4 (1989), p. 1007-1060.

[G-M 3]   N. Ghoussoub, B. Maurey: *Balayage, dentability and optimization on manifolds.* Unpublished lecture notes (1988).

[G-M-S]   N. Ghoussoub, B. Maurey, W. Schachermayer: *Pluriharmonically dentable complex Banach spaces.* J. Reine. und Angew. Math. **402** (1989), p. 76-127.

[H 1]   H. Hofer: *Variational and topological methods in partially ordered Hilbert spaces.* Math. Ann. **261** (1982), p. 493-514.

[H 2]   H. Hofer: *A geometric description of the neighborhood of a critical point given by the mountain pass theorem.* J. Lond. Math. Soc. **31** (1985) p. 566–570.

[K-P-R]   N. J. Kalton, N. T. Peck, J. W. Roberts: *An F-space Sampler,* London Math. Society Lecture Notes **89**, Cambridge University Press (1985).

[K 1]   N. J. Kalton: *Plurisubharmonic functions on quasi-Banach spaces,* Studia Mathematica, TLXXXIV (1986), p. 297-324.

[K 2]   N. J. Kalton: *Differentiability properties of vector-valued functions,* Lecture Notes in Math., Springer-Verlag **1221** (1985), pp. 141-181.

[Ku]   C. Kuratowski: *Topology.* Volume II. Academic Press, New York and London (1968).

[Kr]   M. Krasnoselski: *Topological Models in the Theory of Nonlinear Integral Equations.* Pergamon, Oxford (1965).

[L-S]   A. Lazer, S. Solimini: *Nontrivial solutions of operator equations and Morse indices of critical points of Min-Max type.* Non Linear Analysis, Methods & Applications, **12**, No. 8 (1988), p. 761–775.

[L]   P. L Lions: *Solutions of Hartree-Fock equations for Coulomb systems.* Comm. Math. Physics, **109** (1987), p. 33-97.

[L-Sc]   L. Ljusternik, L. Schnirelmann: *Méthodes topologiques dans les problèmes variationels.* Hermann, Paris (1934).

[Lu-S]   D. Lupo, S. Solimini: *A note on a resonance problem.* Proc. Royal. Soc. Edinburgh. **102 A** (1986), p. 1-7.

[M]   S. Mizohata: *The Theory of Partial Differential Equations.* Cambridge University Press (1973).

[Mi]   C. Miranda: *Un'Osservazione sul teorema di Brouwer.* Boll. Un. Mat. Ital. ser.II, Anno III, n.1, **19** (1940), p. 5-7.

[M-P]   A. Marino, G. Prodi: *Metodi perturbativi nella teoria di Morse.* Boll. Un. Mat. Ital. (vol. II) Suppl. Fasc. 3 (1975), p. 1-32.

[M-W]   J. Mawhin, M. Willem: *Critical point theory and Hamiltonian systems.* Applied Mathematical Sciences, **74**, Springer-Verlag (1989).

[Mo]   G. Mostow: *Equivariant embeddings in euclidian space.* Ann of Math. **65** (1957), p. 432-446.

[Ne]   J. Neveu: *Discrete parameter martingales.* North Holland (1975).

[N 1]   L. Nirenberg: *Topics in non-linear Functional Analysis.* Lecture Notes—Courant Institute, New York (1974).

[N 2]   L. Nirenberg: *Variational Methods in non-linear problems.* Lecture Notes in Math, **1365**, Springer-Verlag (1989), p. 100-119.

[Pa 1]   R. Palais: *Lusternik–Schnirelman Theory on Banach manifolds.* Topology, **5** (1966) p. 115–132.

[Pa 2]   R. Palais: *Slices and equivariant embeddings.* Seminar on transformation groups, Ann. Math. Studies, Princeton University Press. N.Y. (A. Borel, editor) (1960).

[Ph]   R. R. Phelps: *Convex functions, monotone operators and differentiability.* Lecture Notes in Mathematics, **1364**, Springer-Verlag (1989).

[P]   D. Preiss: *Fréchet derivatives of Lipschitz functions.* J. Func.
           Anal. **91** (1990) p. 312–345.

[P-S 1]   P. Pucci, J. Serrin: *A mountain pass theorem.* J. Diff. Eq.
           60 (1985), p. 142–149.

[P-S 2]   P. Pucci, J. Serrin: *Extension of the mountain pass theorem.*
           J. Funct. Anal. **59** (1984), p. 185–210.

[P-S 3]   P. Pucci, J. Serrin: *The structure of the critical set in the
           mountain pass theorem.* Trans. A.M.S. **91**, No. 1 (1987),
           p. 115–132.

[R 1]   P. H. Rabinowitz: *Minimax methods in critical point theory
           with applications to differential equations.* C.B.M.S., A.M.S.,
           No. 65 (1986).

[R 2]   P. H. Rabinowitz: *Variational methods for non-linear eigen-
           value problems.* Prodi ed. Cremonese, Roma (1974), p. 141–
           195.

[Ro]   D. Robinson: PhD dissertation, The University of British
           Columbia. In preparation (1993).

[S-U]   P. Sacks, K. Uhlenbeck: *On the existence of minimal immer-
           sions of 2-spheres.* Ann. of Math. **113** (1981), p. 1-24.

[Sm]   S. Smale: *An infinite dimensional version of Sard's theorem.*
           Amer. J. Math. **17** (1964), p. 307-315.

[So 1]   S. Solimini: *Morse index estimates in Min-Max theorems.*
           Manuscripta Mathematica, **63** (1989), p. 421-433.

[So 2]   S. Solimini: *On the solvability of some elliptic partial differ-
           ential equation with linear part at resonance.* J. Math. Anal.
           Appl. **117**, (1986), p. 138-152.

[So 3]   S. Solimini: *Notes on min-Max theorems with applications to
           some asymptotically linear elliptic problems.* Lecture notes,
           (1990)

[S]   C. Stegall: *Optimization of functions on certain subsets of
           Banach spaces.* Math. Ann. **236** (1978), p. 171-176.

[St]   M. Struwe: *Variational methods and their applications to
           non-linear partial differential equations and Hamiltonian sys-
           tems.* Springer-Verlag (1990).

[Su]   T. Suzuki: *Generation of positive nonradial solutions for semilinear elliptic equations on annului: a variational approach.* Preprint (1991).

[Sz]   A. Szulkin: *Lusternik-Schnirelmann theory on C^1-manifolds.* A.I.H.P. Analyse non linéaire, **5**, No. 2, (1988), p. 119-139

[T 1]   G. Tarantello: *On nonhomogeneous elliptic equations involving the critical Sobolev exponent.* A.I.H.P. Analyse non linéaire, **9**, No. 3 (1992) p. 264–281.

[T 2]   G. Tarantello: *Nodal solutions of semilinear elliptic equations with critical exponent.* C.R. Acad. Sci. Paris Serie I Math. **313**, No. 7 (1991) p. 441–445.

[T 3]   G. Tarantello: *Remarks on forced equations of the double pendulum type.* Trans. A.M.S. **326**, No. 1 (1991) p. 441–452.

[T 4]   G. Tarantello: *Morse index concentration for elliptic problems with Sobolev exponent.* Research report, Carnegie Mellon University, (1991).

[V]   C. Viterbo: *Indice de Morse des points critiques obtenus par minimax.* A.I.H.P. Analyse non linéaire, **5**, No. 3 (1988), p. 221–225.

Index

$\mathcal{A}$-martingale, 244
absolute neighborhood extensor, 78
admissibility, 244
admissible class, 22
admissible cone, 24, 34
admissible space, 24, 34
admissible triples, 239
almost critical sequence, 6
Amann-Zehnder, 182
Ambrosetti-Rabinowitz, 66, 89, 95
analytic martingales, 42
Aoki-Rolewicz, 42
asymptotically linear non-resonant
    boundary value problem, 61, 176
asymptotically linear resonant
    problem of Landesman-Lazer
    type, 178
asymptotically linear strongly
    resonant elliptic problem, 61, 73,
    181
Aubin, T., 165

Bahri, 182, 203
Bahri-Lions, 182, 203
Baire category theorem, 23, 38
Bishop–Phelps theorem, 33
blow-up of singularities, 155, 237
blow-up technique, 156
Bochner integral, 244
bornology β, 26
    β-derivative, 26
    β-differentiable, 26
    β-subdifferentiability, 27
    $C_\beta^1(X)$, 26
    classical β-solution, 29
    viscosity β-(sub, super) solution,
    29, 200
Borsuk-Ulam theorem, 141, 240
Borwein-Preiss, 15, 33
Bourgain, 50
Brezis-Nirenberg, 33, 40, 113, 165,
    233, 237

Brouwer's fixed point theorem, 63,
    97, 241
bump function, 24

Caratheodory function, 12, 13, 41,
    235
Cauchy problem, 66, 131, 170
Cauchy-Schwarz, 20, 74
Cerami-Solimini-Struwe, 151, 165
Chang, K.C., 113
classifying space, 202
coercive, 4, 58
Coffman, 146
cohomological family of dimension
    n, 167
cohomological index, 203
cohomotopic classes, 167
cohomotopic family of dimension n,
    167
concentration compactness method,
    237
connected manifold, 53
connected (A, B connected through
    C), 120
connected set, pathwise, 76, 115
constrained minimization, 5
Coti Zelati-Ekeland-Lions, 223
Coulombic N-body Hamiltonian, 17,
    216
Crandall-Lions, 33
critical exponent, 8
critical level, 67, 69, 80
critical point (non-degenerate), 67,
    119, 125, 127, 167, 169
critical set, 4, 114
critical value, 4, 53

deformation lemma, 54
degree theory, 238
Deville-Godefroy-Zizler, 33
Dirichlet problem, 8
Doob's inequality, 47, 244